Polymer Networks

Principles of their Formation, Structure and Properties

Polymer Networks
Principles of their Formation, Structure and Properties

Edited by

R.F.T. Stepto
Professor of Polymer Science
University of Manchester Institute of
Science and Technology

BLACKIE ACADEMIC & PROFESSIONAL
An Imprint of Chapman & Hall
London · Weinheim · New York · Tokyo · Melbourne · Madras

Published by
Blackie Academic & Professional, an imprint of Thomson Science,
2–6 Boundary Row, London SE1 8HN, UK

Thomson Science, 2–6 Boundary Row, London SE1 8HN, UK

Thomson Science, 115 Fifth Avenue, New York, NY 10003, USA

Thomson Science, Suite 750, 400 Market Street, Philadelphia, PA 19106, USA

Thomson Science, Pappelallee 3, 69469 Weinheim, Germany

First edition 1998

Thomson Science is a division of International Thomson Publishing I(T)P®

Typeset in 10/12pt Times by Academic & Technical, Bedminster, Bristol
Printed in Great Britain by T.J. International Ltd, Padstow, Cornwall

ISBN 0 7514 0480 2

A catalogue record for this book is available from the British Library

Library of Congress Catalog Card Number: 97-76988

∞ Printed on acid-free text paper, manufactured in accordance with ANSI/NISO Z39.48-1992 (Permanence of Paper).

To J.

Preface

Over the past two decades, polymer networks has emerged as a coherent subject area. It is now realized that, in comparison with those of linear polymers, the properties of polymer networks depend to a much larger extent on methods and conditions of preparation. That is, properties depend not only on the chemical structures of the individual polymer chains, but on how those chains connect together to form a network. The inter-relations between formation conditions and network connectivities and properties, coupled with the transformation of liquid reactants of starting materials to a solid during network formation, underlie and define the coherence and breadth of the subject.

This book is the first to treat, in a comprehensive fashion, the fundamental aspects of the science relating to the formation, structure and properties of polymer networks. It seeks to bring together the various scientific strands which constitute the present-day subject and to show how they relate to one another. Hopefully, it will provide a first step towards a deeper and broader-based understanding of polymer network materials and their properties.

I thank sincerely the authors of the individual chapters for their willingness to participate in the book and for their excellent contributions in their fields of expertise. In addition, my particular thanks are extended to Simon Ross-Murphy and John Stanford for their detailed comments on the introductory chapter.

Robert Stepto
University of Manchester Institute of Science and Technology
August, 1997

Contributors

Y. Du	Department of Chemistry, State University of New York, Albany, NY 12222, USA
K. Dušek	Institute of Macromolecular Chemistry, Academy of Sciences of the Czech Republic, 162 06 Prague 6, Czech Republic
B. Erman	Polymer Research Centre, School of Engineering, Bogazici University, *and* TUBITAK Advanced Polymeric Materials Centre, Bebek 80815, Istanbul, Turkey
H.L. Frisch	Department of Chemistry, State University of New York, Albany, NY 12222, USA
M. Ilavsky	Institute of Macromolecular Chemistry, Academy of Sciences of the Czech Republic, 162 06 Prague 6, Czech Republic, *and* Department of Polymer Physics, Faculty of Mathematics and Physics, Charles University, 1800 Prague 8, Czech Republic
J.E. Mark	Department of Chemistry and Polymer Research Center, University of Cincinnati, Cincinnati, OH 45221-0172, USA
S.B. Ross-Murphy	Biopolymers Group, Division of Life Sciences, King's College London, Campden Hill Road, Kensington, London W8 7AH, UK
M. Schultz	Fachbereich Physik, Martin-Luther Universität, 06099 Halle, Germany

J.L. Stanford Polymer Science and Technology Group,
Manchester Materials Science Centre,
UMIST and University of Manchester,
Manchester M1 7HS, UK

R.F.T. Stepto Polymer Science and Technology Group,
Manchester Materials Science Centre,
UMIST and University of Manchester,
Manchester M1 7HS, UK

R.J.J. Williams Institute of Materials Science & Technology,
University of Mar del Plata and National
Research Council, Av. J.B. Justo 4302,
(7600) Mar del Plata, Argentina

Contents

1 Introduction

R.F.T. STEPTO

1.1 Networks, polymer networks and materials

In atomic or molecular terms, network structures are widespread. They are the basis of all solid materials. It is a network structure which permeates the macroscopic volume of a solid material and endows it with a mechanical coherence, thus defining its equilibrium shape. Without a macroscopic atomic or molecular network structure, molecules or atoms are relatively free to undergo translational movements, resulting in fluid, liquid or gaseous, behaviour.

In most inorganic salts, electrostatic interionic forces hold the lattice or network of anions and cations together. When the effects of these are weakened, through the action of temperature or a solvent, the solid state is lost. Diamond is probably the most generally known example of a covalent network, in which each atom is covalently and tetrahedrally bonded to its neighbours. In fact, a diamond crystal is a *macromolecular*, three-dimensional network. Graphite, however, has hexagonal, covalently bonded layers, with relatively weak (van der Waals) forces between the layers. A sample of graphite is thus constituted of layers of two-dimensional, macromolecular networks held together by weaker, secondary forces. The differences in the properties of diamond (hard) and graphite (easily cleaved) are well known and are attributable to their different atomic, network structures.

Treatments of atomic-based solids and relationships between their chemistry, bonding, structures and properties can be found in many textbooks of physical chemistry and materials science, e.g. [1]. In addition to crystalline solids, there are amorphous glassy (vitreous) solids, which are formed by rapid cooling, leading to a rapid increase in viscosity, causing, in turn, static, random atomic arrangements rather than equilibrium, crystalline arrangements.

Organic compounds of low molar mass form molecular-based networks which can be glassy or crystalline at sufficiently low temperatures, or if the molecules contain groups having sufficiently strong interactions (e.g. those forming hydrogen bonds). The intramolecular interactions are usually due to weaker, secondary forces and, generally, solids of organic compounds do not have noteworthy mechanical properties. At ambient temperatures, their applications are often as functional materials, such as micelles and liquid crystals, in which the solid-like molecularly ordered regions are of microscopic rather than macroscopic extents.

In contrast, polymer networks are molecular-based networks whose network structures depend entirely on covalent bonding or on physical intermolecular interactions between *macromolecules*. The former, covalent type of structure can give materials samples which are single macromolecules, as are diamond crystals. However, in contrast to diamond, each pair of adjacent junction points in the networks are not separated by only one covalent bond but by linear (sub)chains of several or many covalent bonds. An example of a junction point in a covalently bonded polymer network is the central carbon atom from pentaerythritol ($C(CH_2OH)_4$). Its $-OH$ groups will have reacted to form, for example, four polyester or polyurethane subchains which connect with adjacent junction points (Chapter 2). If the subchains are sufficiently long and their chemical structure is such as to allow rotational degrees of freedom under the conditions of use, then a soft, flexible, elastomeric material results. Stiff subchains give hard, glassy materials. The flexibility of a material relates to subchain mobility, i.e. to the number of different spatial configurations (conformations) the subchains can adopt, and the reversible extensibility and elasticity relate to the chain mobility *and* the anchoring of the chains by the junction points into a network structure.

The latter type of network structure, formed by physical interactions between linear molecules (Chapter 9), usually has junction zones, rather than junction points, constituted of ordered (often helically structured intertwined) chains or merely topologically entangled chains. The former, ordered type of junction zones can be based on hydrogen bond interactions and gives, under normal conditions, permanent networks. The latter type only endows elastomeric, i.e. network properties, under fast deformations of limited extents, during which the individual macromolecules do not have the opportunity to disentangle within the time-scales of the deformations. Such limited elasticity occurs, for example, in thermoplastics constituted of linear macromolecules, such as polyethylene and poly(dimethyl siloxane).

Distinct from some networks formed from small molecules, polymer networks never have macroscopically crystalline structures. The very presence of junction points means that the subchains between them cannot completely align into crystalline order. The structures and properties of polymer networks are not determined by the interactions between atoms or molecules which constitute the junction points but by the structures and lengths of the network subchains which connect them. Thus, as for linear polymers, the physical properties of polymer networks are dependent on the structures of the polymer chains of which they are constituted, overlaid, however, with a dependence on network connectivity. For example, for a given chain structure, the lengths of the subchains between junction points can be varied by changing the amounts of the different reactants used to prepare a network, or by changing the conditions of preparation. Hence, two polymer networks having chains of similar chemical structures can have markedly different properties because they have different lengths of chains between junction points. Such changes in properties may be exemplified by the sulfur crosslinking of natural rubber, where about 1% sulfur produces soft elastomers and about 30% sulfur forms ebonite, a hard, structural material [2]. In the more highly crosslinked material, the concentration of junction points is high, hence, the lengths of the subchains between them are shorter and there are more subchains per unit volume. In the elastomeric state, because a macroscopic deformation deforms each subchain, the initial modulus of a polymer network is directly proportional to the concentration of such chains (see [3] and Chapters 2 and 7). Hence, a higher modulus results from a higher concentration of chains (or junction points). In addition, shorter chains have a lower total extensibility and this is reflected in a lower, reversible extensibility of a network. Elastomers with low junction-point concentrations can extend uniaxially by, typically, up to 1000%.

The preceding general considerations of the mechanical properties of polymer networks being determined by subchain structure, length and density holds true for all covalent and physical, elastomeric networks. At lower temperatures, glassy solids are formed, as for linear polymers. However, the glass–rubber transition temperature (T_g) is also related to junction-point concentration, with a higher concentration restricting chain motion more and, hence, increasing T_g (Chapters 2, 4 and 5). In addition, the glass–rubber transition becomes broader and less distinct as the concentration of junction points increases.

A combination of both covalent and physical network structures can occur in some, so-called thermoplastic elastomers (Chapter 5). They can be, for example, polyurethane- or polyurea-based. At lower temperatures, physical junction points are formed by associations of intrinsically rigid subchains to give so-called hard-segment domains. As the temperature increases, the associations weaken, the concentration of junction zones decreases and the modulus is reduced.

A characteristic property of permanent elastomeric networks is their ability to swell in suitable solvents. As a polymer network is structurally a solid it cannot dissolve, except by loss of its macroscopically coherent structure. Thus, if a network is immersed in a solvent which is miscible with the polymer chains of the network, the solvent is imbibed and the network swells. Increases in volume of, typically, 500% or more may be achieved by networks of low junction-point concentrations.

Although a network may be deformed, it is a solid and does not flow; it has an infinite steady-shear viscosity. This property has important consequences for the preparation of polymer network materials. Once an incipient network structure has formed, reaction to complete the network formation must take place without flow.

Part of the fascination of the behaviour of polymers are the large changes in physical properties and the wide varieties of mechanical behaviours displayed at easily accessible temperatures and under easily accessible conditions. Because of its macromolecular chains and the changes in their thermal movements and interactions, both with temperature and in the presence of small molecules (plasticizer or solvent), the same polymer can exist as a partially crystalline solid, an amorphous glassy solid, a viscoelastic solid and as a liquid. With polymer networks, such changes of behaviour can also occur. However, they have to conform to the overall network connectivity. That is, a macroscopic sample of a network material also behaves as a single molecule.

1.2 Development of the technology and preparative chemistry of polymer networks

Natural polymer network materials, such as natural rubber, gelatin and wool, have been known and used for many thousands of years [4]. Such materials have of course been modified empirically over the centuries for various applications. For example, in 1823 Macintosh produced a waterproof fabric using natural rubber [5] and natural rubber was crosslinked, or vulcanized, by Goodyear in 1839 using sulfur [2], to give a material with permanent elasticity.

The first, completely synthetic polymer material was a network polymer, Bakelite, or phenol-formaldehyde resin, developed by Baekeland from 1905 to 1909 [5]. This was followed in 1928 by urea-formaldehyde resins [5, 6] and in 1933 by alkyd or polyester resins [6]. In all cases, the chemical reactions involved had been discovered some decades prior to the dates given.

The macromolecular hypothesis was propounded by Staudinger only in 1920 [7]. Hence, the preceding historical examples show that the chemistry and technology related to polymer networks were well established prior to knowledge of the molecular structures of networks and the scientific understanding of their properties. For present purposes, it is useful to consider the examples as typifying various types of polymer network materials. Natural rubber is a physical network which is converted into a covalent network when it is crosslinked. The crosslinking of pre-formed primary, macromolecules, often with sulfur or peroxides, is now a route to many elastomeric network materials [2, 6], for example, polyisobutylene, butadiene-styrene rubber, silicone elastomers.

Gelatin is a classic example of a physical network. It is based on molecules of the linear protein collagen. After denaturation the molecules form a network by partially reassociating in junction zones composed of triple-helical collagen-like sequences ([4] and Chapter 9). Many other physical networks are now known and studied, for example poly(vinyl chloride) gels, polyethylene gels and many types of polysaccharide-based gels (Chapter 9).

The formaldehyde-based resins and the polyester resins are examples of the formation of covalent polymer networks by non-linear polymerizations. In such polymerizations, at least one of the reactants must have more than two reactive groups. That is, one reactant must be polyfunctional rather than difunctional (see [7] and Chapter 2). With only difunctional reactants, linear polymers are formed. Thus, hydrated formaldehyde ($HO–CH_2–OH$) has effectively two reactive –OH groups, phenol is effectively trifunctional as the hydrogens in the *ortho* and *para* positions to the phenolic –OH group react with the –OH groups of the hydrated formaldehyde, and urea ($NH_2–CO–NH_2$) is effectively tetrafunctional because all the –H atoms can react [6]. Phthalic anhydride (difunctional in the acid form) and glycerol (trifunctional in –OH) form the basic ingredients of alkyd resins – polyester networks with trifunctional junction points. Modification by the incorporation of difunctional and monofunctional carboxylic acids allows the concentration of junction points to be changed and properties varied.

Up to the present day, the use of many different chemical reactions to produce a wide variety of networks by non-linear polymerizations has developed, just as in the synthesis of linear polymers. For example, difunctional or polyfunctional isocyanates are reacted with polymeric and oligomeric diols, triols or tetrols, separately or in admixture, to form polyurethane networks, and with diamines and triamines to form polyurea networks (see, e.g., Chapters 2, 5 and 8). Again, junction-point concentration can be varied by using polyols or polyamines of different molar masses. Epoxy resins involve curing (network-forming) reactions between epoxy groups

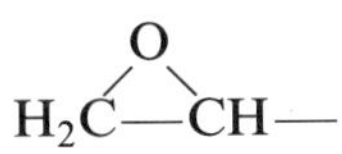

and, often, amine hardeners [8].

In addition to the formation of networks by non-linear polycondensations and polyadditions, non-linear chain polymerizations with unsaturated monomers are also used, in particular those employing mixtures of vinyl and divinyl monomers (Chapter 3). The vinyl monomer is effectively a difunctional reagent and the divinyl monomer a tetrafunctional reagent. The basis of many ion-exchange resins are networks formed by reacting together styrene (difunctional) and *p*-divinyl benzene and subsequently modifying the polymer with ionic groups [9].

The remaining two historical examples of network materials, wool and the waterproofed fabric of Macintosh, are examples of composite materials. Wool fibres have complicated structures composed of keratins, proteins with disulfide crosslinks at cystine units [4]. The thermal lability of the crosslinks is important for the setting of wool (hair) and textiles. The fibrils within a fibre are partially crystalline and are in a cystine-rich, network matrix. The waterproofed fabric is an example of a laminate composite involving a polymer network. The use of networks to make structural, composite materials is now widespread [6, 9]; for example, phenol-formaldehyde resins with webs of cloth or glass, or with wood, as in plywood; unsaturated polyester resins (copolymeric polyester and polystyrene networks) with glass fibres to give fibreglass. Networks are also frequently used with organic and inorganic fillers to produce tough, composite materials. Another type of composite material is the interpenetrating polymer network (IPN), made by preparing two networks, or one network and a linear or branched polymer, in the same volume (Chapter 6).

Obviously, the applications of polymer network materials are multitudinous and it is neither possible nor appropriate to list them all here. Their uses in laminates and composites, as elastomers and structural materials, and as ion-exchange resins, have been indicated. In addition, they find applications, for example [4, 6, 9], as surface coatings (e.g. polyurethane lacquers and polyurethane and alkyd paints), adhesives and sealants (e.g. Araldite epoxy resins, silicone elastomers), cellular materials (e.g. polyurethane, polyurea and polyethylene foams), and as hydrogels and contact lenses (hydrophilic networks).

Products made of network materials have to be manufactured by reactive processing (see Chapters 4 and 5). The reason is that any flow which is required to occur must do so before the point of incipient network (or solid) formation, the gel point. This point is obviously technologically and scientifically important. Reactants for surface coatings are applied by spray or by brush in liquid form and form networks *in situ* by solvent evaporation and atmospheric oxidation. Solvent casting is used for some physical networks (e.g. gelatin). In other cases, the (low viscosity) reactants are mixed as required and react, either *in situ* (compression moulding (CM)) or as they flow into their final position (resin-transfer moulding (RTM) and reaction injection moulding (RIM)). Extrusion and injection moulding are also used (see [4, 10] and Chapter 5).

1.3 Development of the science of polymer networks

The first reported scientific experiments relating to polymer networks concerned the elasticity of natural rubber [3, 7]. Very early experiments by Gough (1805) and Joule (1859), interpreted in terms of the thermodynamics of elastic deformation (Lord Kelvin, 1857), highlighted the exceptional effects of temperature on its stress–strain behaviour and led indirectly to many investigations on the thermodynamics of rubber-like elasticity. The changes on deformation were eventually established as mainly entropic in nature (see [3]), and this behaviour was eventually linked to the deformations of the constituent chains of the network (Meyer, von Susich and Valkó, 1932). Quantitative theories of the stress–strain behaviour of elastomeric networks, the so-called classical theories of rubber-elasticity (see Chapter 7), were developed by Kuhn (1936), Wall (1942), Flory and Rehner (1943), James and Guth (1943) and Treloar (1943). The macroscopic stretching of a network and the resulting deformation of its constituent chains were shown to lead to a reduction in the conformational entropies of those chains and, hence, to a retractive, elastic force. It was also realized (Mullins, 1959) that topological entanglements could effectively produce more junction points and chains, resulting in an increase in the modulus, equivalent to the retractive force per unit area per unit strain.

Independently of the preceding work on the stress–strain behaviour of elastomers, and following earlier work by Flory on the statistics of linear polymerizations (see [7]), advances in polymerization statistics of network-forming polymerizations were made in the early 1940s by Flory [11, 12] and by Stockmayer [13, 14]. These defined, quantitatively, in terms of chemical conversions, the gel point as that point in a polymerization where molecular continuity through the reacting volume first appears. Both non-linear polymerizations and the crosslinking of polymer chains were treated (cf. Chapters 2 and 3). In addition, the growth of a network past the gel point was treated.

The previously mentioned, early work on stress–strain behaviour used mainly traditional elastomers, i.e. those formed by crosslinking existing linear polymer chains. In the 1970s, endlinking non-linear polymerizations were employed to produce so-called model networks, with molecular structures better defined than those of traditional elastomers (see [15]). From the mid-1970s, the classical elasticity theories were further developed (see Chapter 7) by Ronca and Allegra (1975), Flory (1976), Flory and Erman (1982) and Erman and Monnerie (1989). These later theories have resolved the difference between some earlier theories which assumed either 'affine' or 'phantom' molecular chain behaviour in relation to the macroscopic strain. In addition, Edwards and collaborators (1968–86) developed the slip-link theory to account quantitatively for the effects of entanglements. It has been possible, using mainly endlinked networks and the later theories, to improve the quantitative interpretation of stress–strain behaviour (see Chapter 7). However, implicit in all the theories are postulated connections between the molecular deformations of *individual* chains and the macroscopic deformation of a network. Since 1982 [16], neutron scattering has been used to elucidate this connection, but the results are not simply interpreted. In addition, it has usually been assumed that the endlinking polymerizations give perfect networks, with no inelastic chains, so that the initial modulus can be predicted from the structures of the reactants – an assumption that will, eventually, have to be removed (see Chapter 2).

The early Flory–Stockmayer expressions for the gel point have been improved (see Chapter 2) by Frisch (1955), Kilb (1958) and Stepto and collaborators (1974 onwards) to account for intramolecular reaction, leading to improved gel-point predictions. The improvements have made use of the

Jacobson–Stockmayer theory (1950) of ring–chain equilibrium in linear polymerizations to evaluate probabilities of intramolecular reaction. In addition (see Chapter 2), statistical polymerization theories have been developed to account for intramolecular reaction (1962 onwards, Gordon, Dušek and Burchard and collaborators, cascade theory; 1974 onwards, Stepto and Stanford and collaborators, rate theory) and for the possible unequal reactivities of chemically similar reactive groups, due to steric hindrance etc. (Gordon, Dušek and Burchard and collaborators, cascade theory). More recently, Dušek and Somvársky (1985) have developed a combination of kinetics and cascade theories (see Chapter 3) to account for molecular growth in chain polymerizations. Finally, Monte-Carlo simulations are starting to be developed which give more complete descriptions of the molecular species formed in non-linear and crosslinking, network-forming polymerizations (Eichinger and collaborators, 1984 onwards [17]; Somvársky and Dušek, 1994 [18]; Dutton, Stepto and Taylor, 1996 [19]).

The preceding developments in the theory and simulation of network formation mean that connectivities of chains within networks can be described more accurately. Accordingly, the use of such theories in interpreting the mechanical properties of model, endlinked networks by Dušek and collaborators (1977 onwards) and Stepto and collaborators (1979 onwards) have shown that such networks do not have perfect network structures. This conclusion has important consequences for future interpretations of elastomeric properties in terms of the aforementioned theories of rubber elasticity (see Chapter 2).

The viscoelastic properties of polymer networks are at least as important as the elastic properties. A feature which makes them distinct from those of linear polymers is the existence of very long (essentially infinite) relaxation times due to the macroscopic network connectivity (see Ferry [20] and Chapter 8). The effects of entanglements are also important. The latter are presently interpreted in terms of the tube model (Edwards, 1967 [21] and de Gennes, 1971 [22]). In addition, the critical regime around the gel point, where a viscoelastic, liquid reaction-mixture turns into a viscoelastic, solid-like gel is significant, both scientifically and technologically. Following Stauffer [23], it can be interpreted in terms of critical phenomena. In addition, Chambon and Winter [24] have proposed the occurrence of self-similar, viscoelastic behaviour over all distance scales as the rheological definition of the gel point. The loss tangent, $\tan\delta$, is then independent of frequency. The correlation of this definition with the Flory–Stockmayer definition in terms of molecular connectivity is the subject of current research (see Chapter 8).

Viscoelastic properties in the critical, gel regime are important for characterizing physical as well as covalent networks. Physical networks often exist in a swollen (gel) state so that viscous contributions to mechanical behaviour are significant. Studies of well-characterized physical networks have been pioneered by Burchard and by Ross-Murphy (see [25] and Chapter 9). Importantly, Clark and Ross-Murphy [26] have developed a model for elastic behaviour which enables the modulus, at least at small deformations, to be interpreted in terms similar to those conventionally used for covalent networks.

The formation of polymer network materials by non-linear polymerizations has its own characteristics and it has become important to develop chemical-engineering-based parameterizations of the process. Often, in the preparation of networks of high junction-point concentrations, the T_g of the forming network starts to exceed the reaction temperature and diffusion control of the chemical reaction starts to occur. In addition, in network formation, the point of infinite viscosity (the gel point) occurs before complete reaction. Both phenomena mean that the rate of dissipation of the heat of reaction is reduced and temperature increases. Changes in T_g and viscosity are now characterized in terms of the conversion-time-temperature (CTT) and the time-temperature-transformation (TTT) diagrams of, respectively, Williams and collaborators (1982) [27] and Gilham and collaborators (1983) [28] (see Chapter 4). The use of such diagrams has been particularly important for rapid-reacting thermosetting resins formed by CM and RTM and polyurethane and polyurea materials formed by RIM. The RIM process, developed in the 1960s and early 1970s, also requires detailed correlations to be made between reactant flow into the mould, viscosity changes and conversions of reactive groups (see Stanford, Ryan and Elwell [10], Macosko [29] and Chapter 5).

In Section 1.2, the formation of complex (composite) network materials was mentioned. Systematic studies and theoretical interpretations of several types of such materials have increased greatly in recent years. IPNs were introduced by Millar (1960) [30] and developed by Shibayama and Suzuki [31], Lipatov and Seregeva [32], H.L. Frisch, Klempner and K.C. Frisch [33] and Sperling and Friedman [34]. A theoretical understanding of their formation and properties is emerging (see Chapter 6). Newer topics are the control of phase separation in rubber-toughened network composites (Pascault and collaborators [35]). Also, Dušek [36], Rozenberg [37] and Bowman [38] are investigating heterogeneities developing during network formation. The *in situ* production of inorganic fillers during network formation has been developed by Mark and Wen [39], and liquid-crystal networks are being investigated (Mormann and Zimmermann [40]).

There are important areas of research into gels. Hydrogels have numerous applications, as soft contact lenses (invented by Wichterle, 1959), as membranes and in drug delivery [4]. The close control of their structure and properties is important. Interesting phenomena, related to the rapid swelling of gels and phase transitions in gels have been reported by Tanaka and collaborators (1984) [41]. The exploitation of these and related phenomena for thermomechanical and electromechanical gel devices is actively being explored, e.g. [42, 43]. Finally, the formation and properties of microgels are being explored and studied systematically (e.g. Funke [44]; Graham and Hayes [45]). Microgels, colloidal-sized particles, have been known for some years and form when intramolecular reaction dominates in localized spatial regions of network-forming polymerizations. They can arise due to back-biting reactions (see Chapter 3) or at sufficient dilutions of reaction systems of reactants of relatively low functionalities.

1.4 Polymer networks: principles of their formation, structure and properties – scope of the book

It is evident that the science of polymer networks no longer involves just rubber elasticity, gelation and network-formation statistics, as it did in 1953 when Flory wrote his definitive book [7]. To the author's knowledge, the present book is the first to take the reader from the science of polymer networks of that time to a structured account of the much broader-based subject of the present day. By design, and also because of the current breadth of the subject, only principles of general applicability are covered. To this end, considerations are limited in most chapters to results and theories for model systems, giving spatially homogeneous network formation and networks although, in Chapters 5 and 6 on reactive processing and IPNs, considerations of phase-separated materials are of prime importance. Subjects in relatively early stages of development, such as heterogeneities occurring during network formation, microgels and liquid-crystal networks, are not treated; nor are composite network materials, of which the properties often relate more to those of composites than those of networks.

The book leads the reader from network formation through to structure and properties. The close connections which exist in polymer network materials between reactant structures and formation conditions and network properties are emphasized wherever possible. In addition, theory is linked with experiment. Apart from Chapter 9, all the chapters focus principally on covalent networks. This emphasis does not signify that physical networks are less important. Rather, from a general scientific and technological viewpoint, the opposite is true. However, it is a reflection of the fact that the chemical complexity of many physical networks has resulted in a relatively slow progress towards general laws.

Chapter 2 (Stepto) considers the statistics of formation of networks through non-linear polycondensations and polyadditions, covering gelation, intramolecular reaction and how reactant structures and formation conditions determine the modulus of a network. Chapter 3 (Dušek) considers the statistics of network formation via the crosslinking of primary chains and chain copolymerizations. Cyclization and diffusion control are discussed. Connections between the network formation by the polycondensations and polyadditions of Chapter 2 and the network formation by the crosslinking of primary chains and chain copolymerizations of Chapter 3 are indicated in both chapters. Chapter 4 (Williams) describes the characterization, parameterization and understanding of network formation through transformation diagrams, describing changes in materials states related to T_g, viscosity, chemical conversion and heat of reaction.

In Chapters 5 and 6 close connections between formation, structure and properties in network materials are evident. Chapter 5 (Stanford) describes the use of reactive processing to prepare network materials, with particular emphasis on RIM. The significance of polymerization statistics, of the thermodynamics of phase separation and of transformation diagrams, and the links which exist between reactant structures and formation conditions and the morphologies and properties of the resulting materials properties are demonstrated. Chapter 6 (Frisch, Du and Schulz) treats IPN materials and details the different classes of IPNs and their methods of formation. The theory of the principles of IPN formation, including phase separation, and the properties of IPN materials in relation to formation conditions, the structures of their component polymers and phase morphology are described.

Chapters 7, 8 and 9 treat the elastic and viscoelastic properties of polymer networks as they exist. Chapter 7 (Mark and Erman) gives a comprehensive account of molecular theories of rubberlike elasticity, from their early beginnings to their present state of development and also their application to experimental results. Chapter 8 (Ilavsky) treats the viscoelastic behaviour of network materials, presenting a general description of the viscoelastic properties of polymers and how those of network and linear polymers are related. It also gives an account of viscoelastic behaviour in the critical, gel regime, relating theory with experiment. Chapter 9 (Ross-Murphy) deals with physical polymer networks. Different classes of this type of network are defined and both synthetic and biopolymer gels are considered. The formation of physical networks and models for interpreting their properties are discussed.

1.5 Terminology and nomenclature

The continuing developments in the subject of polymer networks from its early beginnings to the present time mean that no standard, accepted terminology and nomenclature exist. IUPAC and those involved in polymer networks and related subjects have recently attempted to start to rectify this situation, and a document, the *Source-Based Nomenclature for Non-Linear Macromolecules and Macromolecular Assemblies*, has been published [46].

For many of the quantities used in the present book, no standard symbols yet exist. Hence, the symbols and abbreviations in the individual chapters are those favoured by the individual authors, and a separate glossary of symbols and abbreviations has been compiled for each chapter. However, terms used in the text have, as far as possible, been made consistent with the aforementioned IUPAC document [46]. Particularly noteworthy, for example, are the terms 'crosslink' and 'junction point'. The former now has a more restricted meaning, signifying a branch point formed by reactions involving sites or groups on *existing* macromolecules or by interactions between *existing* macromolecules. Because of the importance of clear nomenclature and terminology, particularly in a developing subject, and because of its relatively recent publication, the text of the IUPAC document is included as an appendix to the present book.

Glossary of symbols and abbreviations

T_g	glass–rubber transition temperature
$\tan\delta$	loss tangent
CM	compression moulding
IPN	interpenetrating network
RIM	reaction injection moulding
RTM	resin-transfer moulding

References

1. Glasstone, S. (1955) *Textbook of Physical Chemistry*, 2nd edn, Macmillan, London.
2. Fisher, H.L. (1957) *Chemistry of Natural and Synthetic Rubbers,* Reinhold Publishing Corporation, New York.
3. Treloar, L.R.G. (1975) *The Physics of Rubber Elasticity*, 3rd edn, Clarendon Press, Oxford.
4. *Encyclopedia of Polymer Science and Engineering* (1990) (ed. board H.F. Mark, N.M. Bikales, C.G. Overberger and G. Menges), John Wiley and Sons, New York.
5. Hardie, D.W.F. and Davidson Pratt, J. (1966) *A History of the Modern British Chemical Industry*, Pergamon Press, Oxford.
6. Golding, B. (1959) *Polymers and Resins. Their Chemistry and Chemical Engineering*, van Nostrand Reinhold, New York.
7. Flory, P.J. (1953) *Principles of Polymer Chemistry*, Cornell University Press, Ithaca.
8. Skeist, I. (1958) *Epoxy Resins*, Reinhold Publishing Corporation, New York.
9. Winding, C.C. and Hiatt, G.D. (1961) *Polymeric Materials*, McGraw-Hill, New York.
10. Stanford, J.L., Ryan, A.J. and Elwell, M.J.A. (1997) in *Processing of Polymers* (ed. H. Meijer), VCH, Weinheim, Chapter 9.
11. Flory, P.J. (1941) *J. Amer. Chem. Soc.*, **63**, 3083, 3091, 3097.
12. Flory, P.J. (1946) *Chem. Revs.*, **39,** 137.
13. Stockmayer, W.H. (1943) *J. Chem. Phys.*, **11**, 45
14. Stockmayer, W.H. (1944) *J. Chem. Phys.*, **12**, 125.
15. Mark, J.E. (1982), *Adv. Polymer Sci.*, **44**, 1.
16. Beltzung, M., Picot, C., Rempp, P. and Hertz, J., (1982) *Macromolecules*, **15**, 1594.
17. Leung, Y.-K. and Eichinger, B.E. (1984) *J. Chem. Phys.*, **80**, 3877, 3885.
18. Somvársky, J. and Dušek, K. (1994) *Polymer Bull.*, **33**, 369, 377.
19. Dutton, S., Stepto, R.F.T. and Taylor, D.J.R. (1996) *Angew. Makromol. Chem.*, **240**, 39.
20. Ferry, J.D. (1980) *Viscoelastic Properties of Polymers,* 3rd edn, John Wiley and Sons, New York.
21. Edwards, S.F. (1967) *Proc. Phys. Soc.*, **92**, 9.
22. de Gennes, P.-G. (1971) *J. Chem. Phys.*, **55**, 572.
23. Stauffer, D. (1976) *J. Chem. Soc., Faraday Trans. 2*, **72**, 1354.

24. Winter, H.H. and Chambon, F. (1986) *J. Rheol.*, **31**, 683.
25. Burchard, W. and Ross-Murphy, S.B. (eds) (1990) *Physical Networks – Polymers and Gels*, Elsevier Applied Science, London.
26. Clark, A.H. and Ross-Murphy, S.B. (1987) *Adv. Polymer Sci.*, **83**, 57.
27. Adabbo, H.E. and Williams, R.J.J. (1982) *J. Appl. Polymer Sci.*, **27**, 1327.
28. Enns, J.B. and Gilham, J.K. (1983) *J. Appl. Polymer Sci.*, **28**, 2567.
29. Macosko, C.W. (1989) *RIM: Fundamentals of Reaction Injection Moulding*, Hanser, Munich.
30. Millar, J.R. (1960) *J. Chem. Soc.*, 1311.
31. Shibayama, K. and Suzuki, Y. (1967) *Rubber Chem. Tech.*, **40**, 476.
32. Lipatov, Yu. S. and Sergeeva, C.M. (1967) *Russ. Chem. Rev.*, **45**, 1, 63.
33. Frisch, H.L., Klempner, D. and Frisch, K.C. (1969) *J. Polymer Sci., Polymer Phys. Ed.*, **7**, 11, 775.
34. Sperling, L.H. and Friedman, D.W. (1969) *J. Polymer Sci. A-Z*, **7**, 425.
35. Pascault, J.-P. (1995) *Macromol. Symp.*, **93**, 45.
36. Dušek, K. (1986) *Adv. Polymer Sci.*, **78**, 4.
37. Rozenberg, B.A. and Irzhak, V.I. (1995) *Macromol. Symp.*, **93**, 227.
38. Bowman, C.N. and Ansett, K.S. (1995) *Macromol. Symp.*, **93**, 269.
39. Mark, J.E. and Wen, J. (1995) *Macromol. Symp.*, **93**, 89.
40. Mormann, W. and Zimmermann, J. (1995) *Macromol. Symp.*, **93**, 97.
41. Hirokawa, Y. and Tanaka, T. (1984) *J. Chem. Phys.*, **81**, 6379.
42. Kishi, R., Ichijo, H. and Hirasa, O. (1993) *J. Intellig. Mat. Syst. & Struct.*, **4**, 534.
43. Osada, Y. and Ross-Murphy, S.B. (1993) *Scientific American*, **268**, 82.
44. Funke, W. (1988) *J. Coat. Tech.*, **60**, 69.
45. Graham, N.B. and Hayes, C.M.G. (1995) *Macromol. Symp.*, **93**, 293.
46. Kahovec, J., Kratochvil, P., Jenkins, A.D., Mita, I., Paperov, I.M., Sperling, L.H. and Stepto, R.F.T. (1997) *Pure. Appl. Chem.*, **69**, 2511.

2 Non-linear polymerization, gelation and network formation, structure and properties

R.F.T. STEPTO

2.1 Introduction

This chapter considers non-linear polymerizations through the reaction of functional groups, the condition for gelation, or incipient network formation, and the structures, or chemical connectivities, and resulting properties of the networks formed at the ends of polymerizations. The story is not yet complete. However, there are important general relationships between (i) reactant structures (molar mass, chemical functionalities f, chain stiffness) and reaction conditions (dilution and ratio of reactive groups) and (ii) the gel point, network structure and network properties, that can be defined. Such relationships enable more accurate predictions of gel points and network properties, such as rubbery modulus and T_g, to be made, directly from the initial, reaction formulations used. They also provide a deeper molecular understanding of the gel point and of network behaviour.

The chemical reactions used are often the conventional condensation and addition reactions met in linear polycondensations and polyadditions to form, for example, polyesters, polyamides, polyurethanes or polysiloxanes. The difference is that at least *one* of the reactants carries more than two reactive groups, that is, $f > 2$. This property leads to branched molecular structures and, under conditions which can be defined precisely for idealized polymerizations, to network formation, namely, macroscopic molecular growth limited only by the size of the reaction volume. The values of f of the reactants are usually less than about 10, although not necessarily so. Often the reactive groups are at the ends of chains or subchains, giving rise to the term 'endlinking' (polymerization).

Endlinking may be contrasted with crosslinking, referring to reactions linking together existing, primary polymer chains to form branched molecules and eventually networks. Such processes are treated in Chapter 3. The principal distinction in comparison with the present types of non-linear polymerization is the effectively much higher values of f, in terms of numbers of pendant reactive groups or reactive sites (e.g. double bonds) on the primary chains. High values of f lead to gelation at low conversions of the reactive groups or sites and often to relatively high amounts of cyclization (intramolecular reaction). Endlinking and crosslinking processes are illustrated in Figure 2.1. Each crosslink chemically produces two trifunctional junction points, which, because of

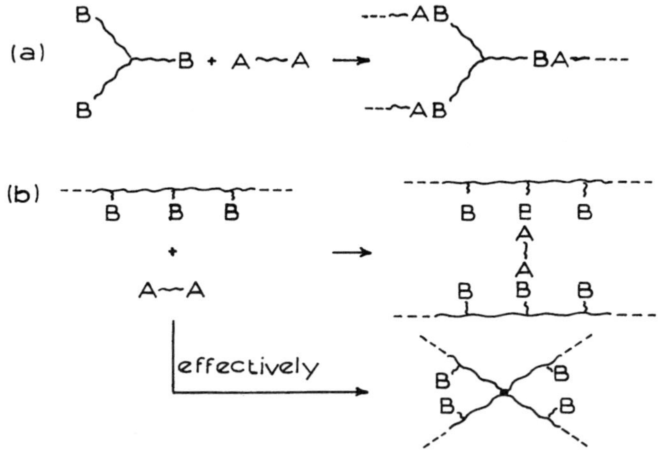

Figure 2.1 (a) Endlinking via an $RA_2 + R'B_3$ polymerization; (b) crosslinking – one crosslink usually gives effectively one tetrafunctional junction point [1]. (Reprinted with permission.)

the short length of the crosslinking molecule compared with chemical chain lengths between pairs of reacted groups along the primary polymer chains, are considered as one tetrafunctional junction [1].

Network formation by non-linear free-radical chain polymerizations (Chapter 3) is distinct from non-linear polycondensations and polyadditions. First, propagation often leads to species of high functionalities being formed, which in turn can lead to significant cyclization and possibly microgel formations. The Trommsdorff effect can also become important.

Gelation and network formation from endlinking non-linear polycondensations and polyadditions can result in elastomeric networks (Chapter 7), in vitrification (Chapter 4) and also in microgel formation (see later and [2]). However, many polymerization formulations exist in which random (intermolecular) reaction between reactive groups means that reaction proceeds homogeneously throughout the volume of the polymerizing mixture. Such polymerizations have gel points which may be predicted with reasonable accuracy and may be used to form model (but not perfect) elastomeric networks, whose properties can be predicted and interpreted in molecular terms with more certainty than those prepared via the crosslinking of polymer chains. The present chapter emphasizes such polymerizations.

The types of network materials which result from the polymerizations to be considered can range from elastomers of low modulus to glassy thermosets, depending, as mentioned, on reactant structures and reaction conditions. Reactants of high molar masses, low functionalities (values of f) and flexible chains lead generally to materials of low modulus as they result in network structures with low concentrations of junction points and flexible, elastic chains between them. Reactants of lower molar masses, higher functionalities and stiff chains generally lead to materials of high modulus, often glasses at ambient temperatures, because of the high concentrations of junction points and the stiff chains between them. Thus, the basic science to be developed is key to understanding the formation, structure and properties of, for example, silicone elastomers and polysulfide sealants, on the one hand, and phenol-formaldehyde and epoxy resins on the other hand.

In addition, *en route* to forming a final network material, the gel point (the point of incipient network formation) is traversed [1, 3]. This point is important both fundamentally and technologically. The fundamental importance is discussed in detail in following sections. Technologically, after the appearance of infinite species, the viscosity of a polymerizing mixture becomes infinite and it starts to develop an elastic modulus. From the gel point onwards, a liquid-to-solid transformation starts to occur with the solid (network) fraction increasing as the reaction proceeds. Thus, in processing, liquid flow is no longer possible after the gel point. Such phenomena are of prime importance in the reactive processing of network-forming polymerizations, as discussed in detail in Chapters 4 and 5. In reaction injection moulding and in the reactive processing of thermosets generally [4–6], any transport into a mould with simultaneous reaction of the reactants has to occur before the gel point.

2.2 Idealized gelation

2.2.1 Molecular basis of gelation

During the formation of covalent networks, via endlinking or crosslinking, there are innumerable molecular structures which form and eventually join together to give the completed network. The structures become innumerable because as highly branched molecules grow they eventually acquire 'infinite' numbers of reactive groups or sites which can react further. The term 'infinite' implies limited only by the macroscopic amounts of reactants used in the polymerization.

The growth with degree of polymerization of the number of reactive groups on a molecule is illustrated in Figure 2.2 with respect to an $RA_2 + R'B_4$ polymerization. Generalizing the figure, it can be seen that, for an $RA_2 + R'B_f$ polymerization, a molecule with n branch points has $n(f-2)+2$ reactive groups [1, 3]. Thus, for $f > 2$, a non-linear polymerization, the number of reactive groups increases with the number of branch points. The increase, linked with the assumption of equal reactivity of like groups, means that more complex molecules react in preference to simpler molecules, leading to infinite or network species before complete reaction.

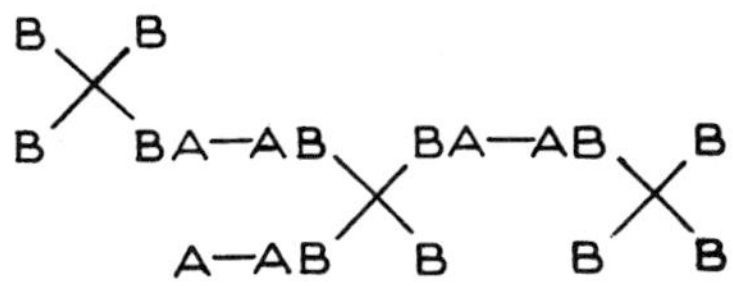

Figure 2.2 A molecule, in an $RA_2 + R'B_4$ polymerization, with three $R'B_4$ units. It has eight reactive groups [1]. (Reprinted with permission.)

The point in a non-linear polymerization where infinite, or unlimited, species first occur is the gel point.

2.2.2 Flory–Stockmayer (F-S) gel points

The molecular changes underlying the phenomenon of gelation in polymerizations were first described quantitatively by Flory [3, 7, 8] and by Stockmayer [9, 10]. The gel point is that point in a polymerization where continuing structures first occur with unit probability. The statistical considerations are illustrated in Figure 2.3 for an $RA_4 + R'B_3$ polymerization. Imagine A^1 is a randomly chosen A group. A^2 is a group which is statistically equivalent to A^1 and could have been chosen instead [1]. The chain from A^1 to A^2 defines the basic repeat structure of continuing chains. The probability that a chain continues from A^1 to A^2 becomes equal to unity at the gel point. That is, from that point onwards in the polymerization some infinite structures must exist.

For an $RA_{fa} + R'B_{fb}$ polymerization the F-S gel point is simply [7–10]

$$(f_a - 1)p_a(f_b - 1)p_b = 1 \qquad (2.1)$$

where p_a and p_b are the extents of reaction of A and B groups, namely, the fractions of the two types of groups which have reacted. The assumptions leading to equation (2.1) are that intramolecular reaction does not occur and that all A and all B groups have the same probabilities of being reacted, namely p_a and p_b. In the sense of a polycondensation or polyaddition, proceeding by random (intermolecular) reactions of pairs of groups, such polymerizations may be termed *random polymerizations* [11, 12]. The factor $(f_a - 1)$ in equation (2.1) is the number of paths which can lead from a randomly chosen A group on an RA_{fa} unit to an adjoining $R'B_{fb}$ unit and

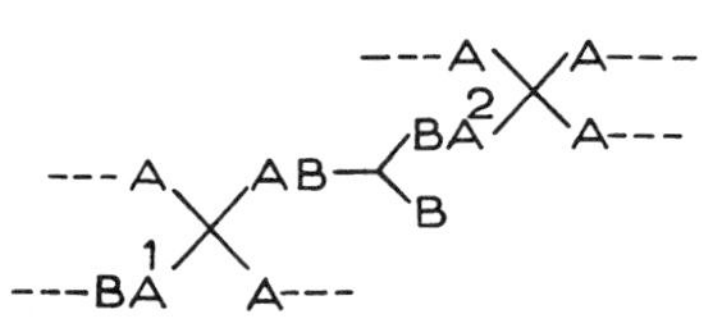

Figure 2.3 Depiction of the repeat structures ($A^1 \rightarrow A^2$) of continuing chains in an $RA_2 + R'B_3$ polymerization [1]. (Reprinted with permission.)

p_a is the probability per path. Similarly, $(f_b - 1)$ is the number of paths which can lead from the $R'B_{fb}$ unit and p_b is the probability per path.

Subject to the same assumptions, Stockmayer [13] treated the general $\sum_i RA_{fai} + \sum_j R'B_{fbj}$ polymerization of mixtures of reactants of different functionalities, with some reactants bearing A groups and some reactants bearing B groups, and A groups reacting only with B groups. The reactant mixture is N_{a1} moles of $RA_{fa1}, \ldots, N_{ai}$ moles of $RA_{fai}, \ldots$ reacting with N_{b1} moles of $R'B_{fb1}, \ldots, N_{bj}$ moles of $R'B_{fbj}, \ldots$ Following a generalization of the statistical reasoning leading to equation (2.1), the gel-point condition becomes [13, 14]

$$(f_{aw} - 1)p_a \cdot (f_{bw} - 1)p_b = 1 \tag{2.2}$$

p_a and p_b are overall extents of A and B groups and f_{aw} and f_{bw} the so-called mass-average (weight-average) functionalities of the A- and B-bearing reactants, with

$$f_{aw} = \frac{\sum_i N_{ai} f_{ai}^2}{\sum_i N_{ai} f_{ai}} = \frac{\sum_i N_{ai} f_{ai} \cdot f_{ai}}{\sum_i N_{ai} f_{ai}} \tag{2.3}$$

and

$$f_{bw} = \frac{\sum_j N_{bj} f_{bj}^2}{\sum_j N_{bj} f_{bj}} = \frac{\sum_j N_{bj} f_{bj} \cdot f_{bj}}{\sum_j N_{bj} f_{bj}} \tag{2.4}$$

In fact, the terminology 'mass-average functionality' is misleading. f_{aw} and f_{bw} are really functionalities averaged according to numbers of functional groups, and are so expressed in the second equalities in equations (2.3) and (2.4). In the derivation of equation (2.2), when randomly choosing an A group, for example, the reactant to which it belongs contains f_{ai} groups with probability [14]

$$\frac{N_{ai} f_{ai}}{\sum_i N_{ai} f_{ai}}$$

Not all mixtures of $\sum_i RA_{fai}$ and $\sum_j R'B_{fbj}$ reactants will gel. If r_a is the initial molar ratio of A groups to B groups, then

$$r_a = \frac{n_{a0}}{n_{b0}} = \frac{p_b}{p_a} \tag{2.5}$$

The latter equation follows because one A group reacts with one B group, so that $n_{a0} \cdot p_a = n_{b0} \cdot p_b$. Hence, for $r_a > 1$ the maximum value of $p_b = 1$ and that of $p_a = 1/r_a$. Substitution of these two limiting conditions in equation (2.2) shows that gelation can only occur for [3, 15]

$$(f_{aw} - 1)(f_{bw} - 1) \geq r_a \geq (f_{aw} - 1)^{-1}(f_{bw} - 1)^{-1} \tag{2.6}$$

The gel-point condition, equation (2.2), includes equation (2.1) as a special case. It also encompasses the gel-point expressions derived separately and more laboriously by Flory [3, 7] for $RA_2 + R'A_3 + R''B_3$ polymerizations and the crosslinking of polymer chains. In the former case, $f_{bw} = f_b = 3$ and f_{aw} is evaluated from the relative amounts of RA_2 and $R'A_3$ reactants used. In the latter case, the crosslinking molecules may be considered difunctional (equivalent to RA_2 reactants) and the chains to be crosslinked often bear one reactive group or site (e.g. double bond) per structural unit. Thus, $f_w(=f_{bw}) = x_w$, where x_w is the mass-average degree of polymerization (dp). Hence, directly from equation (2.2),

$$(x_w - 1)\rho = 1 \tag{2.7}$$

at gel, where ρ is the fraction of units crosslinked. The term ρ is statistically equivalent to $p_a p_b$, although the chemistry of crosslinking reactions usually means that crosslinker molecules always react at both ends. Equation (2.7) is in fact the same as equation (3.4). That equation is derived using cascade theory and, in the idealized case of random polymerizations, cascade theory and F-S statistics yield equivalent results. Equations (2.7) and (2.2) emphasize the fact that gelation occurs very early in reactions involving reactants of high functionalities.

Another basic type of network-forming polymerization is the $\sum_i RA_{fai}$ self-polymerization, where A groups react with each other as, for example, in the preparation of polysiloxanes. The condition for gelation in this case is

$$(f_w - 1)p_a = 1 \tag{2.8}$$

Often the reactants used for preparing network materials have like functional groups of intrinsic or induced unequal reactivities. Intrinsic unequal reactivity can occur, for example, when both primary and secondary hydroxy groups are present on reactants. Induced unequal reactivities (or the substitution effect) can occur when the similar groups on one molecule depend on how many groups on the molecule have already reacted. This can be, for example, for reasons of increased steric hindrance. The evaluation of gel points in simple polymerizations involving like groups having intrinsic unequal reactivities has been treated by Case [16] and others [14]. The evaluation of gel points for a number of polymerizations having induced unequal reactivities of like functional groups has been carried out by Dušek, Ilavsky and co-workers and others [17–29] using cascade theory, as developed for polymerizations by Gordon [30]. Cascade theory is particularly suitable for treating induced unequal reactivities.

Recently, gelation in other types of non-linear polymerizations has received attention. Rolfes and Stepto [15] have evaluated gel-point expressions for polymerizations of the general type $\sum_i RA_{fai} + \sum_j RB_{fbj}$ in which A groups react with A *and* B groups, and B groups react *only* with A groups. Such polymerizations occur in the formation of phenol-formaldehyde resins and in the preparation of polysiloxanes through the hydrolysis of chlorosilane monomers. An expression, bridging equation (2.2), representing no A + A reaction, and equation (2.8), representing no A + B reaction, results, namely

$$\frac{p_b p_{ab}(f_{bw} - 1)(f_{aw} - 1)}{1 - p_{aa}(f_{aw} - 1)} = 1 \tag{2.9}$$

where p_{ab} is the fraction of A groups that has reacted with B groups and p_{aa} is the fraction that has reacted with other A groups.

The syntheses of hyperbranched polymers from reactants of the type ARB_{f-1}, where A groups react only with B groups, have recently gained in importance. It was shown many years ago by Flory [3] that, because of the molar imbalance of A and B groups, gelation occurs only at complete reaction of the A groups. The gelation condition is

$$p_b(f - 1) = 1 \tag{2.10}$$

and equation (2.5), which still applies, gives

$$r_a = (f - 1) = p_b/p_a \tag{2.11}$$

Hence, equation (2.10) can only be satisfied when $p_a = 1$. Thus, hyperbranched polymers are not network polymers.

Finally, what may be termed triplet polymerizations are being used to prepare network materials, in particular polyisocyanurates [31] and polycyanurates [32]. The general scheme of the polymerization reaction for bifunctional reactants forming a polycyanurate is shown in Figure 2.4 [33]. The general case of f-functional reactants RA_f polymerizing through reactions of *triplets* of functional groups was studied some years ago by Fukui and Yamabe [34], who showed that

$$p_a 2(f - 1) = 1 \tag{2.12}$$

at gelation. The case shown in Figure 2.4 is for $f = 2$ so that $p_a = 0.5$ at gel.

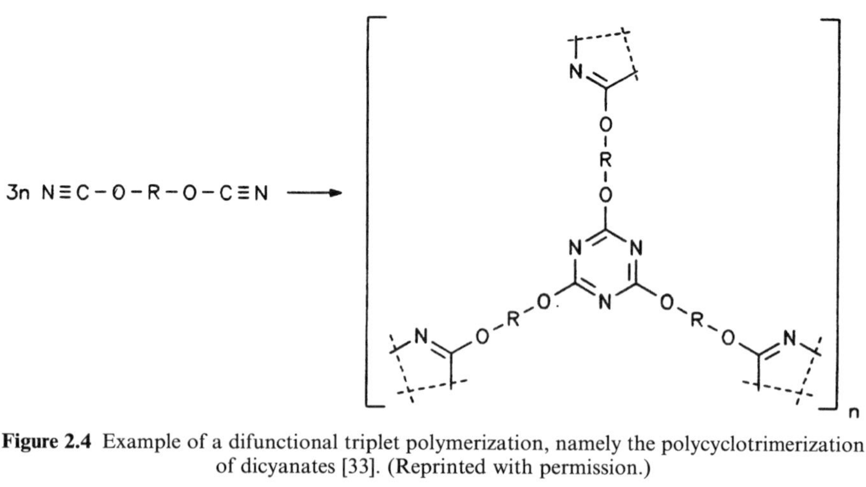

Figure 2.4 Example of a difunctional triplet polymerization, namely the polycyclotrimerization of dicyanates [33]. (Reprinted with permission.)

2.2.3 *Degree-of-polymerization (dp) averages at gelation*

As may be expected, the occurrence of infinite species at the gel point is accompanied by a divergence of the molar mass distribution or, more fundamentally, the dp distribution, in that the mass average and higher averages become infinite. The expression for the number-average dp for the general $\sum_i \mathrm{RA}_{fai} + \sum_j \mathrm{R'B}_{f\mathrm{b}j}$ polymerization is simply derived, independently of the distribution:

$$x_\mathrm{n} = \frac{\text{number of units}}{\text{number of molecules}} = \frac{\sum xN_x}{\sum N_x} \tag{2.13}$$

where N_x is the number of molecules of degree of polymerization x. In the absence of intramolecular reaction, the number of molecules is the initial number (i.e. the number of units) less the number lost (i.e. the number of A groups or B groups reacted $= n_{\mathrm{a}0}\,p_\mathrm{a} = n_{\mathrm{b}0}\,p_\mathrm{b}$). Hence,

$$x_\mathrm{n} = \frac{N_\mathrm{a} + N_\mathrm{b}}{N_\mathrm{a} + N_\mathrm{b} - n_{\mathrm{a}0}\,p_\mathrm{a}} \tag{2.14}$$

Using equation (2.5), equation (2.14) reduces to

$$x_\mathrm{n} = \frac{1}{1 - p_\mathrm{a}/(1/f_\mathrm{an} + 1/f_\mathrm{bn}r)} \tag{2.15}$$

where f_an and f_bn are the number-average functionalities of the A-bearing and B-bearing reactants:

$$f_\mathrm{an} = \frac{\sum N_{\mathrm{a}j} f_{\mathrm{a}i}}{\sum N_{\mathrm{a}i}} = \frac{n_{\mathrm{a}0}}{N_\mathrm{a}} \tag{2.16}$$

and

$$f_\mathrm{bn} = \frac{\sum N_{\mathrm{b}j} f_{\mathrm{b}i}}{\sum N_{\mathrm{b}j}} = \frac{n_{\mathrm{b}0}}{N_\mathrm{b}} \tag{2.17}$$

The expression for x_w, which requires an expression for the distribution of species, is due to Stockmayer [13], namely

$$x_\mathrm{w} = \frac{\sum x^2 N_x}{\sum xN_x} = 1 + \frac{p_\mathrm{a}\,p_\mathrm{b} f_\mathrm{an} f_\mathrm{bn}[p_\mathrm{a}(f_\mathrm{aw} - 1) + 2]}{(p_\mathrm{a} f_\mathrm{an} + p_\mathrm{b} f_\mathrm{bn})[1 - p_\mathrm{a}\,p_\mathrm{b}(f_\mathrm{aw} - 1)(f_\mathrm{bw} - 1)]} \tag{2.18}$$

Comparison with equation (2.2) shows that $x_\mathrm{w} \to \infty$, yet x_n is finite at the gel point.

For a $\sum_i \mathrm{RA}_{fai}$ self-polymerization, the expressions for x_n and x_w are

$$x_\mathrm{n} = \frac{1}{1 - (f_\mathrm{an} \cdot p_\mathrm{a}/2)} \tag{2.19}$$

and

$$x_\mathrm{w} = \frac{1 + p_\mathrm{a}}{1 - p_\mathrm{a}(f_\mathrm{aw} - 1)} \tag{2.20}$$

In this case, the number of molecules lost is one-half of the number of A groups reacted.

An early criterion for gelation proposed by Carothers [35] that $x_n \to \infty$ at gel is unfortunately still often quoted in the literature. It is based on the erroneous precept that all the units in a polymerizing mixture are connected to form an infinite tree at the gel point. Thus, according to equation (2.19), for an RA_f polymerization, gelation occurs when $p_a = 2/f$. On average, this corresponds to just enough groups on each unit reacting for all units to be linked together to give only one (≈ 0) molecule(s). Similarly, in a stoichiometric $RA_{fa} + RB_{fb}$ polymerization mixture ($r = 1$), equation (2.15) would give [1] $p_a = p_b = (1/f_a + 1/f_b)$, again the condition for the polymerizing mixture to connect into a single, infinite tree. These conditions for gelation are inconsistent with the random reaction of groups as they imply that groups so organize themselves that only two per original reactant molecule react.

The distinction between the condition $x_n \to \infty$ and $x_w \to \infty$ in fact illustrates well the phenomenon of gelation [1] (cf. Section 2.2.1). The condition $x_n \to \infty$ means that the average dp per molecule is infinite, because the number of molecules reduces to zero. On the other hand,

$$x_w = \frac{\sum x \cdot xN_x}{\sum xN_n} = \sum x \cdot u_x \tag{2.21}$$

where

$$u_x = \frac{xN_x}{\sum xN_x} \tag{2.22}$$

is the unit fraction of x-mer. Hence, $x_w \to \infty$ means that the average dp per unit is infinite. That is, some units belong to infinite species.

2.2.4 *Perfect network formation*

The classical Flory–Stockmayer treatment of the gel point and the accompanying changes in distributions of species give a basic explanation of the phenomena to which the behaviour and changes in actual polymerizations may be related. The equations in the preceding section relate to changes from the beginning of a polymerization up to the gel point. As discussed in detail by Flory [3], the infinite species which occur from the gel point to complete reaction cannot be enumerated as individual molecules. The sum of reactant units which have joined together to form infinite species is the gel fraction, namely the unit fraction [1] (or approximately mass fraction) of gel, u_g. The gel fraction increases monotonically from zero at the gel point until complete reaction. For stoichiometric formulations, $u_g = 1$ at complete reaction. The complementary quantity, the unit fraction of sol u_s, can be evaluated directly by summing over the units in finite species knowing the dp distribution. For example, for an RA_f self-polymerization,

$$u_s = \sum_{x=1}^{x} u_x = 1 - u_g \tag{2.23}$$

where u_x is the fraction of reactant molecules (or units) in the species of degree of polymerization x. (For an RA_f polymerization $u_s = w_s$, the so-called mass or weight fraction of sol, provided all units have the same mass. The equality neglects any loss of small molecules resulting from the reaction of functional groups. Generally, for any type of polymerization it is u_s, not w_s, which is the more fundamental quantity.) Although x_w overall tends to infinity at the gel point and beyond, the sum in the numerator of equation (2.21) converges provided it is restricted to species with enumerable numbers of unreacted groups, namely $x(f-2)+2$ for degree of polymerization x (cf. Section 2.2.1).

The detailed expression for u_x is

$$u_x = \frac{x(1-p_a)^2}{p_a} \cdot \frac{f[x(f-1)]!}{[x(f-2)+2]!\,x!} \cdot \beta^x \tag{2.24}$$

where

$$\beta = p_a(1-p_a)^{f-2} \tag{2.25}$$

The combinatorial factor relates to the number of distinguishable isomers of molecules of x units with $x(f-2)+2$ unreacted ends. The expression for u_s is

$$u_s = \frac{(1-p_a)^2 p_a^*}{(1-p_a^*)^2 p_a} \tag{2.26}$$

where p_a ($0 \leq p_a \leq 1$) is the extent of reaction and p_a^* is the lowest value of p_a which satisfies equation (2.25) for a given value of β.

The key to the difference between species pre-gel and post-gel is β; $d\beta/dp_a = 0$ at $p_a = 1/(f-1)$ defines a maximum value. Thus, $p_a^* \leq 1/(f-1)$, and as $p_a \to 1$, $p_a^* \to 0$. The reversion in the values of β from the gel point onwards means that the distribution of finite species also reverts and conjugate points (p_a and p_a^*) exist, pre-gel and post-gel, which have identical distributions. The behaviour can be illustrated by combining equations (2.24) and (2.26) to give the unit fraction (u_x') of species within the sol only with

$$u_x' = \frac{u_x}{u_g} = \frac{x(1-p_a^*)^2}{p_a^*} \cdot \frac{f[x(f-1)]!}{[x(f-2)+2]!\,x!} \cdot \beta^x \tag{2.27}$$

Thus, at extent of reaction p_a post-gel, $\beta = p_a(1-p_a)^{f-2} = p_a^*(1-p_a^*)^{f-2}$ and the distribution is the same as that which occurred at p_a^*, pre-gel. Concurrently, equation (2.26) shows that $u_s = 1$ pre-gel ($p_a^* = p_a$), and decreases from the gel point onwards to zero at $p_a = 1$ ($p_a^* = 0$).

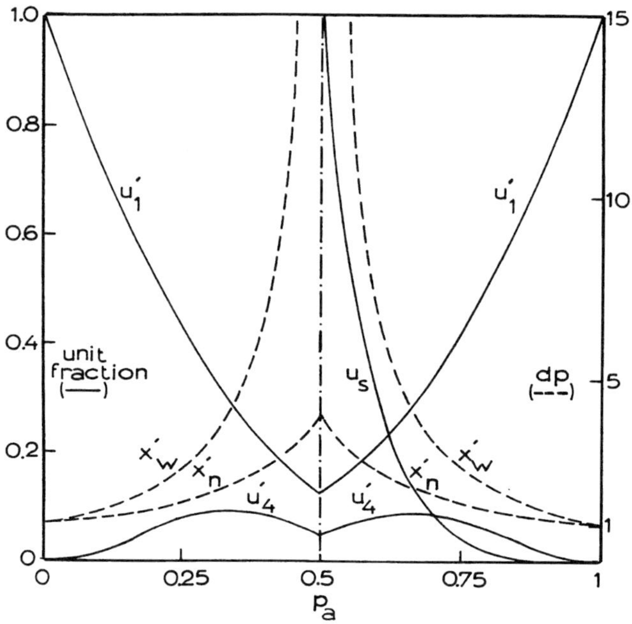

Figure 2.5 Illustrating the reversion of the distribution of sol species after the gel point in an RA_3 polymerization. Curves for u_1' and u_4' follow equation (2.27) and x_n' and x_w' follow equations (2.19) and (2.20) with p_a^* written for p_a. The curve for u_s is given by equation (2.26): (——) unit fractions, left-hand ordinate; (– – –), degree of polymerization, right-hand ordinate [1]. (Reprinted with permission.)

The behaviour is illustrated in Figure 2.5 for an RA_3 polymerization [1, 3]. In this case $p_a^* = (1 - p_a)$ for $p_a > 1/(f - 1) = 0.5$ and the reversion is symmetrical about p_a at gel. For higher values of f, reversion is unsymmetrical about p_a at gel, but the distributions at $p_a > 1/(f - 1)$ map onto those at p_a^*, with p_a and p_a^* evaluated numerically through equation (2.25).

Reversion is a direct result of the unequal probability of growth of molecular species arising from the equal reactivity, or random reaction, of like functional groups, with only intermolecular reaction occurring between finite species. As discussed previously, the larger species have more unreacted groups and, hence, react with higher probabilities. As a polymerization proceeds past the gel point, more and more units are consumed by the gel species and the finite species remaining become less and less, and contain increasing fractions of smaller molecules. For stoichiometric $\sum_i RA_{fai} + \sum_j R'B_{fbj}$ polymerization mixtures and for self-polymerizations, the vanishing sol fraction eventually contains only monomeric reactants and the gel fraction becomes the network which is formed by a polymerization.

For $RA_2 + RB'_{fb}$ polymerizations, expressions of the same forms as in equations (2.19) to (2.27) apply for u_x', x_n', x_w' and u_s, provided these quantities are defined on the basis of the branch units (RB'_{fb}) only, irrespective of how many RA_2 are in a given molecule, and p_a is replaced by $\alpha = p_a p_b$. Free RA_2 monomer is then excluded from considerations. Expressions for complete molecular species, including the counting of RA_2 units, are available, but are more complex and are not often needed. It is the branch-point or complexity distribution which is of prime importance [3].

Flory–Stockmayer theory says nothing concerning the detailed topology of the network, which grows and defines its structure through the random reaction of its reactive groups with other groups on the gel and with groups on sol species. To obtain a perfect network, all reactions (sol–sol, sol–gel and gel–gel) are assumed to yield elastically active chains between junction points in the final network. This assumption is rarely true and is discussed in detail in Section 2.4.

If a perfect network structure is assumed to be formed then the value of the molar mass of the elastically active chain connecting a pair of junction points, M_c, can be calculated directly from the reactant structures, taking account of unreacted groups for nonstoichiometric reaction mixtures. Simplified formulae for such calculations and for the calculations of sol and gel fractions have been given by Miller and Macosko [36, 37]. It is also possible to calculate the same quantities directly from the distribution of molecular species and the fractions of unreacted groups on sol and gel molecules [38].

The detailed relationship between M_c and reactant structures depends on reactant architectures as well as extents of reaction and it is not possible to give a completely general formula. For example, comb-like reactants with pendant reactive groups give loose ends, which are connected to only single branch points, and two types of chains between junction points, those involving sections of the backbone chains and those involving side chains. However, formulae for M_c from most stoichiometric endlinking polymerizations at complete reaction can be derived relatively simply assuming the perfect network structure and relating it to the structures of the reactants from which it is formed [1]. In such networks there are no loose ends. The quantity M_c is related to the molar concentration and number of moles of chains connecting pairs of junction points, n_c and N_c, respectively, by the relationships

$$n_c = \frac{N_c}{W_{net}} \cdot \frac{W_{net}}{V_{net}} = \frac{1}{M_c} \cdot \rho \tag{2.28}$$

where W_{net} and V_{net} are the mass and the volume of network, respectively, and ρ is the density of the network. It is assumed that any sol fraction has been removed. For networks formed from several reactants $1/M_c$ is an average value, denoted $\langle 1/M_c \rangle$, and may be related to reactant structures after factorization of the expression for $\langle 1/M_c \rangle$ in equation (2.28) to give

$$\left\langle \frac{1}{M_c} \right\rangle = \frac{N_c}{N_J} \cdot \frac{N_J}{W_{net}} \tag{2.29}$$

where N_J is the number of junction points in the network.

For a $\sum_i \mathrm{RA}_{fai}$ polymerization

$$\frac{N_\mathrm{c}}{N_\mathrm{J}} = \frac{\sum' N_{\mathrm{a}i} f_{\mathrm{a}i}}{2\sum' N_{\mathrm{a}i}} = \frac{f'_{\mathrm{n,J}}}{2} \tag{2.30}$$

where $\sum'$ denotes summation over chains and junction points, excluding RA_2 units ($f_{\mathrm{a}i} = 2$), and $f'_{\mathrm{n,J}}$ is the number-average functionality of the junction points or the branched reactants (compare equation (2.16)). Secondly,

$$\frac{N_\mathrm{J}}{W_\mathrm{net}} = \frac{\sum' N_{\mathrm{a}i}}{\sum' N_{\mathrm{a}i} M_{\mathrm{a}i}} \cdot \frac{\sum' N_{\mathrm{a}i} M_{\mathrm{a}i}}{\sum N_{\mathrm{a}i} M_{\mathrm{a}i}} \tag{2.31}$$

that is

$$\frac{N_\mathrm{J}}{W_\mathrm{net}} = \frac{1}{M_\mathrm{n,br}} \cdot w_\mathrm{br} \tag{2.32}$$

where $M_\mathrm{n,br}$ is the number-average molar mass of the branched reactants and w_br is the mass fraction of branched reactants. The $M_{\mathrm{a}i}$ in equation (2.31) are the molar masses of the reactant molecules as they occur in the network (i.e. excluding any small molecule lost in the reaction A + A). Combination of equations (2.29), (2.30) and (2.32) shows that

$$\left\langle \frac{1}{M_\mathrm{c}} \right\rangle = \frac{f'_{\mathrm{n,J}}}{2} \cdot \frac{w_\mathrm{br}}{M_\mathrm{n,br}} \tag{2.33}$$

If no linear reactants are used and the junction-point functionalities are all the same ($= f$) then the usual expression

$$\left\langle \frac{1}{M_\mathrm{c}} \right\rangle = \frac{1}{\langle M_\mathrm{c} \rangle} = \frac{f}{2} \cdot \frac{1}{M_\mathrm{n}} \tag{2.34}$$

is obtained, with M_n the number-average molar mass of the reactants. For $\sum_i \mathrm{RA}_{fai} + \sum_j \mathrm{R'B}_{fbj}$ networks, $\langle 1/M_\mathrm{c} \rangle$ is still given by equation (2.33), with $f'_{\mathrm{n,J}}$, $M_\mathrm{n,br}$ and w_br evaluated over both A-bearing and B-bearing species.

Equation (2.30) holds generally for any network structure with no loose ends, so that *concentrations* of chains and junctions, n_c and n_J, are related by

$$n_\mathrm{c} = \frac{f'_{\mathrm{n,J}}}{2} \cdot n_\mathrm{J} \tag{2.35}$$

or, from the generalized form of equation (2.28), and equations (2.29) and (2.32),

$$n_\mathrm{c} = \left\langle \frac{1}{M_\mathrm{c}} \right\rangle \rho = \frac{f'_{\mathrm{n,J}}}{2} \cdot \frac{w_\mathrm{br}}{M_\mathrm{n,br}} \cdot \rho \tag{2.36}$$

showing that

$$n_\mathrm{J} = \frac{w_\mathrm{br}}{M_\mathrm{n,br}} \cdot \rho \tag{2.37}$$

For branched reactants of a single functionality f, equation (2.35) gives

$$n_c = \frac{f}{2} \cdot n_J \tag{2.38}$$

Relationships between concentrations of chains and junction points which assume perfect network structures, such as equations (2.35) and (2.38), are often used when interpreting elastic properties of endlinked networks, as in Chapter 7. All branch points of functionality f are assumed to be elastically active junction points from which $f/2$ elastically active chains emanate. In this case, one may replace the symbols n_c and n_J by $n_{e,c}$ and $n_{e,J}$, denoting concentration of elastically active chains between pairs of junction points and the concentration of elastically active junction points. In practice, deviations from these assumptions occur and the values of $n_{e,c}$ or $n_{e,J}$ deduced from elastomeric properties are rarely those expected from the amount of chemical reaction which has occurred. Such deviations may be due to topological entanglements and chain interactions [39–43], to side reactions, incomplete reaction in endlinking polymerizations (giving loose ends) [44, 45] and, more fundamentally and generally, inelastic chain or loop formation due to the intramolecular reaction of pairs of groups [17–29, 43, 46–73].

It is obviously a key matter in predicting and interpreting the elastomeric properties of networks to be able to calculate the value of $n_{e,J}$ from the reactants and reaction conditions used. This topic is discussed in detail in the following sections in relation to covalent endlinked networks. Calculations of $n_{e,J}$ for networks from crosslinking processes and for physical networks are more difficult (see Chapters 3 and 9).

2.3 Intramolecular reaction and gelation

2.3.1 General characterization of intramolecular reaction

There are two fundamental assumptions of Flory–Stockmayer statistics as applied to linear and to gelling and network-forming polymerizations – the random reaction of pairs of functional groups and the exclusion of intramolecular reaction between finite (sol) species [3].

Random reaction requires that like groups must be intrinsically equally reactive, a condition which can be met by the choice of suitable reactants, and secondly, that all groups are uniformly accessible to all other groups in the polymerization. Thus, diffusion control cannot occur, nor intramolecular reaction. For finite species, on which the number of reactive groups is negligible compared with the total number of groups in a polymerization, random reaction means that intermolecular reaction will dominate, unless there are close spatial relations between pairs of reactive groups on individual molecules. However, in a non-linear polymerization, the number of groups per molecule increases as the polymerization proceeds and, if gelation occurs, then species form which have unlimited numbers of groups. In Flory–Stockmayer statistics the groups on such species still undergo random reaction but because there are so many groups on the gel molecule some of that reaction is intramolecular. The author has evaluated [38] the amount of post-gel intramolecular reaction in stoichiometric RA_f and $RA_2 + R'B_f$ polymerizations according to Flory–Stockmayer statistics. For example, for RA_4 polymerizations, 50% is intramolecular gel–gel reaction and for $RA_2 + R'B_2$ polymerizations the amount is 25%, corresponding, respectively, to the cycle rank and one half of the cycle rank of the networks. It is such reaction which gives the network its final structure. What may be termed the perfect network results for stoichiometric formulations, if the intramolecular reaction all leads to elastic chains (Section 2.4).

The neglect of intramolecular reaction between finite species in polymerizations is a fundamental assumption of Flory–Stockmayer statistics. It has been shown experimentally [74] that for linear polymerizations in bulk, in which each molecule has only two reactive groups, the assumption is usually justified. However, for non-linear polymerizations, the increasing number of reactive groups per molecule together with the spatial correlations between groups on the same molecule mean that intramolecular reaction cannot generally be neglected. This point is illustrated by comparison of number fractions of rings species as functions of extent in Figure 2.6 for linear ($RA_2 + R'B_2$) and non-linear ($RA_2 + R'B_3$) bulk polymerizations [50].

The ring structures formed by intramolecular reaction cause gelation to be delayed, because they reduce the probability of chains continuing to infinity. Thus, the extent of reaction at the gel point for the non-linear polymerization in Figure 2.6 was 0.765 rather than 0.707, as predicted by equation (2.1) for an $RA_2 + R'B_3$ polymerization [75]. In addition, some of the pre-gel and post-gel intramolecular reaction will lead to inelastic loops in the final network. It is the evaluation and prediction of inelastic loops which is the key to the characterization and prediction of network structures and, hence, their physical properties.

The first treatment of the intramolecular reaction occurring during random polymerizations was due to Jacobson and Stockmayer [76] who treated linear ring–chain equilibria. The treatment has been applied extensively by Semlyen and co-workers for the analysis of ring-chain equilibriates and the preparation of macrocyclics [77]. However, as with the data in Figure 2.6, most polymerizations yielding high molar-mass polymers are carried out under essentially irreversible conditions. The balance between intermolecular and intramolecular reaction is then altered and changes as a polymerization proceeds.

The basic parameter characterizing intramolecular reaction, introduced by Jacobson and Stockmayer [76] and by Haward [78] for radical polymerizations, is $P(\mathbf{0})$, the value of the probability density function of the

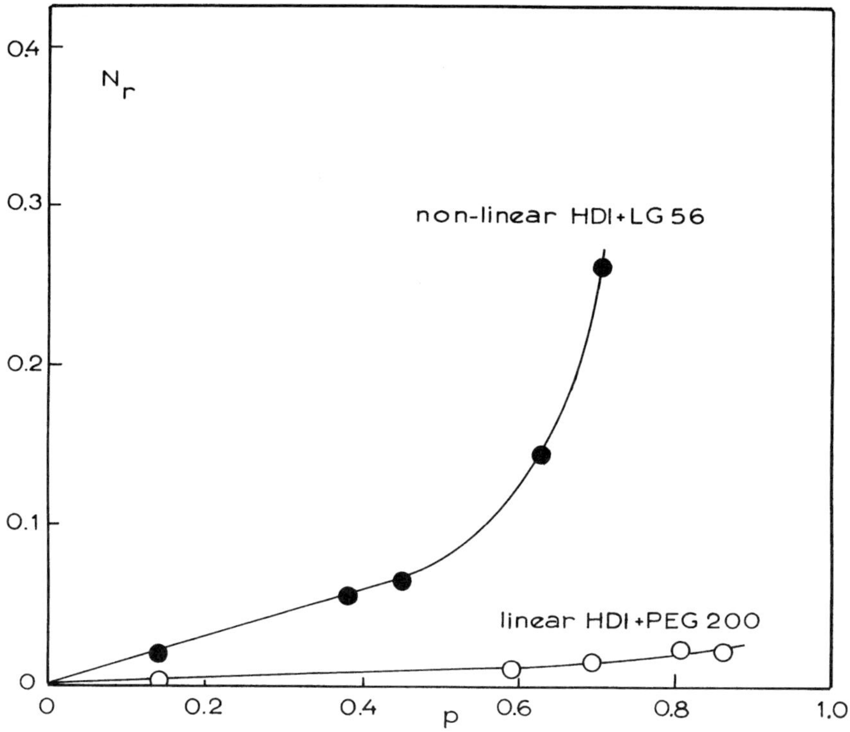

Figure 2.6 Number of ring structures per molecule N_r as a function of extent of reaction p ($= p_a = p_b$) for bulk, linear and non-linear, stoichiometric polyurethane-forming reactants ($r_{NCO} = [NCO]_0/[OH]_0 \approx 1$) [50]: (○) linear, hexamethylene diisocyanate (HDI) + poly(oxyethylene) triol (PEG200) at 70 °C, $[NCO]_0 = 5.111\,\text{mol kg}^{-1}$, $[OH]_0 = 5.188\,\text{mol kg}^{-1}$, number of bonds in the chain forming the smallest ring structure $\nu = 25.2$; (●) non-linear; HDI + poly(oxypropylene) (POP) triol (LG56) at 70 °C, $[NCO]_0 = 0.9073\,\text{mol kg}^{-1}$, $[OH]_0 = 0.9173\,\text{mol kg}^{-1}$, $\nu = 115$. (Reprinted with permission.)

end-to-end vector of a chain $P(\boldsymbol{r})$, for $\boldsymbol{r} = \boldsymbol{0}$, namely, the probability density or concentration of coincident reactive groups. Assuming a Gaussian form for $P(\boldsymbol{0})$, one may write

$$P(\boldsymbol{0}) = \left(\frac{3}{2\pi\langle r_o^2\rangle}\right)^{3/2} \Big/ N_{\text{Av}} \quad (\text{mol vol}^{-1}) \tag{2.39}$$

where N_{Av} is the Avogadro constant and $\langle r_o^2\rangle$ is the mean-square end-to-end distance of the chain. In the context of a polymerization, only certain sizes of ring form, as defined by the functionalities and molar masses of the reactants. This is illustrated in Figure 2.7 for $RA_2 + R'B_2$ and $RA_2 + R'B_4$ polymerizations. If the smallest possible chains forming ring structures contain ν skeletal bonds then (to within about one bond) only rings formed from chains with $\nu, 2\nu, \ldots, i\nu$ skeletal bonds can occur. Thus, assuming

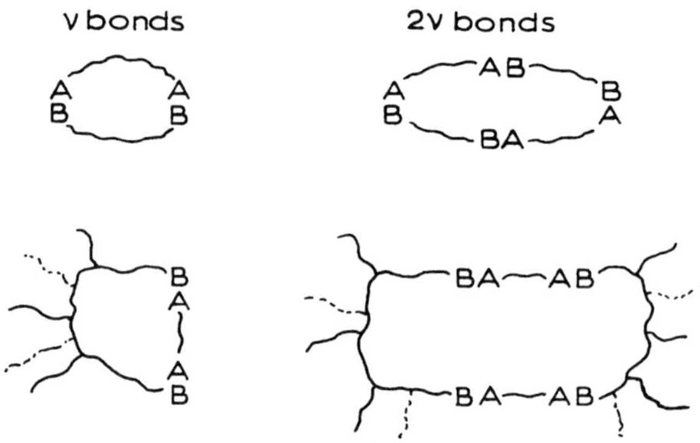

Figure 2.7 Illustrating the two smallest ring structures of ν and 2ν bonds in $RA_2 + R'B_2$ and $RA_2 + R'B_f$ polymerizations [1]. (Reprinted with permission.)

independent Gaussian subchain statistics, the single parameter for the smallest ring structure,

$$P_1 = \left(\frac{3}{2\pi\nu b^2}\right)^{3/2} \Big/ N_{Av} \tag{2.40}$$

is sufficient to characterize intramolecular reaction for all sizes of ring structure. In equation (2.40)

$$\langle r_o^2 \rangle = \nu b^2 \tag{2.41}$$

where b is the effective bond length of the chain of ν bonds. To account for excluded volume, $\langle r_o^2 \rangle$ in equation (2.39) may be replaced by $\langle r^2 \rangle$. However, this is not normally necessary unless diluted reaction systems are of interest and ν is greater than about 100. Chains of $i\nu$ skeletal bonds have a concentration of coincident ends equal to

$$P_i = \left(\frac{3}{2\pi i\nu b^2}\right)^{3/2} \Big/ N_{Av} = P_1 i^{-3/2} \tag{2.42}$$

The competition between the intermolecular and intramolecular reaction of a molecule which occurs during a polymerization is characterized by P_1, by the number and positions of pairs of groups on the molecule which can react intramolecularly, and by the concentrations of groups from other molecules. The situation is illustrated in Figure 2.8. The internal concentrations, namely the total mutual concentration due to all pairs of groups on the given molecule, is c_{int}. It may be expressed by the equations

$$c_{int} = \sum c_{int,i} = \sum_i n_i P_i \tag{2.43}$$

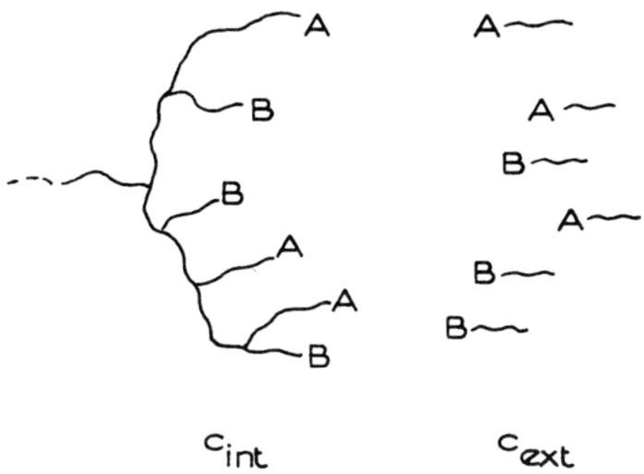

Figure 2.8 Illustrating the competition between intramolecular and intermolecular reaction in an A + B polymerization: c_{int} is the internal concentration of a given molecule (i.e. total mutual concentration due to all pairs of A and B groups on the molecule); c_{ext} is the external concentration due to A and B groups on other molecules [1]. (Reprinted with permission.)

where $c_{int,i}$ is the internal concentration due to all pairs of groups which could form ring structures of size i, and n_i is the number of such pairs. Depending upon how a polymerization theory or model is constructed, c_{int} can also refer to a given group rather than a given molecule. The quantity n_i is then the number of groups which can react intramolecularly with the given group. In either case, the general evaluation of n_i for highly branched and gel species is not feasible and theories of non-linear polymerization, gelation and network formation which include intramolecular reaction essentially make different approximations for the n_i.

Intramolecular reaction is governed by c_{int} and intermolecular reaction by the external concentration c_{ext}, which is the total concentration of A and B groups from the molecules which can react with the B and A groups on the given molecule. To a good approximation c_{ext} may be expressed simply as

$$c_{ext} = c_a + c_b = c_{a0}(1 - p_a) + c_{b0}(1 - p_b) \tag{2.44}$$

where c_{a0} and c_{b0} are the initial concentrations, and c_a and c_b the instantaneous concentrations. Equation (2.44) assumes that intermolecular reaction occurs by the random reactions of pairs of groups (as in Flory–Stockmayer statistics). That is, diffusion effects which would restrict the movement of molecules do not dominate. It also assumes that the number of groups on an individual molecule is not a significant proportion of $c_a + c_b$. This assumption, which has yet to be examined critically, will become less valid as complete reaction is approached.

The combination of equations (2.43) and (2.44) defines the ring-forming parameter

$$\lambda = \frac{c_{int}}{c_{int} + c_{ext}} \tag{2.45}$$

which is the probability of intramolecular versus intermolecular reaction. An alternative parameter is

$$\lambda' = \frac{c_{\text{int}}}{c_{\text{ext}}} = \frac{\lambda}{1 - \lambda} \tag{2.46}$$

The parameters λ and λ' will vary with molecular structure, through c_{int} and, with the overall concentration of reactive groups, through c_{ext}. The variation with molecular structure is reflected through the n_i, defined by the graph topology of the molecule, and through P_1. The latter quantity depends, through $\langle r_o^2 \rangle$, on molar mass ν and chain stiffness b. The dependence on c_{ext} means that λ and λ' increase as a polymerization proceeds and c_{ext} decreases, and that they are larger for diluted systems.

Parameters equivalent to λ and λ' are used in all those theories and models of polymerizations that express intramolecular reaction by taking the random reaction of groups (Flory–Stockmayer statistics) as a basis and modify probabilities of reaction due to the spatial correlations which must exist between reactive groups on the same molecule. For example, equivalent parameters are used in approaches based on differential equations – kinetics theories, cascade theory [20] and rate theory [14] – and on Monte-Carlo simulations [62, 72]. In addition, although the factors influencing intramolecular reaction have been discussed here in the context of endlinking polymerizations, similar considerations apply to the crosslinking of polymer chains (cf. Section 3.3.1).

The interplay of factors which determine λ or λ' may be illustrated by reference to analysis of the data in Figure 2.6. Because of the lower equivalent masses of the reactants and their more flexible chains, ν and b^2 for the linear polymerization are less than those for the non-linear polymerization. Hence, c_{int} per opportunity for intramolecular reaction is larger for the linear polymerization. Again, because of lower equivalent masses, $c_{a0} + c_{b0}$ is larger for the linear polymerization. The ratio of the initial values of λ' is

$$\frac{\lambda_{0,\text{l}}}{\lambda_{0,\text{nl}}} = \frac{P_{1,\text{l}}}{(c_{a0} + c_{b0})_{\text{l}}} \cdot \frac{(c_{a0} + c_{b0})_{\text{nl}}}{P_{1,\text{nl}}} \approx 1.7 \tag{2.47}$$

where l denotes linear and nl non-linear. Thus, if the same number of opportunities per molecule for intramolecular reaction existed in both polymerizations, larger ring fractions would occur in the linear polymerization. That the opposite effect occurs is *directly* attributable to the much larger average number of opportunities per molecule for intramolecular reaction (that is, larger values of n_i in equation (2.43)) in the gelling system. This number is the dominating factor and emphasizes the importance of trying to take account of the complex structures which occur in non-linear polymerizations.

Figure 2.6 shows results from polymerizations in bulk. In both the linear and non-linear cases, dilution of the same reactants increases intramolecular reaction [73, 74], as expected from the decreases in c_{ext}. The resulting ring-fraction data on the linear polymerizations at various dilutions have been interpreted satisfactorily in terms of a variety of theoretical approaches, kinetics theory, cascade theory, rate theory and modified Jacobson–Stockmayer theory [12].

The corresponding non-linear ring-fraction data and associated gel-point data [74] have not been fitted completely satisfactorily using cascade theory [79], rate theory [79] or Monte-Carlo simulations [80], indicating that approximations in the theories, as applied to non-linear polymerizations, can be significant (Section 2.4.2). In this respect, it should be noticed that the use of total ring-fraction data together with gel-point data provides an extreme test of a theory.

It is evident from Figure 2.6 that a decrease in c_{ext} with dilution does not mean, conversely, that intramolecular reaction is absent in polymerizations in bulk. Reactive groups are still diluted by the inert units which connect them. Hence, the maximum values of c_{a0} and c_{b0} are always limited by the molar masses per reactive group (the equivalent masses). The limited values result in compensating effects on the amounts of intramolecular reaction which occur in related polymerizations in bulk, using reactants which differ only in molar masses. Decreases in molar masses mean increases in $c_{a0} + c_{b0}$ and decreases in ν. Hence, both c_{int} and c_{ext} increase and the values of λ' at the start of a reaction can remain approximately constant. Thus, intermolecular reaction will not necessarily be reduced if reactant molar masses are increased.

2.3.2 *General survey of theories and models of polymerizations*

Theories and models of polymerizations mentioned previously – cascade theory [17–28], rate theory [48, 53–58] and Monte-Carlo simulations [62, 66–73, 80, 81] – have been developed to account for intramolecular reaction occurring alongside the random intermolecular reaction of pairs of groups, particularly in non-linear polymerizations. They can also account for the unequal reactivities of like functional groups. Most have been described adequately elsewhere and only brief descriptions of their bases are given here.

Cascade and rate theory consider in detail the reactions between subsets of significant structures in a polymerization and treat only approximately intramolecular reaction in larger species. Cascade theory was developed originally by Gordon [30] and applied particularly by Dušek, Ilavsky and co-workers [17–28, 81, 82] to a variety of polymerizations, including the crosslinking of polymer chains and chain crosslinking (co)polymerizations (Chapter 3). It is based on the direct application of branching theory to define the probability of chains continuing from randomly chosen reactant units. A key approximation made is that only units up to one unit (generation) away from a chosen unit are considered in detail and the relative probabilities of intramolecular reaction to give ring structures of different sizes are essentially equated to P_i (equation (2.42)), neglecting any variation in n_i (equation (2.43)) with i. This so-called spanning-tree approximation means that only two probability generating functions (pgfs) are required for each reactant species. The zeroth-order pgf defines probabilities of different numbers of groups on a chosen unit being intermolecularly reacted, intramolecularly reacted and unreacted, and the first-order pgf defines the corresponding probabilities on units known to be attached to the chosen unit. Unreacted groups turn into intermolecularly and intramolecularly reacted groups as a polymerization proceeds, and this process is described by numerical solutions to a set of simultaneous differential equations describing the rates of change of the pgfs, starting from the initial condition that all groups are unreacted. The number of ring structures is equal to the number of pairs of groups reacted intramolecularly and the gel point is related to the first-order generating functions. Beyond the gel point the first-order pgf is used to define, in terms of an extinction coefficient, the fractions of intermolecularly reacted groups which give chains continuing indefinitely and those which do not. Only those junctions (branched units) which have three or more chains continuing indefinitely are elastically active in the final network. Dušek *et al.* [82] have improved the spanning-tree approximation by using simultaneous kinetics rate equations to trace the detailed interconversion of local structures and then cascade theory to combine these structures. A similar approach has also been used by Samoria, Vallés and Miller [83].

Rate theory was developed by the author and collaborators [11, 12, 14, 48, 53–58, 61]. In the context of a polymerization, it uses F-S statistics to express, with respect to extent of reaction (rather than time), the probabilities of existence of subsets of structures based on the reactants. The rates of interconversion of the structures are expressed by independent analytical expressions. The resulting set of independent (rather than simultaneous) differential equations is extended to account for all the pairs of groups which could react intramolecularly. Fractions of the rates corresponding formally to λ (equation (2.45)) are used to give rates of intramolecular reaction with rates of intermolecular reaction described by $1 - \lambda$. Within the structures in a subset, all the n_i (equation (2.43)) are counted correctly. In comparison with cascade theory, the connectivity of a larger number of reactant units is accounted for locally, but outside the subset all reactions are assumed to be intermolecular. Although the rate equations can in principle be solved analytically, numerical integration is normally used to avoid the derivation of cumbersome expressions. The subset of structures contains unreacted, intermolecularly reacted and intramolecularly reacted groups, and sequences of reacted groups which give continuing chains. The number of pairs of intramolecularly reacted groups gives the number of ring structures and the probability of continuing chains defines the gel point. The gel fraction and the number of elastically active junctions in the gel fraction and in the final network are also counted in terms of numbers of continuing chains from units. Rate theory has so far been applied to model, endlinking polymerizations.

In addition to the cascade and rate-theory approaches, it is possible to write systems of simultaneous kinetics equations to describe the interconversion of molecular species including intramolecular reaction, rather than subsets based on the growth from randomly chosen reactant units. Temple [84] has shown that for a non-linear polymerization, the number of simultaneous equations needed for an accurate representation of N_r is excessive. In addition, intramolecular reaction is underestimated compared with cascade theory and rate theory, for the same initial value of the ring-forming parameter λ (equation (2.45)).

A method of Monte-Carlo (MC) simulation for non-linear polymerizations has been developed by Eichinger and co-workers and has been applied to endlinking polymerizations and crosslinking [66–73, 80]. One reactant is designated a polymer. Rotational-isomeric-state (RIS) statistics can be used to define probabilities of separations of the reactive groups on each molecule and, hence, the effective mutual concentrations of pairs of groups which could eventually become parts of ring structures. The other reactant is designated a crosslinker and its molecules are treated as sticks of defined length. Several thousands ($\geq 10\,000$) of reactant molecules are placed at random in a box and distances between all pairs of A and B groups recorded. A capture sphere is defined, and all A,B pairs whose distances of separation fall within a sphere can react. The pairs are allowed to react sequentially in increasing order of their mutual separations and, as crosslinkers react, they are moved to join the groups on the polymer molecule with which they have reacted. A record is kept of the connectivity of all molecules during the linkage process via a spanning-tree algorithm (SPANFO), which can delineate gel and sol components. Structures, such as smallest loops in the gel, double-loop species and various types of dangling ends are also enumerated. A polymerization is followed by increasing the size of the capture sphere in steps. This allows more pairs of groups to react and the extent of reaction to increase. The molecular connectivities and structures are analysed after each step.

The gel point is located as the mean of: (i) the extent of reaction at which the reduced mass-average degree of polymerization, i.e. the average over all species except the largest one, displays a maximum (lower bound); and (ii) the extent of reaction at which the rate of increase of the overall mass-average degree of polymerization is a maximum (upper bound).

As a reaction proceeds and capture spheres increase beyond a certain volume, the occurrence of gel–gel reaction is suppressed in the simulations, but gel–sol and sol–sol reactions are allowed to continue. The value of the capture-sphere radius chosen for the suppression of gel–gel reaction is $2\langle r_o^2\rangle^{1/2}$, where $\langle r_o^2\rangle$ is, for example, the average mean-square end-to-end distance between two reactant groups on the polymer molecule. The value is chosen arbitrarily but is related to the magnitudes of junction-point fluctuations in networks. It was introduced to reflect the possible reduction, imposed by molecular connectivity, in the mutual reactivity of pairs of groups in the gel, compared with the reactivity of pairs of groups of which at least one group is on a sol molecule. It also leads to more efficient computations. Significantly, however, it means that not all the connectivities which can form are accounted for.

The author and collaborators have developed a new MC algorithm for $RA_{fa} + R'B_{fb}$ polymerizations [62]. It is based on the random reaction of pairs of groups for intermolecular reaction, with intramolecular reaction weighted according to statistics and the size of ring structures being formed, namely P_i of equation (2.42). Distinct from the method of Eichinger *et al.*, the reactants are not embedded in a pseudo space. The algorithm is basically a process for recording the reactions between pairs of groups on the same molecule or different molecules and the molecular connectivities so formed. However, the algorithm counts *all* the ring structures formed and enables polymerizations to be simulated efficiently up to complete reaction. Ring species and loops of all sizes are enumerated and predictions of gel points, and of gel fractions and modulus, as the network is formed, can be made.

2.3.3 Gelation

Gelation has been studied mostly using non-linear polymerizations. As mentioned previously, in the crosslinking of polymer chains, the gel point is hardly significant (equation (2.7)). In non-linear polymerizations, with reactants of low functionalities and reaction mixtures usually, initially, of low viscosity, the gel point occurs well after the start of a polymerization and is manifest by a dramatic increase in viscosity and the appearance of a non-zero elastic shear modulus.

The increase in viscosity and the appearance of an elastic modulus are not only important parameters in the processing of gelling polymerizations (see Section 2.1), they are also the phenomena used to determine the gel time [85, 86]. For fundamental studies, the conversion versus time curve for one type of group is determined and extent(s) of reaction at the gel point can be found by extrapolation or interpolation, together with the use of equation (2.5) for $A + B$ reactions. Other studies assume the chemical kinetics are known and use them to evaluate extents of reaction corresponding to the gel time.

Gelation rarely occurs at the F-S gel points. Delayed gelation always occurs because of intramolecular reaction and delayed or early gelation can occur because of the unequal reactivities of like functional groups. There have been a number of applications of cascade theory to interpret gel points in polymerizations, such as polyepoxide- and polyurethane-forming polymerizations [19, 20, 25, 27, 28, 82]. Such studies can give quantitative interpretations of gel points but, if both unequal reactivities and intramolecular reaction are significant, it is possible to obtain the same gel point with different sets of reactivity and intramolecular-reaction parameters.

Intramolecular reaction is the more fundamental and universal phenomenon and, to elucidate unambiguously its effects on the gel point, it is necessary to choose reactants which have like groups with equal reactivities. Polymerizations with such reactants forming polyesters and polyurethanes have been used, particularly by the author and co-workers [12, 46–48, 52, 54–56, 85, 86] and by Dušek and co-workers [17–21, 23, 24, 26].

Rate theory [48, 57] and Monte-Carlo simulations [72, 73] have been used successfully to interpret gel points in polyurethane-forming polymerizations carried out at various initial dilutions using hexamethylene diisocyanate (HDI) and poly(oxypropylene) (POP) triols and tetrols. Excess reaction at gelation has been shown to be consistent with intramolecular reaction superimposed upon random intermolecular reaction. The amounts of intramolecular reaction found experimentally were shown to be consistent with the spatial correlations between reactive groups being governed by chain conformational statistics, as depicted in equations (2.39) to (2.46).

In order to predict and interpret gel points, it is not necessary to use theories and models of complete polymerizations. Gelation theories have been developed. They focus solely on the gel point and cannot be used to evaluate ring fractions (Figure 2.8) nor network defects (discussed later). They are, from the outset, approximate because they essentially assume a constant value of λ (equation (2.45)), which means that some average value of c_{ext} is used. They have the advantage that they can give analytical expressions for some polymerizations. Extensive computation is not required, as for polymerization theories or Monte-Carlo simulations, and predictions of gel points can be just as accurate. The successive development of several expressions over a number of years [87–91] have led to the so-called Ahmad–Rolfes–Stepto (ARS) theory [92] for $\sum_i \mathrm{RA}_{fai} + \sum_j \mathrm{R'B}_{fbj}$ and $\sum_i \mathrm{RA}_{fai}$ polymerizations, the most accurate and general gelation theory. For the former polymerizations, it evaluates the probability of growth between statistically equivalent points B^1 and B^2, in the molecular structure shown in Figure 2.9. It is necessary to use the complex structure shown in order to define a reasonable approximation to the numerous opportunities for intramolecular reaction which exist. The resulting expression for the gel point, defining unit probability of growth from B^1 to B^2, is

$$\begin{aligned} rp_{\mathrm{ac}}^2(f_\mathrm{a}-1)(f_\mathrm{b}-1) = {} & (1+(f_\mathrm{a}-2)\varphi(1,3/2)\lambda_{\mathrm{a}0} \\ & + (f_\mathrm{b}-2)(f_\mathrm{a}-1)r_\mathrm{a}\varphi(1,3/2)\lambda_{\mathrm{a}0}\,p_{\mathrm{ac}}) \\ & \times (1+(f_\mathrm{b}-2)r_\mathrm{a}\varphi(1,3/2)\lambda_{\mathrm{a}0} \\ & + (f_\mathrm{a}-2)(f_\mathrm{b}-1)r_\mathrm{a}\varphi(1,3/2)\lambda_{\mathrm{a}0}\,p_{\mathrm{ac}}) \end{aligned} \tag{2.48}$$

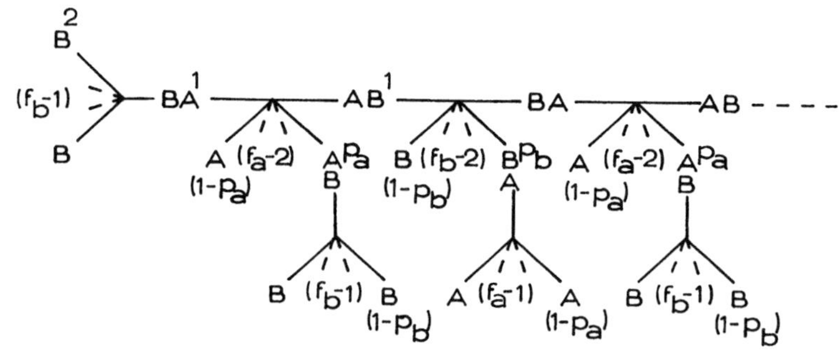

Figure 2.9 Sequence of structural units used to define gelation in an $RA_{fa} + R'B_{fb}$ polymerization in ARS theory. Side chains up to two units long are considered. p_a and p_b are extents of reaction [92].

$$\lambda_{a0} = \frac{P_{ab}}{c_{a0}} \tag{2.49}$$

and

$$P_{ab} = P_1 = \left(\frac{3}{2\pi\langle r^2\rangle}\right)^{3/2} \Big/ N_{Av} \tag{2.50}$$

Here, p_{ac} and p_{bc} are the overall extents of reaction of A and B groups at the gel point. For general mixtures of reactants, f_a and f_b are replaced by f_{aw} and f_{bw}. The ring-forming parameter of the theory is λ_{a0}, defined as the ratio of the concentration of one end of a chain which can form the smallest ring structure around the other end of the same chain (P_{ab}) to the initial concentration of A groups c_{a0}. $\varphi(1, 3/2) = \sum_{i=1}^{\infty} 1^i i^{-3/2} = 2.612$ sums over one opportunity of each size of ring structure. The actual numbers of opportunities are accounted for by the various factors in f_a and f_b in equation (2.48). P_{ab} is equivalent to P_1 of equation (2.40) and is characteristic of the chain forming the smallest ring structure, of mean-square end-to-end distance $\langle r^2\rangle$. It is calculable, independently, from the molar masses, functionalities and chain structures of the reactants (Section 2.2.1 and [1, 59, 61, 92, 93]). When $\lambda_{a0} = 0$ the r.h.s. of equation (2.48) equals 1 and the F-S expression for the gel point is obtained.

Equation (2.48) is quadratic in both p_{ac} and λ_{a0} and may be arranged to solve analytically for either of these quantities [92]. The calculation of p_{ac}, given values of λ_{a0}, allows gel points (values of p_{ac}) to be predicted. This application of equation (2.48) is incorporated into the *Networks Module* of the Polymer Software by Molecular Simulations Incorporated [94]. Alternatively, measured values of p_{ac} can be interpreted in terms of λ_{a0}, derived using equation (2.48).

An illustration of the use of equation (2.48) to interpret gel point data [92, 95] is given in Figure 2.10, for the systems defined in Table 2.1. Two

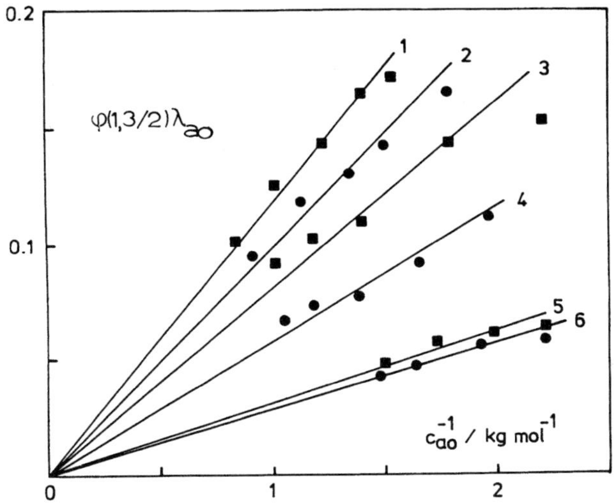

Figure 2.10 Plots of the ARS ring-forming parameter ($\varphi(1,3/2)\lambda_{a0}$) versus initial dilution (c_{a0}^{-1}). Values of $\varphi(1,3/2)\lambda_{a0}$ were evaluated according to equation (2.48) from experimental extents of reaction at gelation [95] for the trifunctional polyester-forming systems 1–6 listed in Table 2.1 ($f_a = 2, f_b = 3, r_a = 1$). Plots were interpreted using equations (2.49) and (2.50) [92].

diacid chlorides (RA_2) and three, star-like polyoxypropylene (POP) triols (LHT240, LHT112 and LG56) ($R'B_3$) were used. Stoichiometric reaction mixtures ($r_a = 1$) were studied and polymerizations were carried out at 60 °C in bulk and in diglyme as solvent. $\varphi(1,3/2)\lambda_{a0}$ was evaluated from measured values of p_{ac} and plotted versus dilution (c_{a0}^{-1}). The F-S gel point is $\lambda_{a0} = 0$ and λ_{a0} (or p_{ac}) increases with dilution. Even in bulk, corresponding to the points at the lowest values of c_{a0}^{-1}, intramolecular reaction occurs, $\lambda_{a0} > 0$ and the gel point is delayed.

The linear plots in Figure 2.10 are consistent with equation (2.48) and their slopes are equal to $\varphi(1,3/2)P_{ab}$ ($= 2.612P_{ab}$), and from equation (2.48)

Table 2.1 Reaction systems (1–6) comprising diacyl (diacid) chlorides and poly(oxypropylene) triols (LHT240, LHT112 and LG56) used for gel-point determinations [95]. Reactions were carried out at 60 °C in bulk and at various dilutions in diglyme: ν, number of skeletal bonds in the chain forming the smallest ring structure; ν_2/ν, fraction of bonds in the chain of ν bonds contributed by the diacid chloride residue; b, effective bond length of the chain of ν bonds, evaluated from the slopes of the plots in Figure 2.10

System	ν	ν_2/ν	b/nm
2. sebacoyl chloride + LHT240	41	0.268	0.382
1. adipoyl chloride + LHT240	37	0.189	0.378
4. sebacoyl chloride + LHT112	70	0.157	0.348
3. adipoyl chloride + LHT112	66	0.106	0.322
6. sebacoyl chloride + LG56	136	0.081	0.319
5. adipoyl chloride + LG56	132	0.053	0.313

should be proportional to $\langle r^2 \rangle^{-3/2}$, characteristic of the chain forming the smallest ring structure. If the chain has ν bonds then, for the star-like triols used, ν is equal to the number of bonds in two arms of a triol and an acid chloride residue [92, 93] and $\langle r^2 \rangle = \nu b^2$ (equation (2.41)), with b the effective bond length. Thus, the slopes in Figure 2.10 decrease as ν, or the combined molar masses of the reactants increases. In addition, Table 2.1 shows that analysis of the slopes gives values of b which decrease as ν_2/ν (the fraction of the chain of ν bonds contributed by the acid chloride residue) decreases. Also the values of b are in reasonable accord with values expected from conformational statistics (compare the values of 0.40 nm and 0.34 nm for infinite, unperturbed poly(methylene) and poly(oxypropylene) chains [92, 93]). Overall, Figure 2.10 and Table 2.1 show that gel-point delay is related to the dilution of reactive groups (c_{a0}^{-1}), to the molar masses of the reactants ν, and to their chain structures b. In addition, there are increases in delays with increased functionality, as shown by studies using triols and tetrols in polyurethane-forming polymerizations [14, 29, 46, 47, 50–52, 54–56, 59–65, 93] (see later; Figure 2.11). The amount of intramolecular reaction will also vary with r_a, although this has still to be studied in detail.

The values of λ_{a0} in Figure 2.10 correspond to values of extent of intramolecular reaction at the gel point,

$$p_{r,c} = p_c - p_c^o \tag{2.51}$$

ranging from about 0.03 to 0.11, depending upon reactants and dilution, with the lowest values occurring for the reactants with the highest molar masses. In equation (2.51), p_c is the experimental value of p_{ac} (or p_{bc}) and p_c^o is the F-S value ($= 1/\sqrt{2}$). In comparison, data for $RA_2 + R'B_3$ stoichiometric, polyurethane-forming polymerizations with ν values of 33 and 61, carried out at different dilutions, have values of $p_{r,c}$ ranging from 0.05 to 0.10. Similar $RA_2 + R'B_4$ polymerizations ($p_c^o = 1/\sqrt{3}$) with values of ν ranging from 29 to 66 give values of $p_{r,c}$ ranging from 0.09 to 0.16. Such amounts of excess reaction at gelation can result in *significant* delays in gel times. In addition, the lowest values quoted come from reactions in bulk, showing that even the natural dilution of reactive groups by inert parts of the reactant molecules can cause intramolecular reaction to occur.

2.4 Network formation, structure and properties

2.4.1 Introduction

The close connection between network structures and modulus G is summarized in the equation

$$G = n_{e,c}RT \tag{2.52}$$

which is valid at low strains (see Chapter 7). The term G is the shear modulus measured at zero frequency and low strains, and $n_{e,c}$ was defined and discussed in Section 2.2.4. In endlinking polymerizations, if each pair of groups reacting leads only to linear chains between junction points, then the perfect network structure is formed. In the absence of physical interactions or topological entanglements between chains, concentrations of elastic chains and junctions ($n_{e,c}$ and $n_{e,J}$) can be calculated from the molar masses and functionalities of the reactants (equations (2.35) to (2.38)). However, if intramolecular reaction forms a significant number of inelastic chains and junctions then $n_{e,c}$ and $n_{e,J}$ will be reduced below the values expected, and materials of lower moduli and probably lower T_gs will result. Thus, the first problem to be addressed is the evaluation of $n_{e,c}$ and $n_{e,J}$ for given reactants and polymerization conditions. Having established the structure of a network in this way, measured moduli can be used to elucidate any effects (e.g. additional modulus or increased T_g) due to interactions or entanglements.

The present section first considers experimental evidence for the close link between amount of intramolecular reaction and the absolute value of modulus and the value of T_g. Also, effects due to chain interactions are illustrated. Second, results using theories of polymerization to predict network defects due to intramolecular reaction are discussed.

2.4.2 Experimental results on formation, structure, modulus and T_g

In order to fully characterize the amount of intramolecular reaction which occurs during the formation of a network one should aim to measure: (i) the variation of ring formation with extent of reaction; (ii) the gel point; and (iii) the variation of sol fraction w_s with extent of reaction.

The measurement of (i) is lengthy, requiring the determination of the change of number-average molar mass with conversion and has only been attempted [75] when linked with gel-point determinations (Figure 2.6), rather than experiments on network preparation. The measurement of (ii) linked with the modulus of the network formed at complete reaction has been used by the author and co-workers [1, 46, 47, 49–52, 54–56]. Limited measurements of (iii) as w_s at complete reaction, linked with modulus, have been used by Dušek, Ilavsky and co-workers [18, 20, 21, 23–26].

The use of sol-fraction data linked with modulus data at complete reaction has disadvantages. Calculations using polymerization theories have to be employed for interpretation, and stoichiometric reaction mixtures often yield negligible sol fractions, except at high dilutions. Hence, the data to be discussed in this section are limited to values of gel points and limiting moduli at complete reaction of stoichiometric polymerizations [47, 50, 52, 54–56]. The so-called, smallest-loop analysis is used [47, 54–56], enabling general relationships between formation, structure and properties

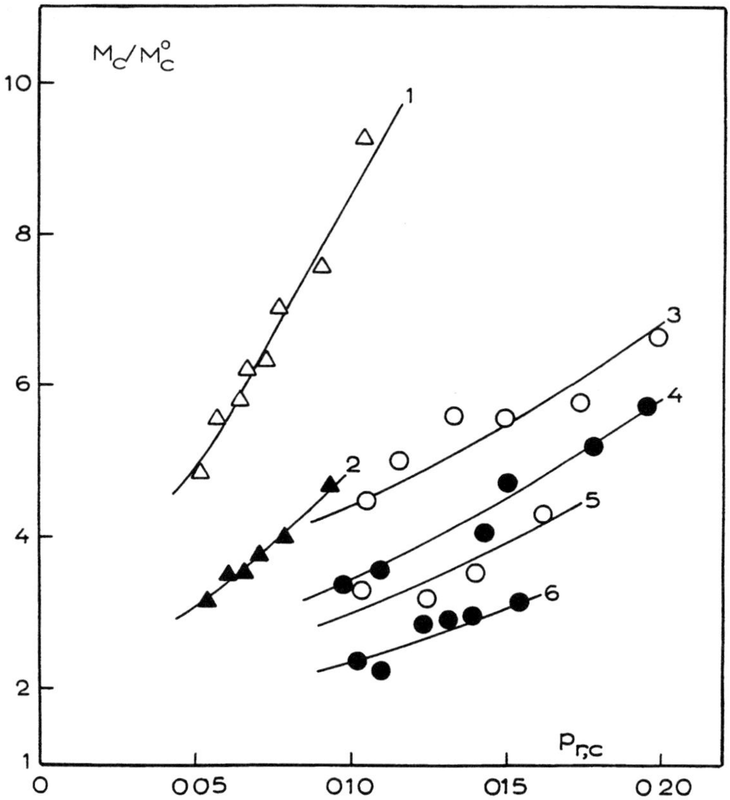

Figure 2.11 M_c from modulus measurements relative to M_c^o (for the perfect networks) versus $p_{r,c}$. Polyurethane networks formed at complete reaction from stoichiometric polymerizations of hexamethylene diisocyanate (HDI) with poly(oxypropylene) POP triols and tetrols, in bulk and at various dilutions in nitrobenzene as solvent. Systems 1 and 2, HDI + POP triols: (1) $M_c^o = 635\,\mathrm{g\,mol^{-1}}$ ($\nu = 33$); (2) $M_c^o = 1168\,\mathrm{g\,mol^{-1}}$ ($\nu = 61$). Systems 3–6, HDI + POP tetrols; (3) $M_c^o = 500\,\mathrm{g\,mol^{-1}}$ ($\nu = 29$); (4) $M_c^o = 586\,\mathrm{g\,mol^{-1}}$ ($\nu = 33$); (5) $M_c^o = 789\,\mathrm{g\,mol^{-1}}$ ($\nu = 44$); (6) $M_c^o = 1200\,\mathrm{g\,mol^{-1}}$ ($\nu = 66$). ν is the number of bonds in chains (of molar mass M_c^o) which can form the smallest loops [63].

to be established without recourse to detailed polymerization theories. (Interpretations using polymerization theories are discussed in the following section.)

Figure 2.11 shows results from polyurethane-forming polymerizations at various initial dilutions using HDI and POP triols and tetrols of various molar masses. The results are plotted as M_c/M_c^o versus $p_{r,c}$. M_c^o is the value of M_c expected for the perfect network structure (Section 2.2.4) and M_c is the quantity derived from measured limiting shear moduli at zero strain. M_c^o is the molar mass of the chain of ν bonds which can form the smallest loop (equation (2.40)). M_c was evaluated from the equations for uniaxial deformation based on Gaussian elasticity theory (Chapter 7),

$$\sigma = G(\Lambda - \Lambda^{-2}) \tag{2.53}$$

and

$$G = ART\rho\varphi_2^{1/3}(V_u/V_F)^{2/3}/M_c \tag{2.54}$$

σ is the applied stress per undeformed area and Λ the experimental deformation ratio. In the absence of free chain ends, A has the values of $(1 - 2/f)$ and 1 for phantom and affine chain behaviours, respectively. In evaluating M_c from G using equation (2.54), A was put equal to 1 as small-strain measurements were used (see Figure 2.13, discussed later). ρ is the density of the dry network, φ_2 the volume fraction of network during measurement ($\varphi_2 = 1$ for unswollen networks having no sol fraction), V_u is the volume of the dry, unstrained network and V_F the volume at formation (assumed to be equivalent to V_o, the volume in the strain-free reference state).

The values of M_c shown in Figure 2.11 are averages from the moduli of the various networks, measured in the dry and equilibrium swollen states. $p_{r,c}$ is the extent of intramolecular reaction at the gel point, as defined by equation (2.51). The points for each reaction system at different values of $p_{r,c}$ come from reactions at different dilutions, with that at the lowest value of $p_{r,c}$ being for the reaction in bulk.

The trends in Figure 2.11 show how the various factors characterizing pre-gel intramolecular reaction to a large extent carry through to determine network structure and modulus. All values of M_c/M_c^o are greater than 1 and also greater than $f/(f-2)$. Hence, the majority of the networks have significantly lower moduli than those of the corresponding perfect networks, even on a phantom basis. In no case is a perfect network formed. Given affine chain behaviour (discussed later), M_c/M_c^o is equal to the reduction in modulus below that expected for the perfect network ($M_c/M_c^o = G^o/G$). The total range of M_c/M_c^o values is about 2.1–9, and for polymerizations in bulk 2.1–5.

There are direct correlations between $p_{r,c}$ and M_c/M_c^o. The former quantity may be taken as a measure of the propensity of a polymerization for intramolecular reaction. Thus, this propensity continues past the gel point and, for a given polymerization system, more inelastic chains are formed in the final network for larger values of $p_{r,c}$. Some of the inelastic chains will form pre-gel and some post-gel (see later). In addition, relatively small values of $p_{r,c}$ can be symptomatic of large reductions in modulus.

More defects occur for trifunctional (curves 1 and 2) compared with tetrafunctional networks (curves 3–6). However, as expected, the delays in gel points are higher for the tetrafunctional polymerizations (equation (2.48)). The major differences between the two functionalities regarding network structure relates to the defects introduced per smallest loop in the two cases. Figure 2.12 shows that *two* elastic junction points (three elastic

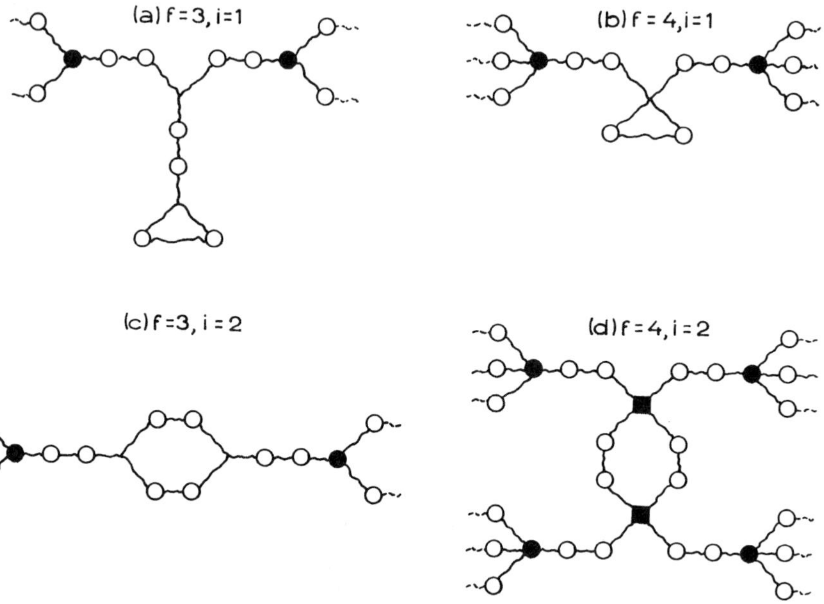

Figure 2.12 Ring structures of the two smallest sizes i in $RA_2 + R'B_f$ networks: (a) $f = 3$, $i = 1$; (b) $f = 4$, $i = 1$; (c) $f = 3$, $i = 2$; (d) $f = 4$, $i = 2$; (○) reacted pairs of groups; (●, ■) elastically active junction points.

chains) are lost per smallest loop ($i = 1$) for $f = 3$, and only *one* elastic junction point (two elastic chains) for $f = 4$. Smaller sizes of loop must give inelastic junctions and chains. Larger loop structures can form parts of elastic chains. In addition, due to their smaller numbers of bonds, smallest loops are the ones most likely to form, per opportunity for intramolecular reaction. The actual distribution of ring sizes is a problem presently under investigation (see Section 2.4.3.2), but the larger values of M_c/M_c^o for $f = 3$ compared with $f = 4$ networks shows that the smallest loops are in fact important for determining network defects.

For a given value of $p_{r,c}$ and functionality, M_c/M_c^o is larger for smaller values of M_c^o (reactants of lower molar mass and lower values of ν). Thus, the smaller is the value of ν, the larger is the proportion of the loops that give elastically inactive junctions. In this respect, it may be noted that $p_{r,c}$ is insensitive to loop size, whereas the value of M_c (reduction in modulus) is expected to be more sensitive to the numbers of smaller loops than the numbers of larger loops. In contrast to the assumption made in most of the existing theories and models of polymerizations, $p_{r,c}$ does not uniquely define M_c/M_c^o. The statistics of post-gel intramolecular reaction depend on detailed spatial correlations which are, therefore, significant in determining the topology of a network.

The reductions in moduli due to intramolecular reaction below these expected for the perfect network structures can be analysed further by converting the values of M_c/M_c^o obtained from modulus measurements to values of extents of intramolecules which have formed inelastic loops by the end of the reaction ($p_{r,e}$). Assuming each inelastic loop or intramolecular reaction of a pair of groups in an $f = 3$ network leads to the loss of two junction points and an $f = 4$ network to the loss of one junction point (conditions which may be satisfied on average, even if larger loops are also significant), the following expressions are obtained [47, 54–56]:

$$p_{r,e} = \frac{1}{6}\left[1 - \frac{1}{A}(AM_c^o/M_c)\right] \quad \text{for } f = 3 \tag{2.55}$$

$$p_{r,e} = \frac{1}{4}\left[1 - \frac{1}{A}(AM_c^o/M_c)\right] \quad \text{for } f = 4 \tag{2.56}$$

The factors 6 and 4, respectively, are equivalent to $2f$ and f, for $f = 3$ and $f = 4$.

Plots of $p_{r,c}$ versus $p_{r,e}$ are shown in Figures 2.13(a)–(c) for $A = 1$ and $(1 - 2/f)$. If all of the reaction which contributes to $p_{r,c}$ also contributes to $p_{r,e}$, again an assumption made in theories of polymerization, then $p_{r,e} \geq p_{r,c}$. It is apparent that this condition is satisfied only if $A = 1$, consistent with the affine chain behaviour expected from the small strain measurements used. On the same basis, the distances of the points above the lines $p_{r,e} = p_{r,c}$ in Figures 2.13(a) and 2.13(b) are the post-gel contributions to $p_{r,e}$. For a given polymerization system, they are larger for the more concentrated reactions (lower values of $p_{r,c}$) indicating that spatial correlations between groups on the gel are then more important. Conversely, as dilution is increased, pre-gel intramolecular reaction yields larger proportions of the inelastic chains. Simple considerations show that, for pre-gel intramolecular reaction in a given polymerization system, the smaller the value of c_{ext} the larger is the proportion of loops formed earlier in a polymerization. Hence, the average loop size is smaller and the loops are more likely to be elastically inactive.

In Figures 2.13(a) and 2.13(b), the trends in $p_{r,e}$, with M_c^o and f, at constant values of $p_{r,c}$, reflect the corresponding trends in M_c in Figure 2.11. Further, extrapolations of the lines in Figure 2.13 to $p_{r,c} = 0$ give limiting values of $p_{r,e} \geq 0$. In other words, even with no pre-gel intramolecular reaction, inelastic chains are formed post-gel. Non-zero intercepts signify the universal occurrence of network imperfections. Some of the gel–gel reaction which must occur to give the final network leads to elastically inactive chains.

The intercepts in $p_{r,e}$ at $p_{r,c} = 0$ can be reconverted, through equations (2.55) and (2.56), into values of M_c/M_c^o in the limit $p_{r,c} = 0$. The reductions in moduli are significant. Thus, the curves in Figure 2.11 do not extrapolate to $M_c/M_c^o = 1$ at $p_{r,c} = 0$ (compare Figures 2.14 and 2.20). The resulting reductions in modulus at $p_{r,c} = 0$ can be analysed further to show that, even in the limit of reactants of infinite molar mass (i.e. no spatial

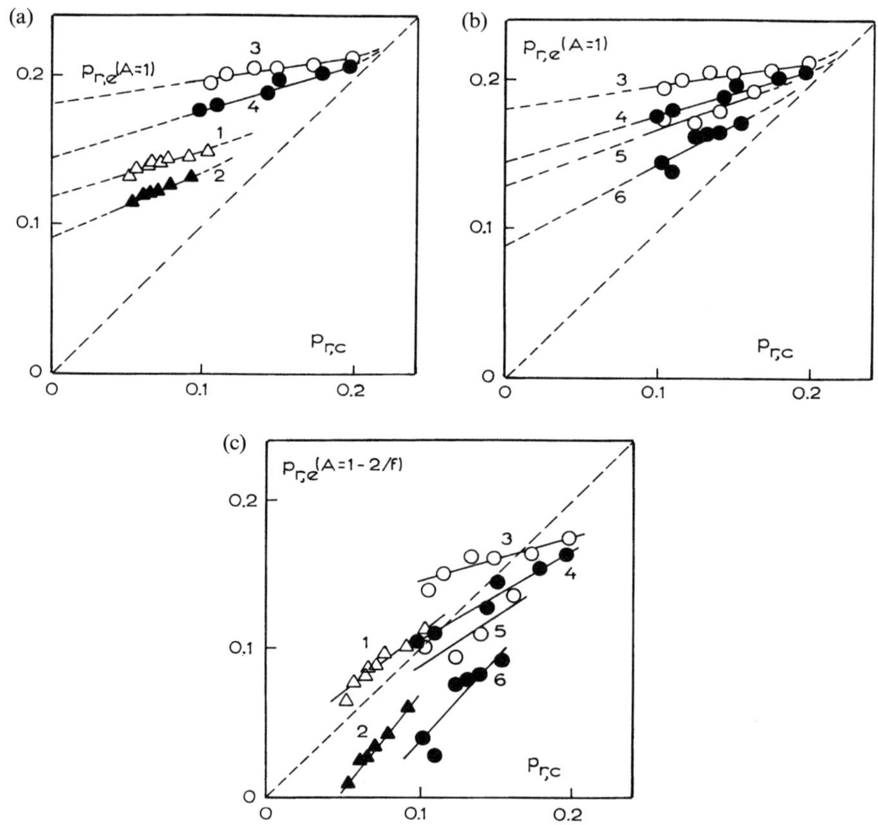

Figure 2.13 Extent of intramolecular reaction at complete reaction ($p_{r,e}$), as derived from modulus measurements versus $p_{r,c}$. Systems as in Figure 2.11 [1]: (a) and (b) $A = 1$; (c) $A = 1 - 2/f$. (Reprinted with permission.)

correlations between reactive groups on the same molecule), some of the random intramolecular reaction between groups on the gel still leads to inelastic chains [47, 56]. Random gel–gel reaction is part of F-S statistics [38], irrespective of the molar masses of the reactants, and does not lead to a perfect network. Obviously, detailed considerations of the spatial correlations governing gel–gel reaction and the structures which result are needed for the quantitative prediction of network structures and properties (see Section 2.4.3.2).

Due to intramolecular reaction, the polyurethane-forming polymerizations just discussed, based on HDI and POP polyols, form networks with moduli lower than those expected for perfect network structures. Related polymerizations, using 4,4′-diphenylmethane diisocyanate (MDI) instead of HDI, give networks showing much smaller reductions in modulus

[24, 50–52]. Results are shown in Figure 2.14, in comparison with the results from system 1 of Figure 2.11. The MDI- and HDI-based systems have similar values of M_c^o. The smaller values of $p_{r,c}$ for the polymerizations with MDI are explicable [50–52] in terms of a stiffer chain structure (larger value of b). The increase in M_c/M_c^o with $p_{r,c}$ means that the effects of intramolecular reaction are still significant. However, the moduli are much larger than would be expected. Calculation [43] shows that the enhanced moduli are consistent with interchain interactions (incipient segmentation) increasing the effective numbers of junction points. The extrapolated value of $M_c/M_c^o = 0.3125$ at $p_{r,c} = 0$ corresponds to one extra (tetrafunctional) junction per elastic chain. For the *bulk* MDI-based polymerization, the compensating effects of intramolecular reaction and interactions combine to give an apparently perfect network. The intercept of the M_c/M_c^o curve at $p_{r,c} = 0$ for the HDI-based system 1 is consistent with the extrapolated value of $p_{r,e}$ at $p_{r,c} = 0$ in Figure 2.13(a) and is the result of only post-gel intramolecular reaction leading to network defects [54–56, 59].

The plots in Figure 2.14 illustrate the difficulty of separating the compensating effects of network imperfections and chain interactions. Topological entanglements in the MDI-based networks are unlikely because the system has a low value of M_c^o and such entanglements are not evident in the HDI-based networks. However, there is much evidence for topological entanglements in networks of high M_c contributing to modulus in amounts relating to the plateau moduli of corresponding (uncrosslinked) linear chains (see Chapter 8 and [39, 41, 42]). Obviously, interpretations of absolute values of moduli have to take account of network imperfections before affine or phantom chain behaviour and the effects of topological entanglements and other interactions can be evaluated. To this end, the preparation of networks at different dilutions and the concomitant measurement of gel points should be of general applicability. A limitation is that methods of determining extents of reaction in the particular reactions used have to be devised. If this is not feasible, then the decrease of the sol (soluble) fraction as a function of time up to complete reaction could be used. However, as mentioned previously, the latter approach relies on the application of theories of polymerization to deduce extents of intramolecular and intermolecular reaction from the differences between experimental and F-S sol fractions. Thus, the interpretation of results rests heavily on the exactness of the theory being used and a critical test of theories is difficult to achieve.

Figure 2.15 illustrates the effects on T_g of network defects caused by intramolecular reaction as represented by α_c $(= p_a p_b$ at gelation) [46, 51]. The values of T_g shown are for dry networks formed from MDI and POP triol, reaction system 7 of Figure 2.14. The measurements of T_g were carried out at 1 Hz, using a torsion pendulum. The two limiting values of T_g correspond to networks with $M_c = M_c^o$ and $M_c = \infty$. Thus, the

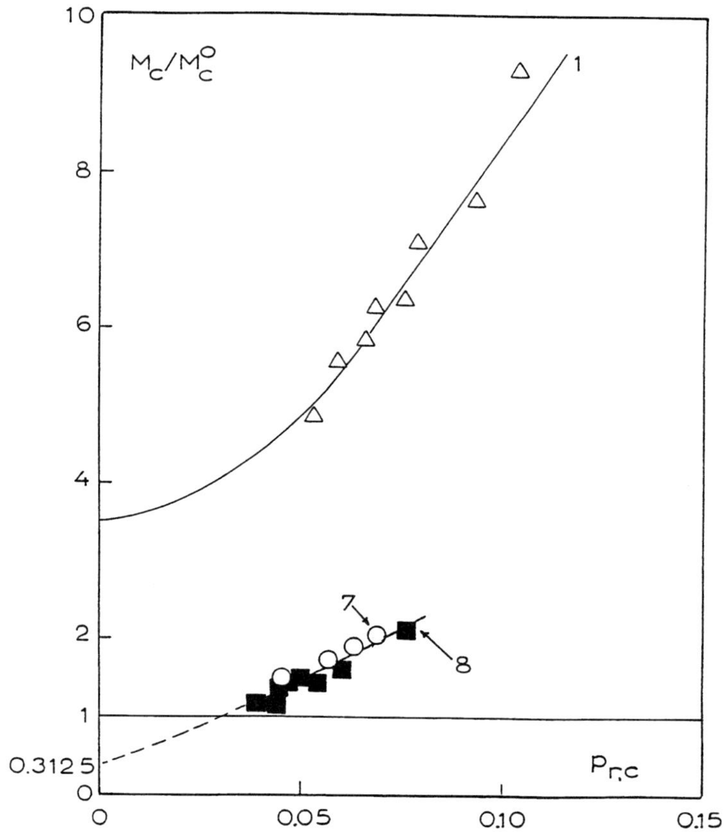

Figure 2.14 Comparison of HDI-based and 4,4′-diphenylmethane diisocyanate (MDI)-based polyurethane networks using M_c^o from modulus measurements relative to M_c^o versus $p_{r,c}$ [1]. System 1, HDI + POP triol (LHT240), system 1 from Figure 2.11; system 7, MDI + POP triol (LHT240), 80 °C, in bulk and in nitrobenzene solution, $M_c^o = 705\,\mathrm{g\,mol^{-1}}$; system 8, MDI + POP triol (LHT240), 70 °C, in bulk and in xylene solution, $M_c^o = 811\,\mathrm{g\,mol^{-1}}$. Curves: system 1 as in Figure 2.11 with extrapolated value of M_c/M_c^o at $p_{r,c} = 0$, consistent with extrapolated value of $p_{r,e}$ at $p_{r,c} = 0$ in Figure 2.13(a). Systems 7 and 8 are discussed in the text and [24, 50–52]. (Reprinted with permission.)

horizontal broken line gives the minimum T_g, that of a linear MDI/POP polymer having a repeat unit of molar mass equal to M_c^o, and the maximum value of T_g at $\alpha_c = 0.5$ was obtained by extrapolation of $(1/M_c, T_g)$ data to $1/M_c^o$. The variation of T_g with α_c (or M_c) is a reflection of the influence of junction-point density on the freedom of segmental motion. The maximum range of T_g values shown, 301–312 K (28–39 °C), reflects the maximum influence for these MDI/POP triol systems.

The effects of conversion and, hence, junction-point density on T_g, which can give similar or much larger variations of T_g are discussed in detail in

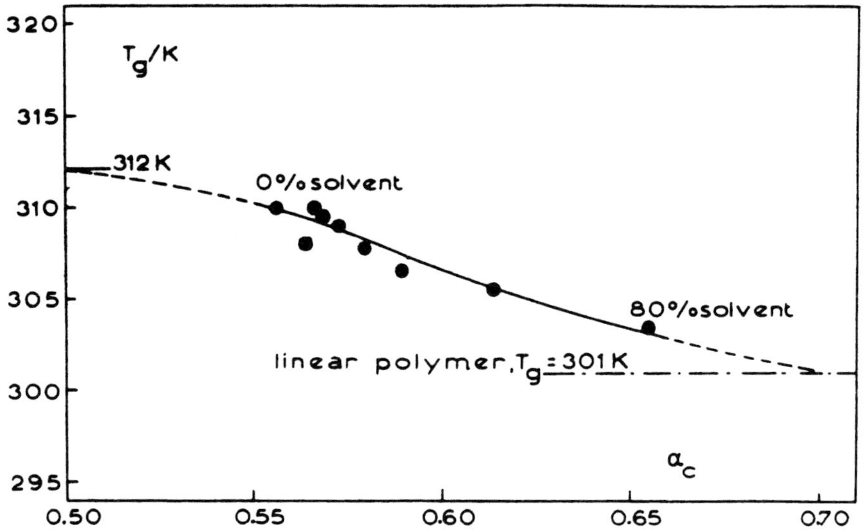

Figure 2.15 T_g versus α_c (●) for dry, trifunctional polyurethane networks from reaction system 7 of Figure 2.14 (MDI + LHT240) with networks prepared at various initial dilutions of reactants. (· — · —) MDI + POP diol with $M_{repeat} = M_c^o = 705\,\mathrm{g\,mol^{-1}}$ [51].

Chapter 4. Relationships between T_g and conversion are important for controlling and predicting the processing behaviour of thermosets and are used in conversion-versus-temperature transformation diagrams.

2.4.3 *Applications of polymerization theories and models*

Cascade theory [17–28], rate theory [48, 53–58] and Monte-Carlo simulations [62, 66–73, 80, 81] (cf. Section 2.3.2) have been used to predict amounts of intramolecular reaction in network-forming polymerizations and, hence, pre-gel ring fractions, gel points, sol fractions and network defects. In addition, the earliest approach, cascade theory, has been applied with some success to interpret gel-point, gel-fraction and limiting-modulus data, often from systems in which like groups have unequal reactivities [19, 20, 25, 27, 28, 82]. In order to give some comparison of the predictions made, the present discussion focuses on interpretations of common data using different theoretical approaches.

2.4.3.1 *Cascade theory*

An example of the interpretation of results using cascade theory by Ilavsky and Dušek [24] is shown in Figure 2.16. It also illustrates the interplay of factors which have to be considered when interpreting results. The polymerizations used a POP triol reacting with MDI at various reactant ratios $r_H = [OH]_o/[NCO]_o$ and dilutions in xylene. The results plotted in

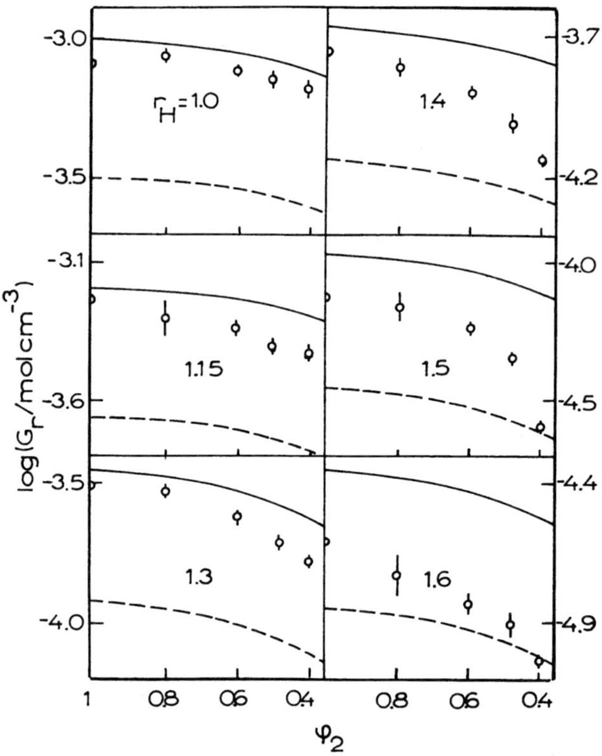

Figure 2.16 Comparison of experimental and calculated dependencies of reduced modulus G_r using cascade theory versus volume fraction of reactants φ_2 for MDI + POP triol polymerizations and networks [24]: $G_r = A\rho/M_c$, $r_H = [OH]_o/[NCO]_o$. Curves calculated using cascade theory, accounting for intramolecular reaction: (——) $A = 1$; (– – –) $A = 1/3$ [1]. (Reprinted with permission.)

Figure 2.14, in comparison with results from polymerization using HDI, are in fact derived from those given in Figure 2.16 at $r_H = 1$.

Figure 2.16 shows log G_r versus φ_2, where the reduced modulus G_r is equal to $An_{e,c} = A\rho/M_c$. The solid and dashed curves give the theoretical predictions from cascade theory for affine and phantom chain behaviour. From the measurements of sol fraction at complete reaction, the effects of cyclization were predicted to be small. With no cyclization, G_r would be independent of φ_2. The experimental values of modulus lie between the predicted values for affine and phantom behaviour.

Ilavsky and Dušek [24] have interpreted the results by assuming phantom chain behaviour and attributing the increases in modulus to pair entanglements between chains. The numbers of pairs of entanglement chains were assumed to be proportional to φ_2^2 and were also calculated using cascade theory. The interpretation appears to have two shortcomings. First, phantom chain behaviour is not expected as limiting moduli at zero strain were measured. Second, concentrations of entanglements should relate to segmentation effects in the final polyurethane networks rather than topological entanglements during formation. An alternative explanation, which avoids these shortcomings and is consistent with the results and also the results on HDI-based networks, is that given in the discussion of Figure 2.14, namely, affine chain behaviour coupled with intramolecular reaction and segmentation effects in the final networks.

2.4.3.2 Monte-Carlo simulations

The MC simulations developed by Eichinger and co-workers [66–73, 80] have applied to several model network-forming endlinking polymerizations and to the crosslinking of polymer chains. In the former applications, data on pre-gel number fractions of rings [80], gel points [72, 73, 80] and modulus at complete reaction [72, 73] have been interpreted. The overall performance of the MC simulations [43, 61, 72] in relation to the experimental results of Figure 2.11 is shown in Figure 2.17, where M_c/M_c^o is again plotted versus $p_{r,c}$ [61]. The results of two MC analyses are shown. Those denoted MC* 5,6 were from Lee and Eichinger's analysis [72], as interpreted by Stepto and Eichinger [43], of the results for systems 5 and 6. They used an early version of the software and a rather exact RIS model of the branched POP tetrol reactants. The results denoted MC 1–2 and MC 3–6 were obtained [61] for systems 1–6 by using a simplified RIS model and a commercial version of the software [94]. The latter results give worse predictions of the gel points. However, importantly, *all* the values of M_c/M_c^o from both MC analyses are too low and do not reflect the experimental differences between $f = 3$ and $f = 4$ networks. There seems to be a serious undercounting of inelastic chains in both the early and commercial versions of the algorithm.

Results of initial calculations using a new MC method developed by the author and collaborators [62] have been published recently. *All* the connectivities in a network are counted and Figures 2.18(a) and (b) show number-fraction distributions of ring structures in $RA_2 + R'B_3$ and $RA_2 + R'B_4$ polymerizations carried out under F-S conditions and for two values of the ring-forming parameter λ_{a0} (see equation (2.49)).

The distributions show several striking features. First, they are broad; ring structures of wide ranges of sizes occur and must somehow be accounted for when considering network defects and their effects on properties. Increasing λ_{a0} markedly increases the proportions of the smaller ring structures, with the smallest dominating. However, large ring structures still cannot be neglected. The maxima, at around $i = 12$ for $RA_2 + R'B_3$ polymerizations and around $i = 8$ for $RA_2 + R'B_4$ polymerizations, are a manifestation of the different numbers of opportunities for forming ring structures of the various sizes, *integrated* over the whole reaction. Indeed,

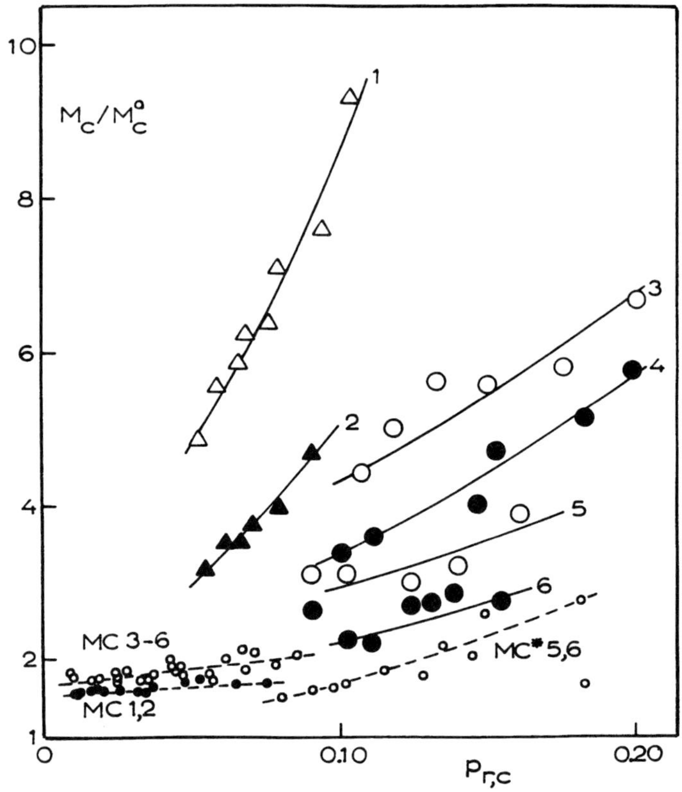

Figure 2.17 Monte-Carlo (MC) values of M_c/M_c^o versus $p_{r,c}$ in comparison with the experimental values from Figure 2.11: (MC) from MSI commercial software, systems 1–6; (MC*) from Lee and Eichinger [72] and Stepto and Eichinger [43] for systems 5 and 6 [61]. (Reprinted with permission.)

for F-S conditions, which show the largest maxima, such intramolecular reaction occurs only post-gel and is determined solely by the numbers of pairs of groups on the gel molecule which can form ring structures of the various sizes. It is possible that the position of the maxima is a function of population size, or the 'finite' size of the gel molecule. Further calculations with different sizes of population are needed to investigate this possibility.

Figure 2.19 shows the calculated and experimental values of M_c/M_c^o versus $p_{r,c}$ for $RA_2 + R'B_3$ polymerizations and networks, systems 1 and 2 of Figure 2.14, have been transferred from Figure 2.17. The same values of λ_{a0} were used in the previous MC 1,2 calculations and the new MC calculations.

The predicted values of gel points $p_{r,c}$, from the new algorithm are improved over those from the commercial software. They can be improved

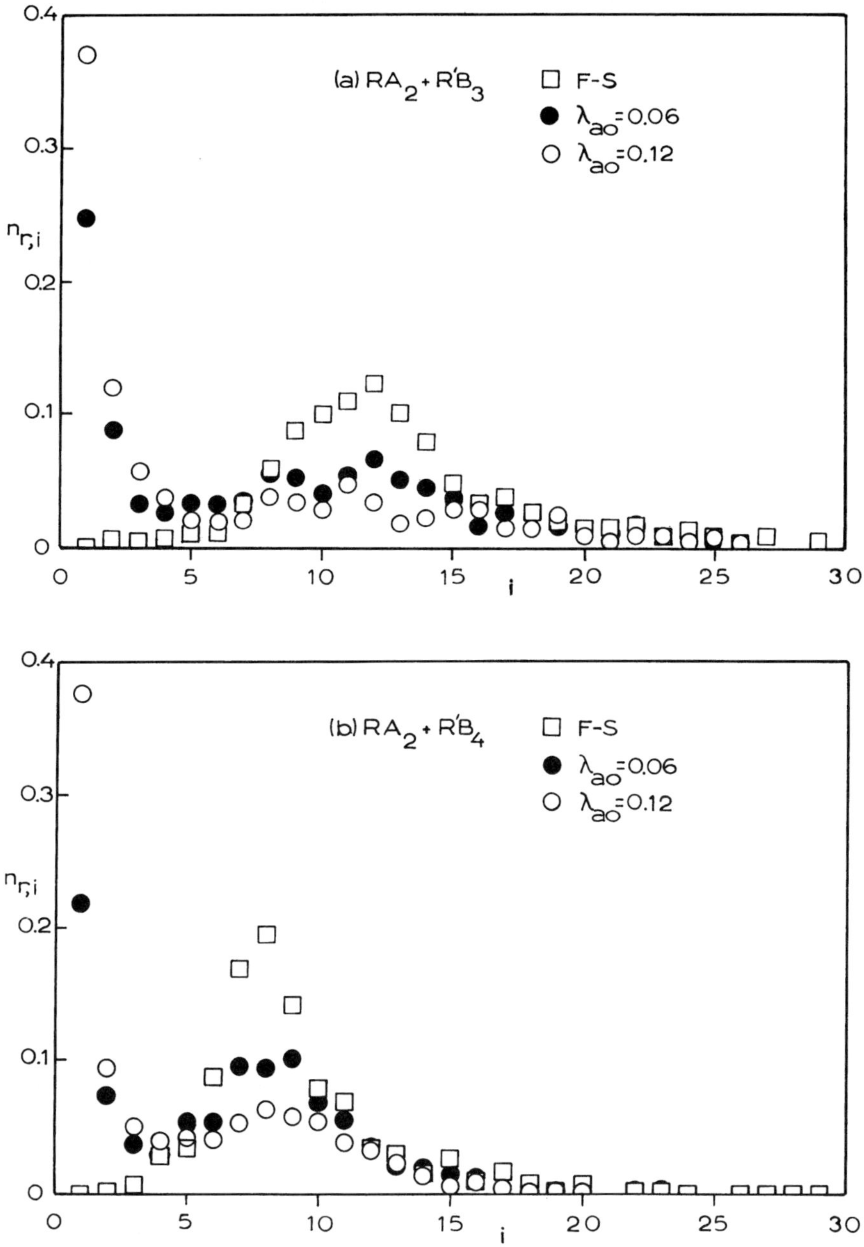

Figure 2.18 Number fraction of ring structures of size i, $n_{r,i}$ versus ring-structure size i: (a) $RA_2 + R'B_3$; (b) $RA_2 + R'B_4$, polymerizations under Flory–Stockmayer (F-S) conditions and with λ_{a0} (equation (2.49)) equal to 0.06 and 0.12 [62].

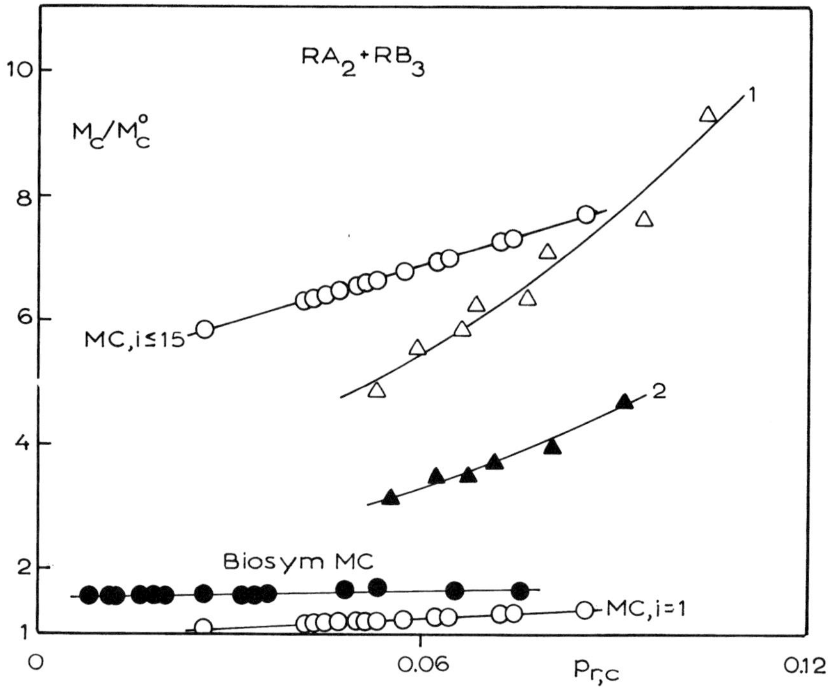

Figure 2.19 Experimental and calculated values of M_c/M_c^o versus experimental and calculated values of $p_{r,c}$ for $RA_2 + R'B_3$ polymerizations. Curves 1 and 2 show experimental results. Biosym MC indicates results using commercial Monte-Carlo software (from Figure 2.17). Curve MC, $i = 1$, reports newer Monte-Carlo results [62] accounting for only smallest ring structures, of size $i = 1$ (Figure 2.12(a)). Curve MC, $i \geq 15$, reports newer MC results [62] accounting for all ring structures up to size $i = 15$ [62].

further by the use of more accurate values of $\langle r^2 \rangle$ in the evaluation λ_{a0} for particular reactions. However, the principal difference between the results from the two MC methods is in the predicted values of M_c/M_c^o. The curve marked $i = 1$ comes from counting only the smallest loops; the curve marked $i \geq 15$ comes from counting ring structures up to size 15 (formed from chains of 15ν bonds), assuming two elastic chains lost per ring structure (as for smallest loops). This assumption gives the maximum loss of elastic chains per ring structure. Indeed, it is possible to predict values of M_c/M_c^o which are much greater than the experimental values by increasing the maximum size of ring structure considered to above $i = 15$. Obviously, more detailed calculations of loss of elastic chains or loss of entropy for ring structures of various types and sizes now need to be explored if accurate predictions of M_c/M_c^o (or modulus) are to be made. Such calculations are in progress [96]. Meanwhile, the results in Figure 2.19 do show the importance of being able to count all ring structures.

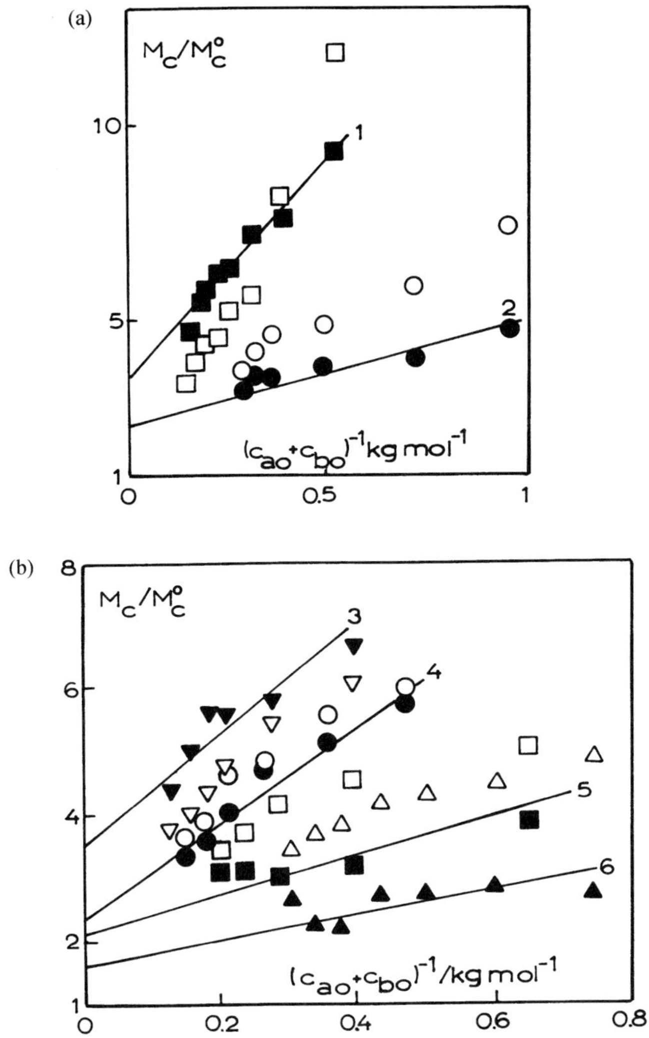

Figure 2.20 Comparison of experimental and calculated values of M_c/M_c^o using rate theory versus $(c_{a0} + c_{b0})^{-1}$ for the HDI-based polymerization and networks of Figure 2.11: (a) HDI + POP triols; (b) HDI + POP tetrols. Solid symbols represent experimental results; open symbols represent calculated values of M_c/M_c^o using values of λ_{a0} *consistent* with the experimental values of $p_{r,c}$ [1]. (Reprinted with permission.)

2.4.3.3 *Rate theory*

First-order rate theory, accounting for smallest loops, has so far been developed for RA_3, RA_4, $RA_2 + R'B_3$ and $RA_2 + R'B_4$ polymerizations [48, 53, 57, 58]. Extensions to other polymerizations are straightforward.

Figures 2.20(a) and (b) show the predictions of modulus obtained [1, 62, 97] for all the data in Figure 2.11. The values of the ring-forming parameter λ_{a0} used were those required to give agreement with the experimental gel points ($p_{r,c}$). Generally, the calculated moduli agree with the experimental values to better than a factor of two. Overall, the prediction of moduli is satisfactory, given that reductions in moduli range from a factor of 2 to a factor of 9. Even though only the smallest loops were counted, the theory seems to be able to predict moduli with reasonable accuracy *provided* that measured gel points are known. The use of ARS theory to predict the gel point coupled with rate theory to predict modulus as illustrated in Figures 2.20(a) and (b) has now been interpreted into commercial software [94]. However, because it counts only smallest ring structures, the theory badly underestimates M_c/M_c^o if values of λ_{a0} consistent with the chain statistics of the reactants are used.

The values calculated using rate theory show approximately the correct relative differences in M_c/M_c^o between trifunctional and tetrafunctional networks. However, they do not show sufficient variation with reactant molar masses (M_c^o or ν). This behaviour again illustrates that the distribution of loop sizes and the resulting different losses of elastic chains for different loop sizes are important. The smaller variation of predicted values of M_c/M_c^o with M_c^o compared with that found experimentally is also a feature of cascade theory. Both rate and cascade theories give M_c/M_c^o versus $p_{r,c}$ as universal curves. In other words, a given ring-forming parameter, which results in a certain delay in gel point, always gives the same reduction in modulus, independent of M_c^o. Such universal behaviour is not observed experimentally, and the recent MC results [62] discussed in the previous section indicate that the different losses of elastic chains due to ring structures of different sizes will be important in explaining this phenomenon.

2.5 Conclusions

Overall, the philosophy of the present chapter has been the prediction and interpretation of the formation, structure and properties of homogeneous, polymer networks formed by non-linear polymerizations. Classical polymerization statistics can be used as a basis for interpreting gel points and the growth of networks during non-linear polymerizations. Intramolecular reaction becomes significant and has to be accounted for if gel points are to be predicted and correctly interpreted.

Intramolecular reaction, which is an integral part of polymerizations involving the reactions of pairs of groups, means that the structures and properties of network polymers depend more closely on the polymerizations by which they are formed than those of linear polymers. The study of gel points linked with the preparation of networks at different dilutions of reactants can be invaluable in highlighting the effects of intramolecular reaction. Because of such reaction, significant reductions in modulus can occur, which can be correlated with gel points. Although gel-point delays can be predicted with reasonable accuracy, the importance of ring structures of various types and sizes on modulus reductions is less well understood.

In addition to modulus being reduced by network defects, it can be increased by chain interactions or entanglements. In this context, the study of networks prepared at different dilutions is also essential if the opposing effects leading to modulus reduction and to modulus enhancement are to be resolved. Although the basis of the elastic behaviour of polymer networks is understood, shortcomings exist in the detailed correlations between macroscopic and molecular deformations (see Chapter 7). To some extent, these are related to the lack of knowledge of network imperfections and the related chain entropy functions, as well as to the cooperative deformation mechanisms of connected network chains.

Acknowledgements

First, I express my sincere gratitude to all the collaborators mentioned in the list of references. The support of the EPSRC, Molecular Simulations (lately Biosym Technologies) Incorporated and ICI Paints is also gratefully acknowledged.

Glossary of symbols and abbreviations

b	$(= \langle r_o^2 \rangle / \nu)$ effective bond length of the chain of ν bonds
c_{ext}	total concentration of A groups and B groups external to a given molecule
c_a, c_b	instantaneous concentrations of A groups and B groups in a polymerization
c_{a0}, c_{b0}	initial concentration of A groups and B groups in a polymerization
c_{int}	total internal concentration for intramolecular reaction arising from all pairs of groups on a given molecule
$c_{\text{int},i}$	internal concentration for intramolecular reaction arising from pairs of groups on a given molecule that can form ring structures of size i
f	chemical functionality of a reactant; functionality of a junction point in a network

f_a, f_b	chemical functionalities of reactants bearing A groups and B groups
f_{ai}, f_{bj}	chemical functionalities of reactants i and j in mixtures of reactants bearing, respectively, A groups and B groups
f_{aw}, f_{bw}	mass-average (chemical) functionalities of mixtures of reactants bearing, respectively, A groups and B groups
f_{an}, f_{bn}	number-average (chemical) functionalities of mixtures of reactants bearing, respectively, A groups and B groups
$f'_{n,J}$	number-average (chemical) functionality of branched reactants in a $\sum_i RA_{fai}$ polymerization
f_w	mass-average (chemical) functionality
n	number (of moles) of branch points in a network
n_{a0}, n_{b0}	initial numbers of (moles of) A groups and B groups in a polymerization
n_c	(molar) concentration of chains between junction points in a network
$n_{e,c}$	(molar) concentration of elastically active chains between junction points in a network
$n_{e,J}$	(molar) concentration of elastically active junction points in a network
n_i	number of pairs of groups on a given molecule that could react to form ring structures of size i
n_J	(molar) concentration of junction points in a network
p	extent of reaction (i.e. fraction of groups reacted) in a stoichiometric A + B polymerization
p_a, p_b	extents of reaction of A groups and B groups
p_{aa}, p_{ab}	fractions of A groups that have reacted, respectively, with A groups and B groups in a mixed A + A/A + B polymerization
p_{ac}, p_{bc}	values of p_a and p_b at the gel point
p_a^*	conjugate value of post-gel p_a that is $<p_{ac}$ and characterizes the distribution of molecular species in the sol
p_c	extent of reaction at the gel point in a stoichiometric A + B polymerization
p_c^o	$(=(f_{aw}-1)^{-1}(f_{bw}-1)^{-1})$ F-S value of p_c
$p_{r,c}$	$(=p_c-p_c^o)$ extent of intramolecular reaction at the gel point
r_a	$(=n_{a0}/n_{b0})$ initial (molar) ratio of A groups to B groups in a polymerization
r_H	$(=[OH]_0/[NCO]_0)$ initial molar ratio of hydroxy groups to isocyanate groups in a polyurethane-forming polymerization
$\langle r_o^2\rangle$	unperturbed, mean-square end-to-end distance of a linear polymer chain
$\langle r^2\rangle$	mean-square end-to-end distance of a linear polymer chain in the presence of excluded volume
u_g	unit fraction of gel (gel fraction)

u_s	unit fraction of sol (sol fraction)
u_x	unit fraction of species of dp = x (x-mer)
u'_x	unit fraction of x-mer in the sol
w_{br}	mass fraction of branched reactants in a polymerization
w_s	mass fraction of sol
x	value of dp
x_n	number-average dp
x_w	mass-average dp
A	front factor in the expression for the shear modulus of an elastomeric network
G	initial shear modulus of an elastomeric network
G^o	initial shear modulus of a perfect, elastomeric network
G_r	($= G/RT$) reduced, initial shear modulus of an elastomeric network
M_{ai}	molar mass of reactant of chemical functionality f_{ai}
M_c	molar mass of (elastically active) chains between junction points in a network
M_c^o	value of M_c for the perfect network structure formed by given reactants
M_n	number-average molar mass
$M_{n,br}$	number-average molar mass of branched reactants in a polymerization
N_a, N_b	initial number of molecules (or moles) of reactants bearing A groups and B groups
N_{Av}	Avogadro constant
N_c	number of (moles) of (elastically active) chains between junction points in a network
N_J	number of (moles) of (elastically active) junction points in a network
N_r	number of ring structures per molecule forming in the course of a polymerization
N_x	number of molecules (or moles) of dp $= x$
P_{ab}	($= (3/(2\pi\langle r^2\rangle))^{3/2})/N_{Av}$) P_1 for polymerizations proceeding via A + B reactions
P_i	($= (3/(2\pi i\nu b^2)^{3/2})/N_{Av}$) molar concentration of coincident ends of a linear chain or sub-chain forming a ring structure of size i, i.e. a chain with $i\nu$ skeletal bonds
$P(\mathbf{0})$	probability density of the end-to-end vector ($\boldsymbol{r}$) of a linear polymer chain at $\boldsymbol{r} = \mathbf{0}$
T_g	glass–rubber transition temperature
V_{net}	volume of a network
V_o	strain-free, reference volume of a network
V_u	volume of a dry, unstrained network
V_F	volume of a network at formation

W_{net}	mass of a network
α	$(= p_a p_b)$ product of the extents of reaction in an $A + B$ polymerization
α_c	$(= p_{ac} p_{bc})$ product of the extents of reaction at the gel point in an $A + B$ polymerization
λ	$(= c_{int}/(c_{int} + c_{ext}))$ ring-forming parameter
λ'	$(= c_{int}/c_{ext})$ ring-forming parameter
λ_{a0}	$(= P_{ab}/c_{a0})$ ring-forming parameter
ν	number of skeletal bonds in the chain forming the (average) smallest ring structure in a polymerization
ρ	fraction of units crosslinked; density of a dry network
σ	stress per undeformed area of a network
φ_2	volume fraction of network in a swollen network sample; volume fraction of reactants in a network-forming polymerization
Λ	experimental deformation ratio
dp	degree of polymerization
pgf	probability generating function
ARB_{f-1}	reactant having one reactive A group and $(f - 1)$ reactive B groups
ARS	Ahmad–Rolfes–Stepto
F-S	Flory–Stockmayer
HDI	hexamethylene diisocyanate
MC	Monte Carlo
MDI	4,4′-diphenylmethane diisocyanate
POP	poly(oxypropylene)
RA_{fa}, $R'B_{fb}$	reactants having, respectively, f_a reactive A groups and f_b reactive B groups
$\sum_i RA_{fai}$	mixture of reactants, $RA_{fa1}, RA_{fa2}, \ldots, RA_{fai}$, bearing, respectively, $f_{a1}, f_{a2}, \ldots, f_{ai}$ reactive A groups
$\sum_i R'B_{fbj}$	mixture of reactants, $R'B_{fb1}, R'B_{fb2}, \ldots, R'B_{fbj}$, bearing, respectively, $f_{b1}, f_{b2}, \ldots, f_{bj}$ reactive B groups
RIS	rotational isomeric state

References

1. Stepto, R.F.T. (1992) in *Comprehensive Polymer Science, First Supplement* (eds S.L. Aggarwal and S. Russo), Pergamon Press, Oxford, chapter 10.
2. Graham, N.B. and Hayes, C.M.G. (1995) *Macromol. Symp.*, **93**, 293.
3. Flory, P.J. (1953) *Principles of Polymer Chemistry*, Cornell University Press, Ithaca.
4. May, C.A. (ed.) (1983) *Chemorheology of Thermosetting Polymers*, American Chemical Society, Washington, DC.
5. Macosko, C.W. (1989) *RIM: Fundamentals of Reaction Injection Moulding*, Hauser, Munich.

6. Stanford, J.L., Ryan, A.J. and Elwell, M.J.A. (1997) in *Processing of Polymers* (ed. H. Meijer), VCH, Weinheim, chapter 9.
7. Flory, P.J. (1941) *J. Amer. Chem. Soc.*, **63**, 3083, 3091, 3097.
8. Flory, P.J. (1946) *Chem. Revs.*, **39**, 137.
9. Stockmayer, W.H. (1943) *J. Chem. Phys.*, **11**, 45.
10. Stockmayer, W.H. (1944) *J. Chem. Phys.*, **12**, 125.
11. Stanford, J.L. and Stepto, R.F.T. (1975) *J. Chem. Soc., Faraday Trans. 1*, **71**, 1292.
12. Stanford, J.L., Stepto, R.F.T. and Waywell, D.R. (1975) *J. Chem. Soc., Faraday Trans. 1*, **71**, 1308.
13. Stockmayer, W.H. (1952) *J. Polymer. Sci.*, **9**, 69; (1953) **11**, 424.
14. Stepto, R.F.T. (1982) in *Developments in Polymerisation-3* (ed. R.N. Haward), Elsevier Applied Science, London, chapter 3.
15. Rolfes, H. and Stepto, R.F.T. (1992) *Makromol. Chem., Theory. Simul.*, **1**, 245.
16. Case, L.C. (1957) *J. Polym. Sci.*, **26**, 333.
17. Dušek, K. and Vojta, V. (1977) *Br. Polym. J.*, **9**, 164.
18. Dušek, K., Hadhoud, M. and Ilavsky, M. (1977) *Br. Polym. J.*, **9**, 172.
19. Matejka, L. and Dušek, K. (1980) *Polym. Bull. (Berlin)*, **3**, 489.
20. Dušek, K. (1982) *Rubber Chem. Technol.*, **55**, 1.
21. Dušek, K. and Ilavsky, M. (1982) in *Elastomers and Rubber Elasticity* (eds J.E. Mark and J. Lal), American Chemical Society, Washington, DC, chapter 21.
22. Ilavsky, M. and Dušek, K. (1982) *Polym. Bull. (Berlin)*, **8**, 359.
23. Ilavsky, M. and Dušek, K. (1983) *Polymer*, **24**, 981.
24. Ilavsky, M. and Dušek, K. (1986) *Macromolecules*, **19**, 2139.
25. Dušek, K. (1986) *Adv. Polym. Sci.*, **78**, 1.
26. Dušek, K. and Ilavsky, M. (1988) in *Biological and Synthetic Polymer Networks* (ed. O. Kramer), Elsevier, Barking, chapter 14.
27. Dušek, K. and MacKnight, W.J. (1988) in *Cross-Linked Polymers, Chemistry, Properties and Applications* (eds R.A. Dickie, S.S. Labana and R.S. Bauer), American Chemical Society, Washington, DC, chapter 1.
28. Zelenka, J., Ilavsky, M., Dobas, I. and Dušek, K. (1991) *Polymer Networks & Blends*, **1**, 93.
29. Ross-Murphy, S.B. and Stepto, R.F.T. (1986) in *Cyclic Polymers* (ed. J.A. Semlyen), Elsevier, Barking, chapter 10.
30. Gordon, M. (1962) *Proc. Roy. Soc. London, Ser. A*, **268**, 240.
31. Sandler, S.R. (1967) *J. Appl. Polymer Sci.*, **11**, 811.
32. Bauer, M., Bauer, J. and Kühn, G. (1986) *Acta Polymerica*, **37**, 715.
33. Bauer, J. and Burchard, W. (1992) *J. Phys. II France*, **2**, 1053.
34. Fukui, K. and Yamabe, T. (1960) *J. Polymer Sci.*, **XLV**, 305.
35. Carothers, W.H. (1936) *Trans. Faraday Soc.*, **32**, 39.
36. Macosko, C.W. and Miller, D.R. (1976) *Macromolecules*, **9**, 199.
37. Miller, D.R. and Macosko, C.W. (1976) *Macromolecules*, **9**, 206.
38. Stepto, R.F.T. (1990) *Polymer Bull. (Berlin)*, **24**, 53.
39. Graessley, W.W. (1974) *Adv. Polymer Sci.*, **16**.
40. Doi, M. and Edwards, S.F. (1978) *J. Chem. Soc., Faraday Trans. 2*, **74**, 1789, 1802, 1818.
41. Heinrich, G., Straube, E. and Helmis, G. (1988) *Adv. Polymer Sci.*, **85**, 34.
42. Kramer, O. (1992) in *Elastomeric Polymer Networks* (eds J.E. Mark and B. Erman), Prentice Hall, Englewood Cliffs, NJ, chapter 17.
43. Stepto, R.F.T. and Eichinger, B.E. (1992) in *Elastomeric Polymer Networks* (eds J.E. Mark and B. Erman), Prentice Hall, Englewood Cliffs, NJ, chapter 18.
44. Gottlieb, M., Macosko, C.W., Benjamin, G.S., Meyers, K.O. and Merrill, E.W. (1981) *Macromolecules*, **14**, 1039.
45. Macosko, C.W. and Saam, J.C. (1985) *Polymer. Prepr., Div. Polymer Chem., Amer. Chem. Soc.*, **26**, 48.
46. Stepto, R.F.T. (1979) *Polymer*, **20**, 1324.
47. Stepto, R.F.T. (1987) in *Advances in Elastomers and Rubber Elasticity* (eds J. Lal and J.E. Mark), Plenum Press, New York, p. 329.

48. Cawse, J.L., Stanford, J.L. and Stepto, R.F.T. (1979) in Proceedings of the 26th IUPAC International Symposium on Macromolecules, Mainz, IUPAC, Oxford, p. 393.
49. Fasina, A.B. and Stepto, R.F.T. (1981) *Makromol. Chem.*, **182**, 2479.
50. Stanford, J.L. and Stepto, R.F.T. (1982) in *Elastomers and Rubber Elasticity* (eds J.E. Mark and J. Lal), American Chemical Society, Washington, DC, chapter 20.
51. Stanford, J.L., Stepto, R.F.T. and Still, R.H. (1982) in *Reaction Injection Molding and Fast Polymerization Reactions* (ed. J.E. Kresta), Plenum Press, New York, p. 31.
52. Stanford, J.L., Stepto, R.F.T. and Still, R.H. (1984) in *Characterization of Highly-Crosslinked Polymers* (eds S.S. Labana and R.A. Dickie), American Chemical Society, Washington, DC, chapter 1.
53. Lloyd, A.C. and Stepto, R.F.T. (1985) *Brit. Polymer J.*, **17**, 190.
54. Stepto, R.F.T. (1988) in *Biological and Synthetic Polymer Networks* (ed. O. Kramer), Elsevier, Barking, chapter 10.
55. Stepto, R.F.T. (1988) in *Cross-linked Polymers, Chemistry, Properties and Applications of Crosslinking Systems* (eds R.A. Dickie, S.S. Labana and R. Bauer), American Chemical Society, Washington, DC, chapter 2.
56. Stepto, R.F.T. (1988) *Acta Polymerica*, **39**, 61.
57. Rolfes, H. and Stepto, R.F.T. (1990) *Makromol. Chem., Macromol. Symp.*, **40**, 61.
58. Rolfes, H. and Stepto, R.F.T. (1991) *Computat. Polymer. Sci.*, **1**, 100.
59. Stepto, R.F.T. (1994) *Progr. Rubber & Plast. Tech.*, **10**, 130.
60. Rolfes, H. and Stepto, R.F.T. (1994) *Polymer Gels & Networks*, **2**, 149.
61. Dutton, S., Rolfes, H. and Stepto, R.F.T. (1994) *Polymer*, **35**, 4521.
62. Dutton, S., Stepto, R.F.T. and Taylor, D.J.R. (1996) *Angew. Makromol. Chem.*, **240**, 39.
63. Ross-Murphy, S.B. and Stepto, R.F.T. (1996) in *Large Ring Molecules* (ed. J.A. Semlyen), Wiley, Chichester, chapter 16.
64. Dutton, S., Stepto, R.F.T. and Taylor, D.J.R. (1997) *Macromolecular Symposia*, **18**, 119.
65. Stepto, R.F.T. (1997) in *Wiley Polymer Networks Group Review Series, Volume 1* (eds K. te Nijenhuis and W. Mijs), John Wiley & Sons Ltd., Chichester, in press.
66. Leung, Y.-K. and Eichinger, B.E. (1984) in *Characterization of Highly-Crosslinked Polymers* (eds S.S. Labana and R.A. Dickie), American Chemical Society, Washington, DC, chapter 2.
67. Leung, Y.-K. and Eichinger, B.E. (1984) *J. Chem. Phys.*, **80**, 3877, 3885.
68. Shy, L.Y., Leung, Y.-K. and Eichinger, B.E. (1985) *Macromolecules*, **18**, 983.
69. Neuberger, N.A. and Eichinger, B.E. (1985) *J. Chem. Phys.*, **83**, 804.
70. Shy, L.Y. and Eichinger, B.E. (1986) *Macromolecules*, **19**, 2787.
71. Galiatsatos, V. and Eichinger, B.E. (1988) *J. Polym. Sci., Polymer Phys. Edit.*, **26**, 595.
72. Lee, K.-J. and Eichinger, B.E. (1990) *Polymer*, **31**, 406.
73. Lee, K.-J. and Eichinger, B.E. (1990) *Polymer*, **31**, 414.
74. Stepto, R.F.T. and Waywell, D.R. (1972) *Makromol. Chem.*, **152**, 263.
75. Stanford, J.L. and Stepto, R.F.T. (1977) *Brit. Polymer J.*, **9**, 124.
76. Jacobson, H. and Stockmayer, W.H. (1950) *J. Chem. Phys.*, **18**, 1600.
77. *Cyclic Polymers* (1986) (ed. J.A. Semlyen), Elsevier, Barking.
78. Haward, R.N. (1950) *Trans. Faraday Soc.*, **46,** 204.
79. Askitopoulos, V. (1981) M.Sc. Thesis, University of Manchester.
80. Shy, L.Y. and Eichinger, B.E. (1985) *Brit. Polymer J.*, **17**, 200.
81. Somvársky, J. and Dušek, K. (1994) *Polymer Bull.*, **33**, 369, 377.
82. Dušek, K., Somvársky, J., Ilavsky, M. and Matejka (1991) *Computat. Polymer Sci.*, **1**, 90.
83. Samoria, C., Vallés and Miller, D.R. (1986) *Makromol. Chem., Macromol. Symp.*, **2**, 69.
84. Temple, W. (1972) *Makromol. Chem.*, **160**, 277.
85. Peters, R.H. and Stepto, R.F.T. (1966) in *The Chemistry of Polymerisation Processes*, Monograph No. 20, Society of Chemical Industry, London, p. 157.
86. Hopkins, W., Peters, R.H. and Stepto, R.F.T. (1974) *Polymer*, **15**, 315.
87. Frisch, H.L. (1955) Proceedings of Polymer Division Meeting, 128th American Chemical Society Meeting, Minneapolis, American Chemical Society, Washington, DC.
88. Harris, F.E. (1955) *J. Chem. Phys.*, **23**, 1518.
89. Kilb, R.W. (1958) *J. Physical Chem.*, **62**, 969.

90. Stepto, R.F.T. (1974) *Faraday Discuss. Chem. Soc.*, **57**, 91.
91. Ahmad, Z. and Stepto, R.F.T. (1980) *Colloid & Polymer Sci.*, **258**, 663.
92. Rolfes, H. and Stepto, R.F.T. (1993) *Makromol. Chem., Macromol. Symp.*, **76**, 1.
93. Stepto, R.F.T. and Taylor, D.J.R. (1996) *Polymer Gels & Networks*, **4**, 405.
94. *Networks Module*, Molecular Simulations Inc., San Diego.
95. Smith, R.S. and Stepto, R.F.T. (1974) *Makromol. Chem.*, **175**, 2365.
96. Stepto, R.F.T. and Taylor, D.J.R., unpublished results.
97. Rolfes, H. and Stepto, R.F.T. (1993) *Makromol. Chem., Macromol. Symp.*, **65**, 233.

3 Network formation involving polyfunctional polymer chains

K. DUŠEK

3.1 Introduction

In the preceding chapter, network formation from monomers or other precursors of low functionality, varying typically between two and six, by *non-chain reactions* have been dealt with. Some of the components were of low molecular weight, some others were represented by bi-, tri- or tetrafunctional telechelic polymers. The competition between intermolecular and intramolecular reactions was an important issue.

Significant groups of polymer networks are also prepared from high-molecular-weight reactants of high functionality. The groups involve the *crosslinking of primary chains*, often called *vulcanization*, as well as *chain polymerizations*. In chain (co)polymerizations, the networks are formed usually from small monomer molecules. However, long primary chains carrying pendant double bonds are formed during the fast propagation step and, because of this, these processes have some similar features with vulcanization. In contrast to non-chain reactions, the neighbouring-group effect and excluded-volume effect are expected to play more important roles. From the point of view of reaction mechanism, vulcanization can usually be considered a non-chain reaction, whereas crosslinking polymerization involves initiation, propagation, termination and chain transfer reactions and a simple recursive approach to branching cannot be applied. Also cyclization reactions are expected to play a more important role particularly at the beginning of polymerizations.

This chapter describes special features of network formation by crosslinking of primary chains and free-radical chain (co)polymerization and gives basic relationships determining the onset of gelation, development of sol fraction and elastically active network chains. Theoretically, the handling of distributions (degree-of-polymerization and functionality distributions of starting components) is a common feature. It is very convenient to describe these distributions by *generating functions*. The formalism of the generating functions in relation to network formation has been described in several reviews ([1, 2] and references therein). Here, this formalism will be briefly explained and commented on in Appendices 3A and 3B.

The existing network formation theories are:

1. the mean-field, statistical generation from units (fragments) in different reaction states;
2. the generation of structures by kinetics differential equations (Smoluchowski coagulation equations);
3. computer generation in space.

The first two groups are discussed here. Group (1) theories are usually sufficient for the description of the crosslinking of primary chains. The application of group (2) theories is necessary for initiated and other complex reaction occurring during crosslinking chain (co)polymerizations. However, at present there is no reliable theoretical method for the description of chain crosslinking (co)polymerizations that take into account *all* the main chemical and physical factors controlling structure growth.

3.2 Crosslinking of primary chains

The crosslinking (vulcanization) of primary chains is one of the oldest methods used in the manufacture of crosslinked polymers – vulcanized rubbers. Sulfur or sulfur-containing compounds, peroxides or ionizing radiation are used for vulcanization. In addition, functional copolymers are often crosslinked with a bifunctional or multifunctional crosslinking agent. For example, hydroxyethyl acrylate copolymers can be crosslinked with diisocyanates or glycidyl methacrylate copolymers with diamines.

The reactions can be considered as non-chain reactions. In vulcanization, the functionality is usually equal or proportional to the degree of polymerization and the functionality distribution is thus equal to the degree-of-polymerization distribution. In functional copolymers, the distribution of units in the primary chains can be random, or non-random. The type of functionality distribution affects gelation and network build-up. Some of the vulcanization reactions can be accompanied by chain degradation. Vulcanization by ionizing radiation is an example. Theoretically, combined crosslinking and degradation is much more complicated than simple crosslinking because the distribution of 'primary chains' is changing as a result of degradation (see [3, 4]).

Cyclization has been found to be relatively weak in a number of systems formed by non-chain reactions: for polyurethanes, only a few per cent of bonds are wasted in ring-forming reactions. However, in vulcanizing systems it may amount to 10–30%. The extent of cyclization depends on chain conformations and dilution.

3.2.1 General theoretical treatment of network formation

The crosslinking of primary chains in the absence of cyclization has been a subject of a number of theoretical studies (see [5]). The combinatorial approach of Flory has been extended to various complicated systems (see, e.g., [6–8]). Subsequently, the crosslinking of primary chains was dealt with using the cascade formalism [9–12], or the 'recursive method' [13]. Computer simulation in space was developed by Galiatsatos and Eichinger [14, 15]. Here, we will explain the application of the theory of branching processes (cascade theory) which can very easily handle various degree-of-polymerization distributions of primary chains.

The simplest case will be considered. An arbitrary distribution of degree of polymerization of primary chains is considered to be randomly crosslinked by the formation of bonds between any two segments of the chains. The reactions are exclusively intermolecular before the gel point. For circuit closing beyond the gel point, the Flory mean-field assumption of crosslinking of any two segments in the gel is used. All segments have the same reactivity; no neighbouring-group effect exists. The distribution of degree of polymerization of the primary chains is given by the number fraction n_i or weight fraction w_i of i-mer. Ring-free cases will be considered and the effect of cyclization will be commented on.

The number-fraction distribution of i-mers and number of bonds issuing from the crosslinked segments are given by the following *generating function* (gf):

$$F_{0n}(Z, z) = \sum_{i=1}^{\infty} n_i Z^i (1 - \alpha + \alpha z)^i = \sum_{i=1}^{\infty} n_i Z^i \sum_{j=1}^{i} (1 - \alpha)^{i-j} \alpha^j z^j \tag{3.1}$$

In this gf, the exponent i of Z identifies the number of segments in a primary chain and the exponent j $(0 \leq j \leq i)$ of z, the number of bonds issuing from segments of such a primary chain; α is the fraction of crosslinked units. The gf is normalized, so that $F_{0n}(Z, z) = 1$ for $Z = z = 1$ (defining a probability generating function). A primary chain with bonds extending to topologically neighbouring units is shown in Figure 3.1.

As shown in Appendix 3A, the gf for the number of *additional* bonds issuing from primary chains already bound to another primary chain is obtained by the differentiation of $F_{0n}(Z, z)$ with respect to z and renormalization:

$$F(Z, z) = \sum_{i=1}^{\infty} w_i (1 - \alpha + \alpha z)^{i-1} \tag{3.2}$$

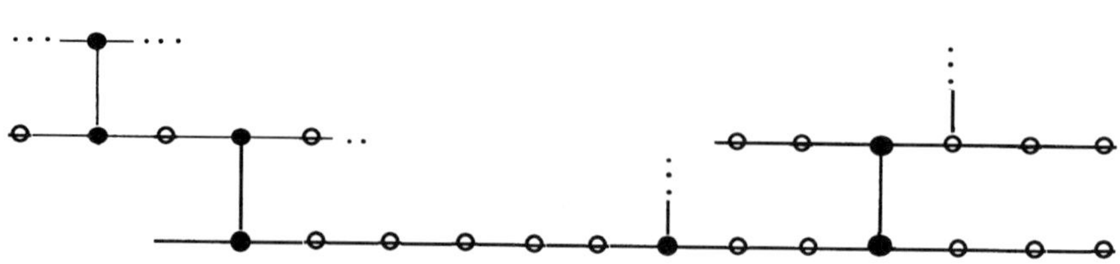

Figure 3.1 Network build-up with primary chains as building units: (○) uncrosslinked monomer unit; (●) crosslinked monomer unit.

Infinite continuation of the structure is possible if the average number of additional bonds is ≥ 1. The average number of additional bonds is given by the value of the derivative of $F(Z, z)$ for $Z = z = 1$,

$$F'(1,1) \equiv \left[\frac{\partial F(Z,z)}{\partial z}\right]_{Z=z=1}$$
$$F'(1,1) = \sum_{i=1}^{\infty} w_i(i-1)\alpha = 1 \tag{3.3}$$

Therefore, the critical fraction of crosslinked units at the gel point α_c is given by

$$\alpha_c = \frac{1}{\overline{P_{0w}} - 1} \tag{3.4}$$

The critical value is thus determined by the *weight-average degree of polymerization* of primary chains ($\overline{P_{0w}}$) irrespective of the type of their distribution, because the chance of participation of a primary chain in structure propagation is proportional to the number of its segments.

In the *pre-gel stage*, crosslinking manifests itself by an increase in the degrees of polymerization (molecular weights) of branched molecules. The number-average degree of polymerization $\overline{P_n}$ is given by the ratio of the average number of units per primary chain to the number of molecules. In the ring-free case, the number of branched molecules is given by the number of primary molecules minus the number of bonds connecting them. The number of bonds per primary chain is given by $F'_{0n}(1,1)/2$, where $F'_{0n}(1,1)$ is the value of the derivative of $F_{0n}(Z, z)$ with respect to z for $Z = z = 1$. Thus,

$$\overline{P_n} = \frac{\overline{P_{0n}}}{1 - F'_{0n}(1,1)/2} \tag{3.5}$$

or

$$\overline{P_n} = \frac{\overline{P_{0n}}}{1 - \alpha \overline{P_{0n}}/2} \tag{3.6}$$

This relationship is also independent of the type of distribution of primary chains. The weight-fraction gf giving the weight-average degree of polymerization $\overline{P_w}$ reads

$$W(Z) = \sum_{i=1}^{\infty} w_i Z^i (1 - \alpha + \alpha u)^i \tag{3.7}$$

where u is given by the recursive equation

$$u = \sum_{i=1}^{\infty} w_i Z^i (1 - \alpha + \alpha u)^{i-1} \tag{3.8}$$

The value of $\overline{P_{\mathrm{w}}}$ is obtained by differentiation of $W(Z)$ with respect to Z and by putting $Z = 1$:

$$\overline{P_{\mathrm{w}}} = \frac{\overline{P_{0\mathrm{w}}}(1+\alpha)}{1-\alpha(\overline{P_{0\mathrm{w}}}-1)} \tag{3.9}$$

The denominator becomes equal to 0 at the gel point and $\overline{P_{\mathrm{w}}}$ diverges whereas $\overline{P_{\mathrm{n}}}$ is still finite.

In the *post-gel stage*, some of the bonds issuing from segments of a chain have finite and some infinite continuation. Finite continuation means that the issuing branch (subtree) is finite; through bonds with infinite continuation, the chain is connected to the (infinite) gel. A bond has a finite continuation with probability v defined by

$$v = F(1, v) = \sum_{i=1}^{\infty} w_i(1-\alpha+\alpha v)^{i-1} \tag{3.10}$$

This equation has one trivial root, $v = 1$, and another root in the interval $0 \leq v \leq 1$, which is the desired one. The value of v is determined by α and the type of distribution; it is obtained by the numerical solution of equation (3.10).

The extinction probability, v, is a key quantity for the calculation of post-gel parameters, such as sol fraction, w_{s}, molecular weight averages of the sol, fractions of crosslinked units in the sol and gel, fractions of material in dangling chains and elastically active network chains (EANC), average degrees of polymerization of dangling chains and EANCs, and number of EANCs per monomer unit, N_{e}, or their concentration per unit volume, ν_{e}. Here, we will show only the derivations of w_{s} and ν_{e}. More details are available elsewhere [1, 2, 9, 16].

The soluble fraction (sol) is by definition composed of units from which only bonds of finite continuation extend. This condition is introduced by weighting the probability that a unit has reacted and a bond has been formed (α) by the probability v that this bond has finite continuation. Thus, the weight fraction of the sol is given by

$$w_{\mathrm{s}} = 1 - w_{\mathrm{g}} = \sum_{i=1}^{\infty} w_i(1-\alpha+\alpha v)^{i} \tag{3.11}$$

From equations (3.10) and (3.11), it follows that

$$w_{\mathrm{s}} = v(1-\alpha+\alpha v) \tag{3.12}$$

so that the sum over the distribution is eliminated. However, v is still dependent on the type of distribution.

Sometimes, the experimental sol fraction is used for determination of the fraction of crosslinked units, α, using the above equation because a direct analytical determination of α is difficult. However, one has to be careful

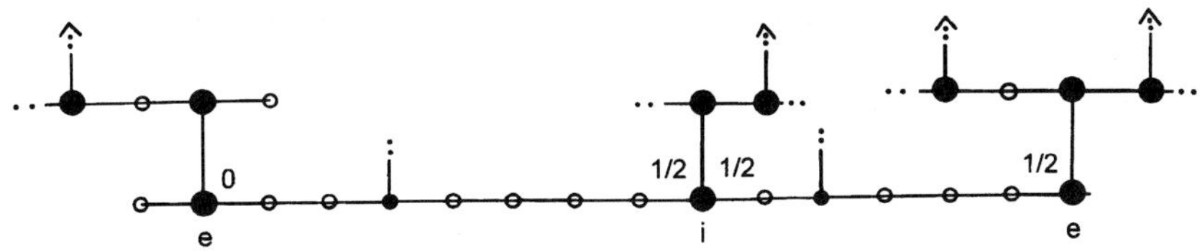

Figure 3.2 Specification of the number of elastically active network chains (EANC): (O) uncrosslinked monomer unit; (•) crosslinked monomer unit; (●) active crosslinked unit (ACU); (e) end ACU; (i) inner ACU. Contributions to the number of EANCs indicated as 0 and $\frac{1}{2}$.

because w_s can be affected by cyclization. Also, it depends on the molecular weight distribution of primary chains.

The calculation of the number or concentration of EANCs is based on considerations of the primary chains as building units. Each primary chain contains j crosslinked units from which k bonds issue with *infinite* continuation. Such a crosslinked unit is called an *active crosslinked unit* (ACU). The distribution for the number of ACUs per primary chain is described by the gf $T(Z)$:

$$T(Z) = \sum_{i=1}^{\infty} n_i \{1 - \alpha + \alpha[v + (1 - v)Z]\}^i = \sum_{k=0}^{\infty} t_k Z^k \tag{3.13}$$

where the coefficient t_k is the probability of finding a primary chain with k ACUs. An EANC is represented by a sequence of units between two ACUs from which at least three paths issue to infinity. In the case of crosslinking of primary chains, each ACU with $k \geq 2$ has two end ACUs and $(k - 2)$ inner ACUs (Figure 3.2). Each inner ACU contributes $2 \times \frac{1}{2}$ to the number of EANCs. The contribution of an end ACU is conditional. The end ACU contributes by $\frac{1}{2}$ if it is bound to an inner ACU of another primary chain; it does not contribute to the number of EANCs if it is connected with an end ACU of the other chain. These contributions to the number of EANCs per primary chain N_e can be expressed through the coefficients, t_k, of the gf $T(Z)$:

$$N_e = \sum_{k=3}^{\infty} (k - 2)t_k + 2\frac{1}{2}t_2 \frac{\sum_{k=3}^{\infty}(k - 2)t_k}{\sum_{k=2}^{\infty}(k - 2)t_k} \tag{3.14}$$

This equation can conveniently be expressed through the values of derivatives of $T(Z)$:

$$N_e = \frac{[T'(1) + T'(0) + 2T(0) - 2][1 + T'(1) - T(0) - 2T'(0)]}{1 - T(0) - T'(0)} \tag{3.15}$$

where

$$T(X) \equiv T(Z = X) \qquad X = 0, 1$$

and

$$T'(X) \equiv \left(\frac{\partial T(Z)}{\partial Z}\right)_{Z=X}$$

The concentration of EANCs per unit volume ν_e is related to N_e as

$$\nu_e = \frac{N_e d}{M_0 \overline{P}_{0n}} \tag{3.16}$$

3.2.2 Explicit degree-of-polymerization distributions of primary chains

The preceding equations were formulated for any arbitrary degree-of-polymerization distribution expressed by sets of fractions $\{n_i\}$ or $\{w_i\}$. In reality, a distribution is a discrete or continuous distribution that has been derived from the polymerization mechanism or is empirical. Examples of distribution are listed in Appendix 3B. Of these, the Schulz–Zimm distribution function covers a wide range of actual distributions.

It is of advantage to use distributions in the form of gfs (Appendix 3B). It is a matter of routine, to replace all the sums in the equations by number- or weight-fraction gfs. Here, we illustrate the procedure using the example of the most probable distribution. In equation (3.17) the variable Z is replaced by $Z(1-\alpha+\alpha z)$ where Z keeps track of the number of all monomer units and z of the number of bonds issuing from *crosslinked* monomer units. Thus,

$$F_{0n}(Z,z) = N(Z,z) = \frac{(1-q)Z(1-\alpha+\alpha z)}{1-qZ(1-\alpha+\alpha z)} \tag{3.17}$$

where

$$q = 1 - \frac{1}{\overline{P}_n}$$

and

$$F(Z,z) = \frac{(1-q)^2}{[1-qZ(1-\alpha+\alpha z)]^2} \tag{3.18}$$

The relationships for other parameters are derived routinely, e.g.

$$\alpha_c = \frac{1-q}{2q} \tag{3.19}$$

$$v = \frac{(1-q)^2}{[1-q(1-\alpha+\alpha v)]^2} \tag{3.20}$$

from which the trivial root $v = 1$ can be eliminated.

3.2.3 *Functional copolymers*

The network-formation relationships for functional copolymers, containing a fraction a of reactive units, are determined by the distribution of the occurrence of functional monomer units. If this distribution is random, the corresponding gf reads as

$$F_{0n}(Z, z) = \sum_{i=1}^{\infty} n_i Z^i [1 - a + a(1 - \alpha + \alpha z)]^i \tag{3.21}$$

and the gel-point condition reads

$$\alpha_c = \frac{1}{a(\overline{P_{0w}} - 1)} \tag{3.22}$$

In block copolymers or 1:1 ($a = 1/2$) alternating copolymers, the corresponding equations are

$$F_{0n}(Z, z) = \sum_{i=1}^{\infty} n_i Z^i (1 - \alpha + \alpha z)^{ai} \tag{3.23}$$

$$\alpha_c = \frac{1}{a\overline{P_{0w}} - 1} \tag{3.24}$$

The difference in the values of α_c given by equations (3.23) and (3.24) is significant only for small values of $\overline{P_{0w}}$.

For the crosslinking of copolymers of varying composition (e.g. due to conversion drift)

$$F_{0n}(Z, z) = \sum_{i=1}^{\infty} n_i Z^i [(1 - a_i) + a_i(1 - \alpha + \alpha z)^i] \tag{3.25}$$

and

$$\alpha_c = \frac{1}{(\overline{aP_{0w}}) - 1} \tag{3.26}$$

3.2.4 *Cyclization*

As in the case of network formation using reactants of low functionalities, described in Chapter 2, cyclization competes with intermolecular branching. Generally, cyclization is expected to occur somewhat more frequently during crosslinking because there are more reactive groups or units per molecule. The fraction of bonds lost by intramolecular reaction typically amounts to 10–30% in the initial stages of crosslinking compared with less than 10% during network formation from low-molecular-weight monomers.

The fraction of bonds lost by cyclization is expected to grow somewhat with increasing conversion. Before the gel point, a definition of an intramolecular bond is straightforward: any ring closure does not increase the molecular weight of a molecule. Beyond the gel point, cyclization must be redefined, since closed circuits (cycles) always exist in the gel. In fact, the equilibrium elastic modulus can be related to the *cycle rank* of the gel. One rather asks how cyclization can affect the equilibrium mechanical response of the network. In that sense, the cycles of interest must be defined as *elastically inactive cycles* [8, 16]. If such a cycle is formed, the equilibrium modulus does not change. However, from this point of view only simple loops can be defined as *elastically inactive* (Figure 3.3). In more complicated ring structures (cycles), the chains contribute to the modulus but not by the same factor as other chains for which the correlation through circuits in the gel exists over large distances. As a result, it is not quite clear what cycle rank value should

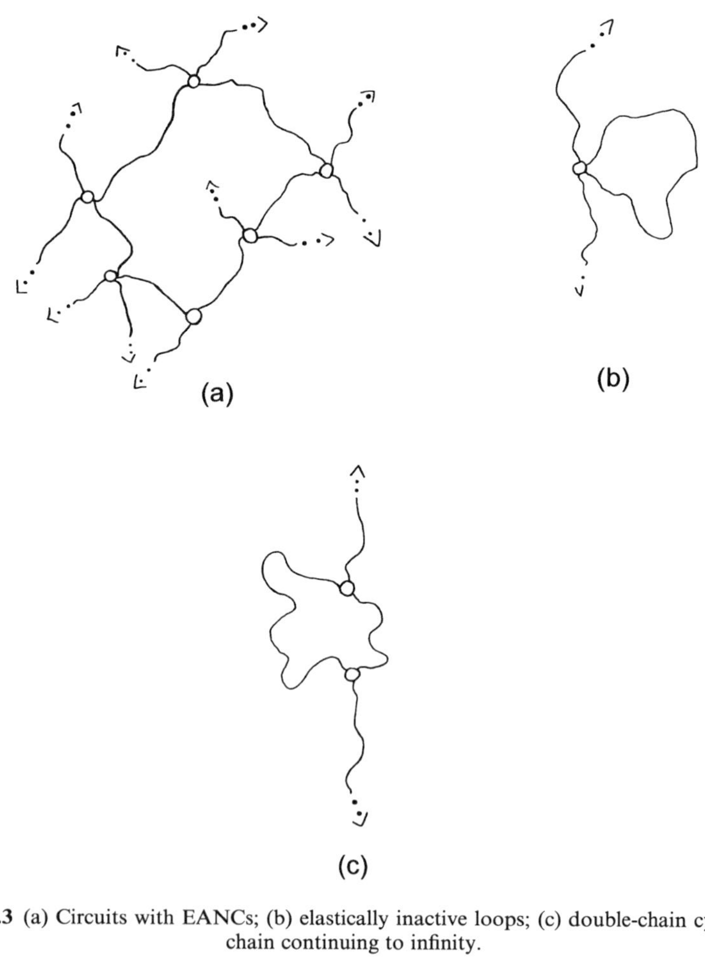

Figure 3.3 (a) Circuits with EANCs; (b) elastically inactive loops; (c) double-chain cycle; (→) chain continuing to infinity.

be considered in the case of phantom network behaviour (see Section 7.4), for which the equilibrium modulus should be proportional to the cycle rank.

Close to the gel point, large cycles can be traced which conform to the definition of elastic inactivity. The segments of the ring-closing sequences can, however, undergo crosslinking at later stages by which these cycles become elastically active. This process has been called the *ring activation process* [10]. The mechanism of formation of elastically inactive rings and of the ring activation process is shown in Figure 3.4.

In view of the preceding difficulties, theoretical modelling of crosslinking in the presence of cyclization is more difficult than in the case of the polymerization of low-molecular-weight monomers, both in the pre-gel and post-gel stages. One can refer to mean-field approaches [10, 17–20] as

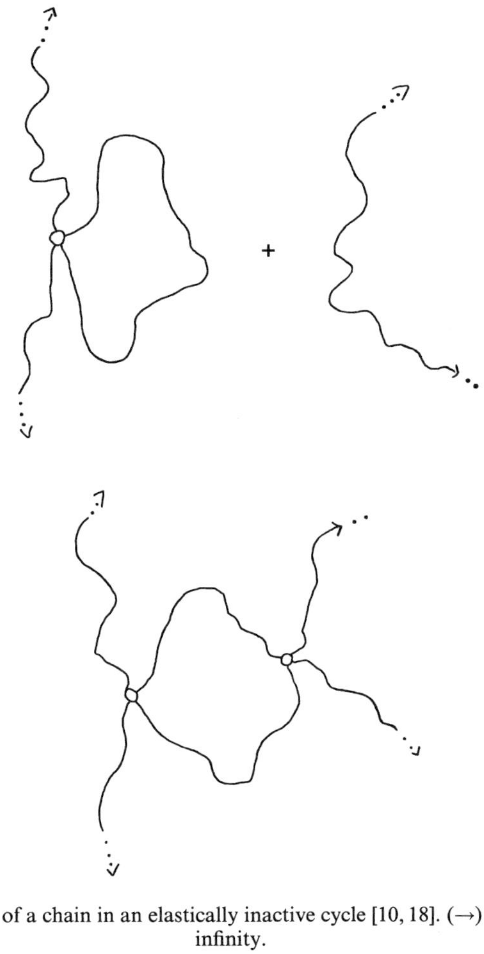

Figure 3.4 Activation of a chain in an elastically inactive cycle [10, 18]. (→) chain continuing to infinity.

well as to Monte-Carlo simulations in space [14, 15]. These studies show that fractions of ring-forming bonds increase with increasing degree of polymerization of primary chains, their flexibility, and dilution of the reaction system. Also, experimental dependences are in agreement with these trends. Determination of the shift in molecular-weight evolution compared with its expected ring-free dependence on conversion and shift of the gel-point conversion are the best characteristics of the intensity of cyclization at present.

3.3 Chain crosslinking (co)polymerization

The term *chain crosslinking (co)polymerization* includes primarily the *free-radical* polymerization of multivinyl monomers or their copolymerization with monovinyl monomers. In a broader sense, other *initiated* growth processes, by which primary chains, traceable in the resulting branched structures, are formed, like living polymerizations, can also be included. The growth of each primary chain is started by an initiation step, proceeds by propagation and is interrupted by chain transfer or termination. However, the features of the crosslinking process and structure of the product depend on whether the propagation step is fast or slow compared with the other steps. In this section, network formation by free-radical polymerization characterized by fast propagation is discussed exclusively.

3.3.1 Network formation mechanism

From the point of view of elementary reaction steps, the free-radical crosslinking copolymerization does not differ from the linear copolymerization of monovinyl compounds and involves *initiation*, *propagation*, *chain transfer* and *termination*, where transfer and termination can occur by several mechanisms. Network formation by the free-radical copolymerization of multivinyl (or, in general, multiunsaturated) monomers also displays special features of linear polymerizations, primarily the occurrence of the Trommsdorff effect (gel effect) due to the diffusion control of termination-by-radical recombination. An increase in the concentration of free-radical species and the resulting increase in the polymerization rate and degrees of polymerization of chains are the main consequences. The onset of diffusion control depends on the viscosity of the polymerizing system and the formation of long chains favours a decrease in the diffusivities of the macroradicals.

In contrast, there exist several main differences arising from the participation of a multi-unsaturated monomer in a polymerization compared with that of a mono-unsaturated monomer [20]:

1. Multi-unsaturated monomer molecules are first incorporated into polymer chains as units with pendant double bond(s) (Figure 3.5).

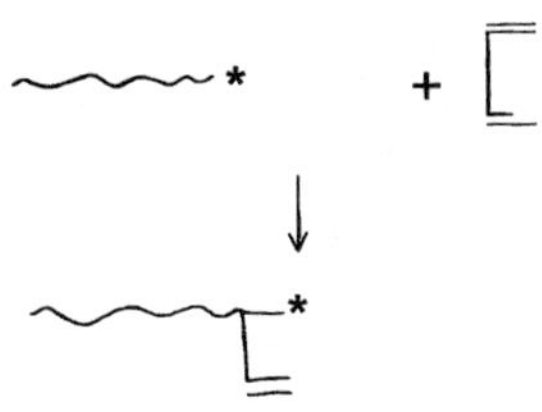

Figure 3.5 Formation of a pendant double bond.

2. Cyclization is much more important: i.e. propagation, transfer and termination reactions can proceed intermolecularly and intramolecularly.
3. The effective reactivities of the pendant double bonds can change, due to chemical reasons, steric hindrance, or thermodynamic excluded-volume effects.
4. The Trommsdorff effect occurs in the early stages of polymerization, as the system approaches the gel point.
5. Different reactivities of double bonds of participating monomers may result in conversion-dependent degrees of crosslinking and possibly in the formation of uncrosslinked polymer, either at the end or at the beginning of network formation.

The enhanced probability of cyclization just at the beginning of a polymerization is explained by a scheme given in Figure 3.6. The active end of the growing macroradical (*) can react (a) with double bonds of the monovinyl or multivinyl monomer, or (b) with pendant double bonds on other, chemically unconnected chains (*intermolecular crosslinking*), or (c) with pendant double bonds on the same chain or primary chains chemically connected with it (*intramolecular crosslinking, cyclization*). The two types of cyclization reactions are sometimes distinguished from each other: ring closing with pendant double bonds on the same chain (c_1) is called *cyclization*

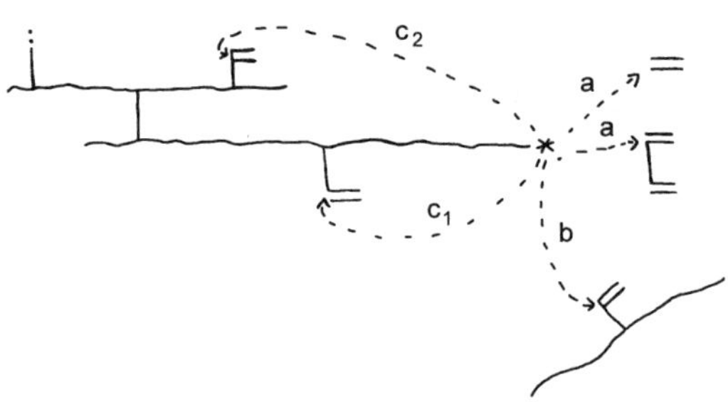

Figure 3.6 Elementary reactions of a growing macroradical during a free-radical chain copolymerization: (a) propagation involving monomers; (b) intermolecular crosslinking; (c_1 and c_2) cyclization, (c_1) involving pendant double bonds of the same primary chain and (c_2) involving double bonds of other connected chains (multiple crosslinking).

and that with double bonds on other chains of the same molecule (c_2) is called *multiple crosslinking*. At the beginning of a polymerization, there exist only a few chains with pendant double bonds in the reacting system, so the probability of intermolecular crosslinking is very low and converges to zero at zero conversion. However, intramolecular reaction (cyclization) can always proceed if the ring closure is allowed by chain flexibility, so the first crosslinks formed are almost exclusively intramolecular. Rings of various sizes are formed, with the smallest one being usually the most probable. If formation of small rings is very probable, as in the case of diallylamines or divinyl formals, only unbranched chains containing small rings are formed (*cyclopolymerization*) [21, 22] and gelation does not occur. Cyclization is not so important when the fraction of the divinyl monomer is low or when the two double bonds are located at the extremities of a telechelic vinyl-terminated polymer chain. The probability of the spatial coincidence of two groups on a polymer chain scales with $N^{-3/2}$, where N is the number of units in the connecting sequence (see Section 2.3.1).

Extensive cyclization is one special feature of free-radical crosslinking copolymerization. A strongly decreased reactivity of the pendant double bonds in the formed molecules is another. This decrease is primarily due to the formation of cycles. Many of the pendant double bonds become entrapped in the internally crosslinked structure. The decrease in reactivity can also be assisted by thermodynamic excluded volume. The relative importance of the two factors can be determined from the fraction of units with pendant double bonds at the beginning of a polymerization; the presence of doubly reacted divinyl units manifests cyclization. Figure 3.7 shows that, even at low conversions, a considerable fraction of the divinyl units has reacted at both double bonds. However, there is a common feature that even at full conversion of monomer, many pendant double bonds survive for a long time. At high concentrations of the divinyl monomer the fraction of units with pendant double bonds *in the polymer* is almost constant (Figure 3.8).

The special features of network formation by free-radical crosslinking copolymerization are determined by the peculiar polymerization behaviour of the pendant double bond. Its *apparent* reactivity is initially enhanced by cyclization but decreased due to steric hindrance (excluded volume) arising as a direct consequence of cyclization. The structures formed are rather compact. The evolution of the structures is illustrated in Figure 3.9. Compact, internally crosslinked molecules are formed at the beginning of a polymerization. The pendant double bonds in the interior cannot take part in further polymerization and only the peripheral double bonds participate in the structure growth. Thus, gelation occurs by the chemical joining of these primary compact molecules. The concept of formation of primary compact (microgel-like) molecules and their chemical 'aggregation' resulting in macroscopic gelation has been corroborated by the existence of bimodal

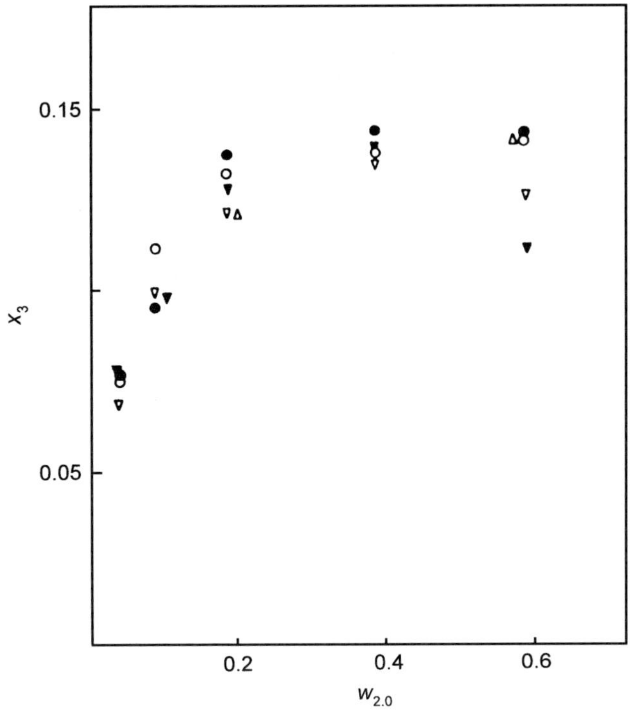

Figure 3.7 Dependence of the fraction of divinyl units with two reacted double bonds, x_3, in soluble (<1% conversion) styrene–ethylene dimethacrylate (ST–EDMA) copolymers on the weight fraction of EDMA in feed, $w_{2,0}$, at different dilutions. Volume fraction of monomers in feed: (○) 1.0, (●) 0.8, (▽) 0.6, (▼) 0.4, (△) 0.2. Redrawn from Ref. 49.

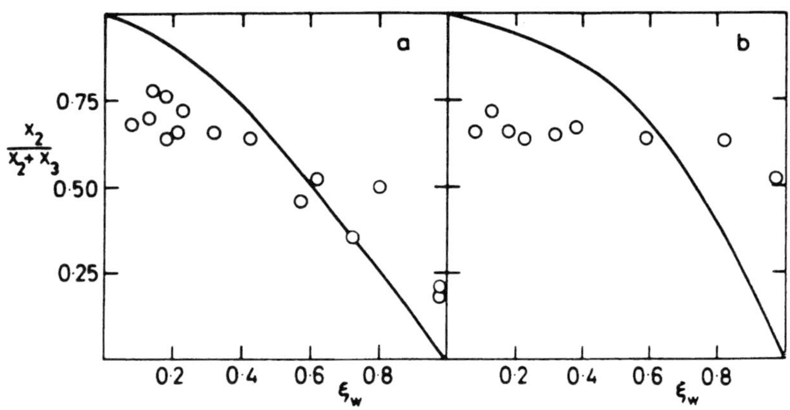

Figure 3.8 Dependence of the relative fraction of units with pendant double bonds, $x_2/(x_2 + x_3)$ on weight conversion of monomers, ξ_w, in copolymers of styrene (ST) with *p*-divinylbenzene (DVB). Concentrations of DVB in the feed: (a) 8% DVB; (b) 15% DVB. Reprinted with permission from Ref. 20.

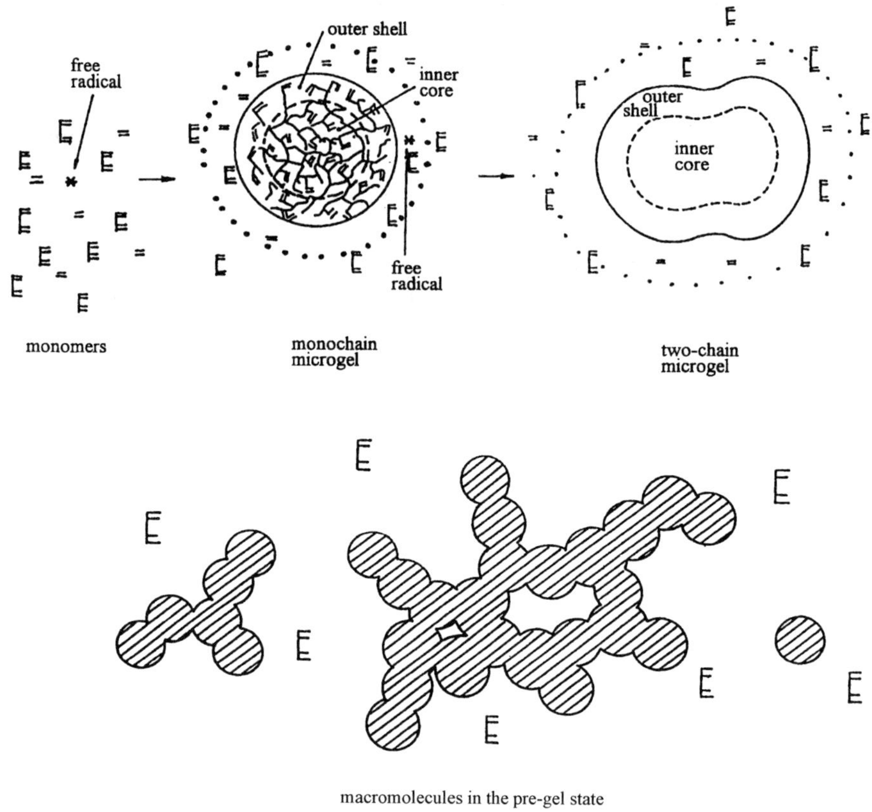

Figure 3.9 Mechanism of network formation by free-radical copolymerization.

molecular weight distributions and transformation of the lower-molecular-weight fractions into a distribution of higher-molecular-weight ones (from size-exclusion chromatography (SEC) and dynamic-light-scattering measurements) [23–25].

The preceding conclusions are based on experimental findings in the range of intermediate and high concentrations of 'small' bisunsaturated monomers. They can be summarized as follows:

- Diffusion control (Trommsdorff effect, propagation limitation etc.) plays a much more important role than in network formation by non-chain reactions or monovinyl polymerizations.
- The fraction of units with two reacted double bonds at the beginning of polymerization is high.
- Above a certain level, the fraction of units with both reacted double bonds does not increase with increasing concentration of the bisunsaturated monomer.

- The efficiency of crosslinking in branching is low and decreases with decreasing conversion.
- The intrinsic viscosity is low and is relatively insensitive to molecular weight.
- The gel point conversion is up to 2–3 orders of magnitude higher compared with the theoretical value for a ring-free system.
- The mobility of a part of the molecular structure, even of the low-conversion pre-gel polymer, is low and decreases with increasing concentration of the bisunsaturated monomer.
- A bimodal distribution of the pre-gel polymer with a relative increase and broadening of the higher-molecular-weight part with increasing conversion.

The picture of network formation according to the scheme in Figure 3.9 has been put forward, for instance, by the groups headed by Kloosterboer (see, e.g., [26]), Okay and co-workers (see, e.g., [27–29]), Lee and co-workers (see, e.g., [23–25, 30]) and Dušek and co-workers (see, e.g., [20, 31, 32]). It should be emphasized again that α,ω-bisunsaturated precursors, such as polyurethane bisacrylates, exhibit almost normal copolymerization behaviour because of the large separation distance between the two double bonds.

3.3.2 Modelling

The theories of all three groups (Section 3.1) have been used. The statistical theories where the network structures are generated from monomer units in different reaction states [33] can be used for modelling only in the very special case of (a) no cyclization, (b) most probable degree-of-polymerization distribution of primary chains, (c) equal and independent reactivity of double bonds and (d) no diffusion control over the whole conversion range. Even if a correction is made for the fractions of bonds wasted in cycles [34], the approximation is still very crude and does not reflect many of the special features of molecular-structure growth.

3.3.2.1 Kinetics theories

Kinetics theories can describe network formation rigorously in the case of ring-free systems and with some approximation when cyclization is involved. Assumptions (b)–(d) in the preceding paragraph can be relaxed and one can formally include cyclization as well as degradation reactions. Translational or segmental diffusion control can be included and diffusion coefficients can be made dependent on the properties of individual molecules or on average properties of their assemblies [35]. The differential equations, where time is the independent variable, are of the type of (generalized) Smoluchowski coagulation equations. The *mass action law*, meaning proportionality of the reaction rate to reactant concentrations or their products, is still valid (mean-field element), but the kernel of the equations (the rate constant) can be made a structure-dependent function, more complex than direct proportionality to the numbers of functional groups on the reacting molecules.

Most theoretical approaches represent an extension of the schemes valid for linear polymerizations [36, 37]. If considered at all, the Trommsdorff effect is taken as nonspecific. This means that the diffusivities of the macroradicals depend only on the average degree of polymerization and not on the individual degrees of polymerization. Molecules and macroradicals are characterized by the number of monomer units x (with possible distinctions between types in copolymers), number of free-radical ends r and number of pendant double bonds l. The change in concentration of such species with time can be considered to be composed of several contributions to the rate v as

$$v = \frac{\mathrm{d}c(x,r,l)}{\mathrm{d}t} = v_{\text{initiation}} + v_{\text{propagation}} + v_{\text{transfer}} + v_{\text{termination}} \tag{3.27}$$

The propagation, transfer and termination reactions are usually considered to be intermolecular but can also occur intramolecularly.

The evolution of the distribution of $c(x,r,l)$ is described by an infinite set of differential equations, assuming the validity of the mass-action law. Thus, the contribution of intermolecular propagation can be considered in the following form:

$$\begin{aligned}\left[\frac{\mathrm{d}c(x,r,l)}{\bar{k}\,\mathrm{d}t}\right]_{\text{propagation}} &= \sum_{i=1}^{r-1}\sum_{j=0}^{r}\sum_{k=1}^{l}(r-j)(k+1)c(x-i,r-j,l-k)c(i,j,k+1)\\ &\quad + j(l-k+1)c(x-i,r-j,l-k)c(i,j,k)\\ &\quad - c(x,r,l)\sum_{i=1,j=0,k=1}^{\infty} jlc(i,j,k) + krc(i,j,k)\end{aligned} \tag{3.28}$$

The population of multiple free radicals ($r > 1$) is usually small and their contribution can be neglected. However, this is not true for the largest molecule – the gel. For the ring-free case, when the reactivities of the macroradical and pendant double bond *are not* a function of the vector $(\mathbf{x}, \mathbf{r}, \mathbf{l})$, the distribution $c(x,r,l)$ can be obtained implicitly, for example, in the form of a generating function. From it, moments of the distribution such as $\overline{P_\mathrm{n}}$, $\overline{P_\mathrm{w}}$, $\overline{P_\mathrm{z}}$, and the condition for gelation can be calculated. Beyond the gel point, the solution gives the properties of the sol [38].

Even simpler is the ring-free treatment of polymerization and copolymerization of a bisunsaturated monomer with independent reactivities of double bonds, when a pendant double bond is of the same reactivity as the double bonds of the monomer. Then, the bridge connecting the two reacted or unreacted double bonds in the bisunsaturated monomer unit can be split. Each of the generated halves bears a dummy functional group y. A distribution of primary chains is generated from the halves (and the monounsaturated monomer in the case of copolymerization) in accord with the kinetics laws of free-radical copolymerization. The species formed (primary chains) still bear the dummy functional group y. Pairs of these are randomly recombined to form the original bridges in the bisunsaturated monomer units. This process is schematically shown in Figure 3.10. For the recombination of primary chains, the statistical method is used. This type of simulation of branched and crosslinked structures has been named the *combined method* and proposed by Dušek [39] and Dušek and Šomvársky [40]. The method also allows one to describe the molecular structure of the gel in terms of primary chains and their branch-point distribution. Due to

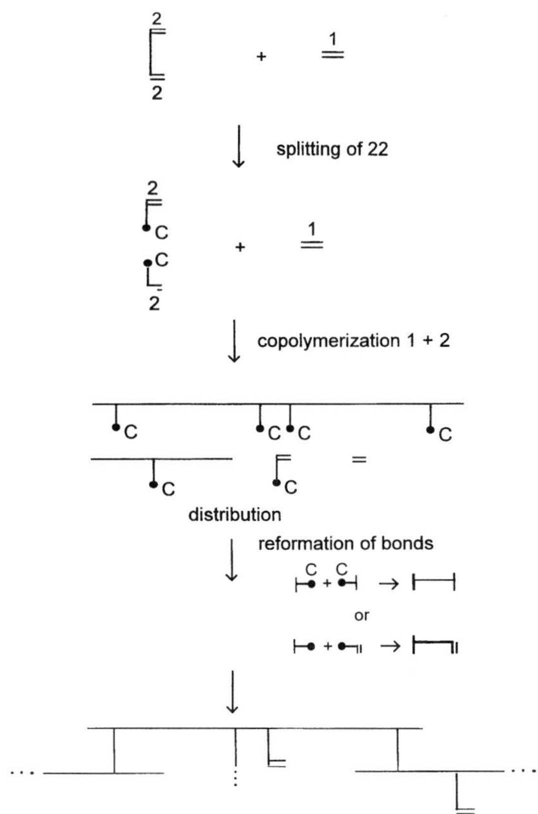

Figure 3.10 Strategy adopted in the combined method for the simulation of network build-up.

Table 3.1 Elementary reactions for the copolymerization of bisunsaturated (1) and mono-unsaturated (2) monomers

No.	Type of reaction	Reaction	Rate constant[a]
1	Dissociation of initiator	$II \rightarrow 2I^*$	k_{II}
2	Initiation	$I^* + M_x \rightarrow IM_x^*$	k_{Ix}
3	Propagation	$\ldots M_x^* + M_y \rightarrow \ldots M_xM_y^*$	k_{xy}
4	Transfer to monomer	$\ldots M_x^* + M_y \rightarrow \ldots M_x + M_y^*$	k_{Axy}
5	Termination by recombination	$\ldots M_x^* + M_y^* \rightarrow \ldots M_xM_y \ldots$	k_{Bxy}
6	Termination by disproportionation	$\ldots M_x^* + M_y^* \rightarrow \ldots M_x + M_y \ldots$	k_{Cxy}

[a] $x, y = 1, 2$.

the neglect of cyclization and of the lowered reactivity of pendant double bonds, the application of this approach is limited to systems involving telechelic bisunsaturated polymers where the separation between the two double bonds is large.

The application of the approach to the copolymerization of bisunsaturated (1) and mono-unsaturated (2) monomers has been presented [41, 42]. The elementary reactions shown in Table 3.1 were considered. The distribution of polymeric species, the macroradicals $(M_1)_m(M_2)_nM_1^*$, $(M_1)_m(M_2)_nM_2^*$, and inactive primary chains $(M_1)_m(M_2)_nM_1$, $(M_1)_m(M_2)_nM_2$, are obtained by summing the contributions from reactions 1–6 in the form of differential equations for generating functions, in which time is the independent variable. The generating functions serve as input information for the branching process of the recombination of primary chains. Standard operations are used to obtain the desired structural parameters. The expressions contain values of the generating functions and of their derivatives. Therefore, one has to find these values at a given time t by solving numerically the equations for the derivatives. An example of calculated dependences is given in Figure 3.11.

If the kernels of the rate equations are more complex (for instance, the kernel is proportional to a power of the number of functional groups on the molecule), the transformation to a generating function is not possible and the moment method is inapplicable. Rather, one has to use a Monte-Carlo simulation (not in space) of the process described by differential equations [43, 44].

Cyclization in the propagation step can be described formally by the equation

$$\left[\frac{dc(x, r, l)}{dt}\right]_{\text{prop. cycl.}} = K_c(x, r, l + 1)c(x, r, l + 1) \qquad (3.29)$$

The 'rate constant' $K_c(x, r, l)$ represents the probability that the free radicals on a molecule can reach pendant double bonds available on the same molecule. In a randomly branched acyclic structure with no excluded-volume effect this probability is determined by the sum over all possible

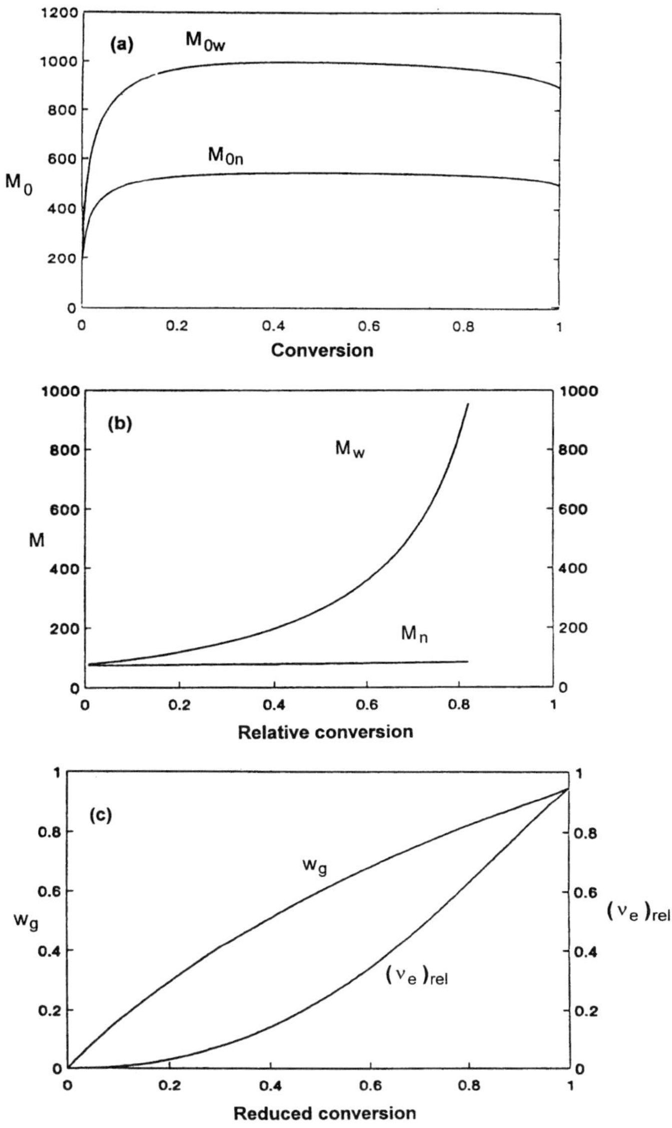

Figure 3.11 Example of the kinetics simulation of network build-up from mono-unsaturated and bisunsaturated monomers using the combined method [42]. Initial composition: number fractions of double bonds, $n_1 = n_2 = 1/2$; molecular weights of monomer fragments, $M_1 = 50$, $M_2 = 60$. Values of kinetics constants defined in text: $k_{II} = 1$, $k_{I1} = k_{I2} = 10^3$; $k_{11} = k_{12} = k_{21} = k_{22} = 10^3$; $k_{A11} = k_{A12} = k_{A21} = k_{A22} = 10^2$; $k_{B11} = k_{B12} = k_{B21} = k_{B22} = 10^4$; $k_{C11} = k_{C12} = k_{C21} = k_{C22} = 10$. (a) Molecular weights of primary chains versus conversion, ξ. (b) Molecular weight of molecules before the gel point versus relative conversion, ξ/ξ_g. (c) Gel fraction, w_g, and relative concentration of EANCs, $(\nu_e)_{rel}$ (relative to the value for $\overline{P_{0n}} \to \infty$) versus reduced conversion, $(\xi - \xi_g)/(1 - \xi_g)$. The critical conversion of double bonds at gel is $\xi_g = 0.1185$.

paths extending from the free radical to the pendant double bonds weighted by $d^{-3/2}$, where d is the separation distance counted in the number repeat units between the free radical and a given pendant double bond (cf. Section 2.3.1). For a cyclic structure, particularly in the gel, the situation is much more complex. One can consider an *Ansatz*

$$K_c(x, r, l) \propto r l^{-\sigma(x,\zeta)} + l r^{-\sigma(x,\zeta)} \tag{3.30}$$

where ζ is the cycle rank of the molecule. Apparently, for large, highly cross-linked (cyclized) structures $l^{-\sigma(x,\zeta)}$ or $r^{-\sigma(x,\zeta)} \rightarrow 0$ and the reaction may stop. The functional form of σ has still to be investigated. Also, such an approach requires one to keep track of ζ of individual molecules, which is impossible. Equation (3.29) can be modified to

$$\left[\frac{\mathrm{d}c(x, r, l, \zeta)}{\mathrm{d}t}\right]_{\text{prop. cycl.}} = K_c(x, r, l+1, \zeta - 1)c(x, r, l+1, \zeta - 1) \tag{3.31}$$

The effect of *limited accessibility* (steric excluded-volume) on intermolecular reactions can also be modelled by a product kernel with exponents different from unity (modification of equation (3.28))

$$\begin{aligned}\left[\frac{\mathrm{d}c(x, r, l)}{k\,\mathrm{d}t}\right]_{\text{propagation}} &= \sum_{i=1}^{r-1}\sum_{j=0}^{r}\sum_{k=1}^{l}(r-j)^{\nu}(k+1)^{\nu}c(x-i, r-j, l-k)c(i, j, k+1)\\ &\quad + j^{\nu}(l-k+1)^{\nu}c(x-i, r-j, l-k)c(i, j, k)\\ &\quad - c(x, r, l)\sum_{i=1, j=0, k=1}^{\infty} j^{\nu}l^{\nu}c(i, j, k) + k^{\nu}r^{\nu}c(i, j, k)\end{aligned} \tag{3.32}$$

If $\nu \leq 1/2$, the system does not gel. Generally, the exponent ν is characteristic of a *reactive surface*: for $\nu = 1$ all groups within the whole volume have an equal chance of reaction: for $\nu = 2/3$ only groups on the surface of a spherical molecule can react. The exponent ν can be a function of x and possibly of cycle rank ζ (cf. also [45]).

As was stressed previously, some or all of the steps in network formation can be controlled by *diffusion*. Reactions before the gel point can be controlled by translational diffusion. The translational diffusion coefficients D_{trans} of macromolecules scale with degree of polymerization (molecular weight) P and concentration C as

$$D_{\text{trans}} \propto P^{-1/3} \qquad \text{translational diffusion, spheres} \tag{3.33}$$

$$D_{\text{trans}} \propto P^{-1}C^{-1} \qquad \text{disentanglement, linear chains} \tag{3.34}$$

$$D_{\text{trans}} \propto P^{-2}C^{-1.75} \qquad \text{reptation, linear chains} \tag{3.35}$$

Thus, one can summarize factors affecting the rate 'constant' for reaction of molecule i with molecule j and the resulting structure evolution as follows:

1. Competition between reactivity control and diffusion control gives

$$\frac{1}{K} = \frac{1}{K_{\mathrm{chem}}} + \frac{1}{K_{\mathrm{D}}} \tag{3.36}$$

This follows from the theory of reaction rates: the slower reaction step determines the overall reaction rate.

2. Chemically controlled reaction with steric excluded-volume effect (reaction between molecule i and j) gives

$$K_{\mathrm{chem},\,ij} = k_{\mathrm{C},\,ij} l_i^{\nu} l_j^{\nu} \tag{3.37}$$

3. Translational and segmental diffusion give

$$K_{\mathrm{D},\,ij} = K_{\mathrm{D},\,\mathrm{trans},\,ij} + K_{\mathrm{D},\,\mathrm{segm},\,ij} \tag{3.38}$$

$K_{\mathrm{D,trans},ij}$ and $K_{\mathrm{D,\,segm},\,ij}$ are interdependent. In the beginning, intermolecular $K_{\mathrm{D,\,segm},\,ij}$ is determined by collisions described by $K_{\mathrm{D,\,trans},\,ij}$. Later, the translational displacement of large molecules becomes so slow that segmental diffusion control dominates. Also, when segmental diffusion becomes extremely slow, e.g. in the glass transition region, translational diffusion also becomes negligible.

4. Collisions controlled by translational diffusion give

$$K_{\mathrm{D},\,\mathrm{trans},\,ij} = k_{\mathrm{D},\,\mathrm{trans}}(D_{\mathrm{trans},\,i} + D_{\mathrm{trans},j}) \tag{3.39}$$

The smaller, more mobile partner determines the reaction rate.

5. Segmental diffusion gives

$$K_{\mathrm{D},\,\mathrm{segm}} = k_{\mathrm{segm}} D_{\mathrm{segm}} \tag{3.40}$$

with D_{segm} determined by average properties (such as viscosity) and displacement from T_{g}.

3.3.2.2 Computer simulation in space

Kinetics (initiated) lattice-percolation experiments by Manneville and DeSeze [46] and Boots *et al.* [47] have revealed a basic difference between a non-chain polymerization and an initiated polymerization with fast propagation. A continuous degree-of-polymerization distribution and more-or-less randomly distributed clusters in space occur in the former case, and large, internally crosslinked clusters immersed in monomer are formed in the latter case (Figure 3.12). The basic deficiency of the lattice percolation, the immobility of structures during the crosslinking process, is not so serious in the case of chain polymerizations because of the high speed of propagation. However, this model is still too crude to introduce cyclization, steric hindrance or excluded-volume effects dependent on the properties of the monomer molecules, composition, dilution [20, 48] and the diffusivities of the species involved in network build-up.

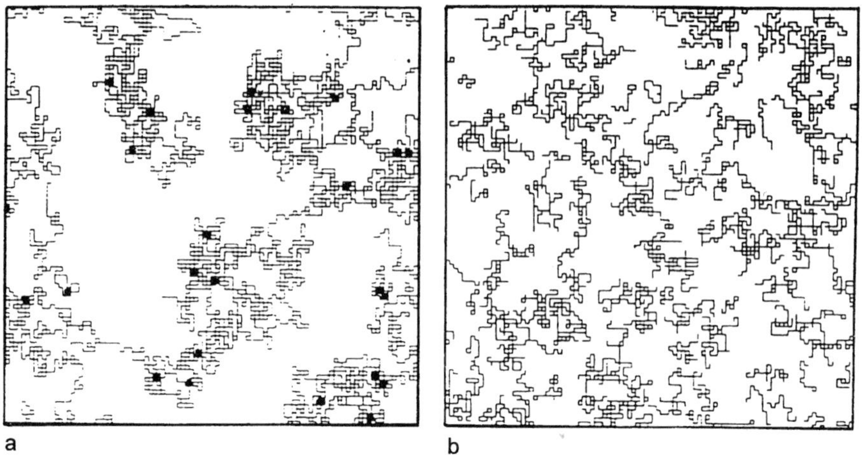

Figure 3.12 Lattice-percolation simulation of network build-up by an initiated reaction. Simulation in two dimensions at a bond conversion of 25%: (a) when a living free radical is trapped, a new free radical starts to grow, a trapped living radical is indicated by a full square; (b) the kinetic length is limited to 20 steps. Reprinted with permission from Ref. 47.

3.3.3 Conclusions

Crosslinking chain (co)polymerization is, from the theoretical point of view, the most complicated way of forming networks. All kinds of non-idealities are involved from the beginning of a polymerization, for example, strong cyclization, excluded-volume effects, steric hindrances and specific diffusion control. No existing theoretical approach can give a reliable prediction of network build-up and the molecular structures of the products formed. Only two kinds of theoretical approaches have any chance of becoming predictive, the generalized Smoluchowski coagulation equations and simulations in space.

Appendix 3A Application of probability generating functions

If the distribution of states of a component (for example, distribution of degrees of polymerization, or distribution of reaction states of an i-mer) are characterized by a set of probabilities of states i, $\{a_i\}$, where a_i is a fraction (number, weight, ...) of the component in the ith state, the corresponding probability generating function (pgf) $F(Z)$ reads as

$$F(Z) = \sum_i a_i Z^i \tag{3.41}$$

where Z is an auxiliary (dummy) variable of the pgf through which operations with the pgfs are performed to get various averages. Since the set

$\{a_i\}$ covers the whole probability space for the variable $\{a_i\}$, it follows that

$$[F(Z)]_{Z=1} = \sum_i a_i = 1 \tag{3.42}$$

By differentiation with respect to the variable Z and putting $Z = 1$, one gets the first moment of the distribution $\{a_i\}$

$$\left[\frac{\partial F(Z)}{\partial Z}\right]_{Z=1} = \sum_i ia_i \tag{3.43}$$

The second derivative gives

$$\left[\frac{\partial^2 F(Z)}{\partial Z^2}\right]_{Z=1} = \sum_i i(i-1)a_i = \sum_i i^2 a_i - \sum_i ia_i \tag{3.44}$$

Also, by integration,

$$\left[\int_0^Z \frac{1}{\zeta} F(\zeta)\,\mathrm{d}\zeta\right]_{Z=1} = \sum_i \frac{a_i}{i} \tag{3.45}$$

The last relationship can be used for the evaluation of lower-moment averages, e.g. by passing from the second- to the first-moment averages.

The formalism of the generating function can be applied, for example, to:

1. A description of branching processes, as explained in the main text. The exponent i of the variable z^i_{xy} identifies the number of bonds of $x \to y$ type issuing from the given building unit. Lower-case z as a variable is reserved for the characterization of the number of bonds linking the building units. The pre-factor characterizes the probability of such an occurrence. Alternatively, the coefficients b_i in $F(z) = \sum_i b_i z^i$ are equal to the probabilities of finding a building unit in the ith reaction state, where a reaction state is characterized by the number of bonds i through which the unit is linked to other units (compare $F_{0n}(Z, z)$ in the main text).
2. Degree-of-polymerization distributions. The variable capital Z is reserved for identification of the building unit or some of its properties, for example, the number of units per primary chain or number of units in a branched molecule. Thus, if n_i and w_i are, respectively, the number and weight fractions of i-mer, then the number-fraction generating function, $N(Z)$, is given by

$$N(Z) = \sum_{i=1}^{\infty} n_i Z^i \tag{3.46}$$

and the weight fraction gf by

$$W(Z) = \sum_{i=1}^{\infty} w_i Z^i = \frac{1}{P_\mathrm{n}} \frac{Z\,\partial N(Z)}{\partial Z} \tag{3.47}$$

The gf, Ξ, for the z fraction defined by

$$z_i = \frac{iw_i}{\sum iw_i} \tag{3.48}$$

is given by

$$\Xi(Z) \equiv \sum_i^{\infty} z_i Z^i = \frac{\sum_i^{\infty} iw_i Z^i}{\sum_i^{\infty} iw_i} = \frac{\sum_i^{\infty} i^2 n_i Z^i}{\sum_i^{\infty} i^2 n_i} \tag{3.49}$$

Likewise, to obtain the number fraction from the weight-fraction distribution, one uses integration (cf. equation (3.45)):

$$\int_0^Z \frac{1}{\zeta} W(\zeta)\,\mathrm{d}\zeta = \sum_1^{\infty} \frac{w_i Z^i}{i} = \frac{1}{P_\mathrm{n}} N(Z) \tag{3.50}$$

or

$$N(Z) = P_\mathrm{n} \int_0^Z \frac{1}{\zeta} W(\zeta)\,\mathrm{d}\zeta \tag{3.51}$$

The generating functions for some distributions are given in Appendix 3B.

Appendix 3B Generating functions for degree-of-polymerization distributions

Most probable (random, Flory) distribution:

$$n_i = (1-q)q^{i-1}, \qquad q = 1 - \frac{1}{\langle P \rangle_n} \tag{3.52}$$

$$N(Z) = \frac{(1-q)Z}{1-qZ} = \sum_i^{\infty} n_i Z^i \tag{3.53}$$

$$W(Z) = \frac{(1-q)^2 Z}{(1-qZ)^2} \tag{3.54}$$

$$\overline{P_\mathrm{n}} = \frac{1}{1-q} \tag{3.55}$$

$$\overline{P_\mathrm{w}} = \frac{1+q}{1-q} \tag{3.56}$$

Poisson distribution:

$$n_i = \frac{\exp(-\nu)\nu^{i-1}}{(i-1)!} \tag{3.57}$$

$$N(Z) = Z \exp[\nu(Z-1)] \tag{3.58}$$

$$W(Z) = \frac{Z(1+\nu Z)}{1+\nu} \exp[\nu(Z-1)] \tag{3.59}$$

$$\overline{P_\text{n}} = \nu + 1 \tag{3.60}$$

$$\overline{P_\text{w}} = \frac{\nu^2 + 3\nu + 1}{\nu + 1} \tag{3.61}$$

Schulz–Zimm distribution:

$$n_i = \left[\frac{(\sigma/\overline{P_\text{n}})^\sigma}{\Gamma(\sigma)}\right] i^{-\sigma} \exp(-(\sigma/\overline{P_\text{n}})i) \tag{3.62}$$

$$N(Z) = \left(\frac{\sigma}{\sigma - \overline{P_\text{n}} \ln Z}\right)^\sigma \tag{3.63}$$

$$W(Z) = \left(\frac{\sigma}{\sigma - \overline{P_\text{n}} \ln Z}\right)^{\sigma+1} \tag{3.64}$$

$$\overline{P_\text{w}} = \langle P \rangle_\text{n} \frac{\sigma + 1}{\sigma} \tag{3.65}$$

Generally, one can derive generating functions for any distribution for which a Laplace transform exists [12].

Acknowledgement

The support of this work by the Grant Agency of the Academy of Sciences of the Czech Republic (Grant No. 4050508) is gratefully acknowledged.

Glossary of symbols and abbreviations

a	fraction of active monomer units in a copolymer
a_i	probability of state i
$c(x, r, l)$	concentration of molecules composed of x segments, r free radicals and l unreacted functional groups
d	density
$\bar{k}$	rate constant
k_{xy}	rate constant for reaction between x and y
l	number of unreacted groups (pendant double bonds)
l_i	number of unreacted functional groups (pendant double bonds) on molecular species i
n_i	number fraction of molecules i
$q = 1 - 1/\overline{P_\text{n}}$	parameter in the most probable distribution
r_i	number of free radicals on molecular species i
t_i	coefficient in the pgf $T(Z)$

u	recursive variable
v, v_i	extinction probability
v	velocity (rate)
w_i	weight fraction of molecular species i
w_g	weight fraction of gel
w_s	weight fraction of sol
x	number of structural units
x_i	number of structural units in molecular species i
z, z_{xy}	variable of pgf (for bond $x \to y$)
z_i	z fraction of molecules i
C	concentration
D	diffusion coefficient
$F_{0n}(Z, z)$	probability generating function for number of bonds
$F'_{0n}(1)$	value of the derivative of $F_{0n}(Z, z)$ for $Z = z = 1$
$F(Z, z)$	probability generating function for number of additional bonds
$F'(1)$	value of the derivative of $F(Z, z)$ for $Z = z = 1$
$K(x, r, l)$	cyclization kernel (rate constant) for molecules of composition x, r, l
K	rate constant
K_D, K_{chem}	rate constants for diffusion and chemically controlled reactions, respectively
M_n	number-average molecular weight
M_w	weight-average molecular weight
M_{0n}	number-average molecular weight of primary chains
M_{0w}	weight-average molecular weight of primary chains
N	number of segments
N_e	number of elastically active network chains
$N(Z)$	number-fraction generating function
P	degree of polymerization
$\overline{P_n}$	number-average degree of polymerization
$\overline{P_{0n}}$	number-average degree of polymerization of primary chains
$\overline{P_w}$	weight-average degree of polymerization
$\overline{P_{0w}}$	weight-average degree of polymerization of primary chains
$\overline{P_z}$	z-average degree of polymerization
$T(Z)$	generating function for the number of bonds with infinite continuation
$T'(1)$	value of the derivative of $T(Z)$ for $Z = 1$
$T'(0)$	value of the derivative of $T(Z)$ for $Z = 0$
T_g	glass-transition temperature
$W(Z)$	weight-fraction generating function
α	fraction of crosslinked units, conversion of functional groups

α_i	conversion of functional groups i
ν	exponent; parameter in Poisson distribution (equation (3.57))
ν_e	concentration of EANCs
$(\nu_e)_{rel}$	relative concentration of primary chains
ξ	conversion of functional groups (double bonds)
σ	exponent; parameter in Schulz–Zimm distribution (equation (3.62))
ζ	cycle rank
$\Xi(Z)$	z-fraction generating function
gf	generating function
pgf	probability generating function
ACU	active crosslinked unit
DVB	divinylbenzene
EANC	elastically active network chains
EDMA	ethylene dimethacrylate
MMA	methyl methacrylate
ST	styrene

References

1. Dušek, K. (1989) in *Telechelic Polymers: Synthesis and Applications* (ed. E.J. Goethals), CRC Press, Boca Raton, pp. 289–360.
2. Dušek, K. (1986) *Adv. Polym. Sci.*, **78**, 1–60.
3. Charlesby, A. (1960) *Atomic Radiation and Polymers*, Pergamon Press, Oxford.
4. Dole, M. (1972) *Radiation Chemistry of Polymers*, Academic Press, New York.
5. Flory, P.J. (1953) *Principles of Polymer Chemistry*, Cornell University Press, Ithaca.
6. Pearson, D.S. and Graessley, W.W. (1980) *Macromolecules*, **13**, 1001.
7. Graessley, W.W. (1975) *Macromolecules*, **8**, 186.
8. Graessley, W.W. (1974) *Adv. Polym. Sci.*, **16**, 1.
9. Dobson, G.R. and Gordon, M. (1965) *J. Chem. Phys.*, **43**, 705.
10. Dušek, K., Gordon, M. and Ross-Murphy, S.B. (1978) *Macromolecules*, **11**, 236.
11. Dušek, K. (1972) *J. Polym. Sci. C*, **39**, 73.
12. Vojta, V. and Dušek, K. (1980) *Br. Polym. J.*, **12**, 1.
13. Miller, D.R. and Macosko, C.W. (1988) *J. Polym. Sci., Polym. Phys. Ed.*, **26**, 1.
14. Galiatsatos, V. and Eichinger, B.E. (1988) *J. Polym. Sci., Polym. Phys. Ed.*, **26**, 595.
15. Taylor, G., Castner, E.S. and Galiatsatos, V. (1995) *J. Chem. Soc., Faraday Trans.*, **91**, 2665.
16. Dušek, K. (1984) *Macromolecules*, **17**, 716.
17. Helfand, E. and Tonelli, A.E. (1974) *Macromolecules*, **7**, 59.
18. Dušek, K. and Vojta, V. (1977) *Br. Polym. J.*, **9**, 164.
19. Dušek, K. and Ilavský, M. (1975) *J. Polym. Sci., Symp.*, **53**, 57.
20. Dušek, K. (1982) in *Developments in Polymerization, 3* (ed. R.N. Haward), Applied Science, Barking, pp. 143–206.
21. Cornfield, G.C. and Butler, G.B. (1972) in *Developments in Polymerization, 3* (ed. R.N. Haward), Applied Science, Barking, pp. 1–54.
22. Sadhir, R.K. and Luck, R.M. (1992) *Expanding Monomers. Synthesis, Characterization, and Application*, CRC Press, Boca Raton.
23. Yang, Y.S. and Lee, J.L. (1987) *Polym. Proc. Eng.*, **5**, 327.
24. Yang, Y.S. and Lee, L.J. (1988) *Polymer*, **29**, 1793.
25. Chiu, Y.Y. and Lee, L.J. (1995) *J. Polym. Sci., Pt. I, Polym. Chem.*, **33**, 257, 269.

26. Kloosterboer, J.G. (1988) *Adv. Polym. Sci.*, **84**, 1–62.
27. Okay, O., Nagash, H.J. and Capek, I. (1995) *Polymer*, **36**, 2413.
28. Okay, O. (1994) *Polymer*, **35**, 797.
29. Okay, O. and Nagash, H.J. (1994) *Polym. Bull.*, **33**, 665.
30. Han, C.D. and Lee, D.-S. (1987) *J. Appl. Polym. Sci.*, **33**, 2859.
31. Dušek, K., Galina, H. and Mikeš, J. (1980) *Polym. Bull.*, **3**, 19.
32. Dušek, K. (1993) *Collect. Czech. Chem. Commun.*, **58**, 2245.
33. Gordon, M. and Malcolm, G.N. (1966) *Proc. Roy. Soc., London, A*, **295**, 29.
34. Landin, D.T. and Macosko, C.W. (1988) *Macromolecules*, **21**, 846.
35. Dušek, K. (1996) *Polym. Gels Networks*, **4**, 383.
36. Tobita, H. and Hamielec, A.E. (1988) *Makromol. Chem., Macromol. Symp.*, **20/21**, 501.
37. Tobita, H. and Hamielec, A.E. (1989) *Macromolecules*, **22**, 3098.
38. Dušek, K. (1979) *Polym. Bull.*, **1**, 523.
39. Dušek, K. (1985) *Br. Polym. J.*, **17**, 185.
40. Dušek, K. and Šomvársky, J. (1985) *Polym. Bull.*, **13**, 313.
41. Dušek, K., Šomvársky, J. and Braam, A.W.J. (1992) Proceedings of 11th Meeting of Polymer Networks Group, San Diego, p. 35.
42. Šomvársky, J. and Dušek, K. (1994) Proceedings of 5th European Polymer Federation Symposium on Polymer Materials, Basel, CAPR 09, p. 2.6.
43. Šomvársky, J. and Dušek, K. (1994) *Polym. Bull.*, **33**, 369.
44. Šomvársky, J. and Dušek, K. (1994) *Polym. Bull.*, **33**, 377.
45. Dušek, K. and Šomvársky, J. (1995) *Faraday Disc. Chem. Soc.*, **101**, 147.
46. Manneville, P. and DeSeze (1981) in *Numerical Methods in the Study of Critical Phenomena* (eds I. Della Dora, J. Demongeot and B. Lacolle), Springer, Berlin.
47. Boots, H.M.J., Kloosterboer, J.G. *et al.* (1985) *Br. Polym. J.*, **17**, 219.
48. Matsumoto, A. (1995) *Adv. Polym. Sci.*, **123**, 41.
49. Dušek, K. and Spěváček, J. (1983) *Polymer*, **21**, 750.

4 Transitions during network formation

R.J.J. WILLIAMS

4.1 Introduction

Network formation is accompanied by several changes in material properties and states. A polymer network generated by the reaction of liquid monomers at constant temperature will exhibit an initial increase in viscosity followed by a change to a gel state. Depending on the nature of the system and the particular reaction (cure) temperature, vitrification may also take place, leading to a glassy state. This transition may occur before or after gelation (causing a liquid–glass or rubber–glass transition, respectively). At high temperatures, degradation of the polymer network, associated with devitrification and revitrification phenomena [1], may be observed. For modified-thermosetting polymers, i.e. rubber-modified epoxies, separation of a new phase rich in modifier takes place during network formation.

This chapter analyses the main transitions occurring during network formation in the framework of transformation diagrams describing changes in material states. Examples illustrating applications to actual systems are discussed.

4.2 Transitions during network formation

4.2.1 Gelation

The sol–gel transition has been analysed in Chapters 2 and 3. It will just be recalled here that gelation (macrogelation) takes place at a critical conversion where a macroscopic macromolecule, confined only by the boundaries of the reaction vessel, is generated. At this point, the weight-average molecular weight $\bar{M}_w$ and the macroscopic viscosity become infinite.

The critical conversion at gelation x_{gel} depends on the number and distribution of reactive sites (functionalities) per molecule, the competition between intermolecular reaction and intramolecular cyclization, and the presence of several other non-idealities such as unequal reactivities of like functionalities and substitution effects.

For systems in which the network structure is a unique function of conversion, x_{gel} is a constant that does not depend on the polymerization condition (i.e. the cure temperature). When there are different reaction paths available

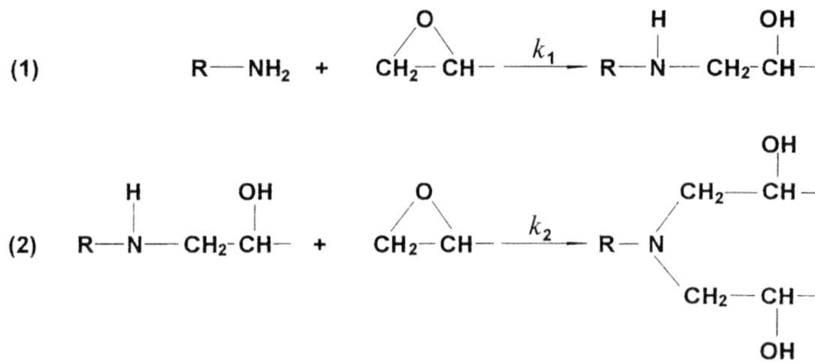

Figure 4.1 Main reactions taking place in an epoxy-amine system.

for network formation, x_{gel} remains constant only if the competition between different reactions does not change significantly with polymerization temperature, i.e. if different paths exhibit similar activation energies.

Let us consider a typical diepoxide–diamine reaction (a system that will be frequently used for illustration purposes in this chapter). The diepoxide is a two-functional monomer whereas the diamine is a four-functional monomer. Figure 4.1 illustrates the main reactions taking place. A third reaction, namely the polyetherification of epoxy groups, may occur in the presence of specific initiators or at high temperatures, once the concentration of amine hydrogens has been sufficiently depleted [2]. It is particularly relevant for formulations containing an excess of epoxy groups over stoichiometry [3].

For a stoichiometric diepoxide–diamine system without substitution effects ($k_1 = k_2$, defined on the basis of the concentration of amine hydrogens), and in the absence of any other non-ideality, the gel conversion is equal to $x_{gel} = 0.577$ (the ideal value for the polycondensation of a two-functional monomer with a four-functional monomer; see equation (2.1)). If $k_2 \ll k_1$, linear chains will be formed first and only when the primary amine groups are almost exhausted will branching through the secondary amine groups take place, eventually leading to gelation. However, the gel conversion for this limiting case is equal to $x_{gel} = 0.618$ [4], i.e. close to the ideal value. This explains why reported x_{gel} values for stoichiometric epoxy–amine systems of different chemical nature are frequently close to 0.60 and constant over the range of cure temperatures.

4.2.2 Vitrification

4.2.2.1 General description

Vitrification corresponds to the transformation from a liquid or rubbery state to a glassy state. The glass-transition temperature T_g indicates the temperature below which a material exhibits properties typical of a glass. At temperatures higher than T_g the molecular mobility is sufficient for the material to attain the properties (enthalpy, specific volume etc.) characteristic of the equilibrium rubbery or liquid state.

As discussed by Choy and Plazek [5], upon cooling, a glass-transition temperature is determined which depends exclusively on the cooling rate. For a given material and a particular cooling rate, the same T_g will always be obtained. Once the material enters the glassy condition, its now non-equilibrium state is determined by its thermal history, such as the particular state attained in its relaxation to equilibrium. The T_g determined by heating is not only a function of the rate of temperature increase but also of the past thermal history. And this is precisely the situation for the case of partially reacted network-forming polymers. The standard procedure is to advance the polymerization reaction for a given time at a particular cure temperature, T_{cure}, to cool the sample and to determine T_g by a subsequent temperature increase at a particular heating rate. Figure 4.2 shows typical thermograms obtained by differential scanning calorimetry (DSC).

From Figure 4.2(a), T_g and the conversion x may be obtained. Regarding the T_g value, it has to be taken into account that: (i) it depends on an arbitrary definition, i.e. onset (as marked in Figure 4.2(a)), mid-point or end value of the transition; (ii) it depends on the nature of the experimental

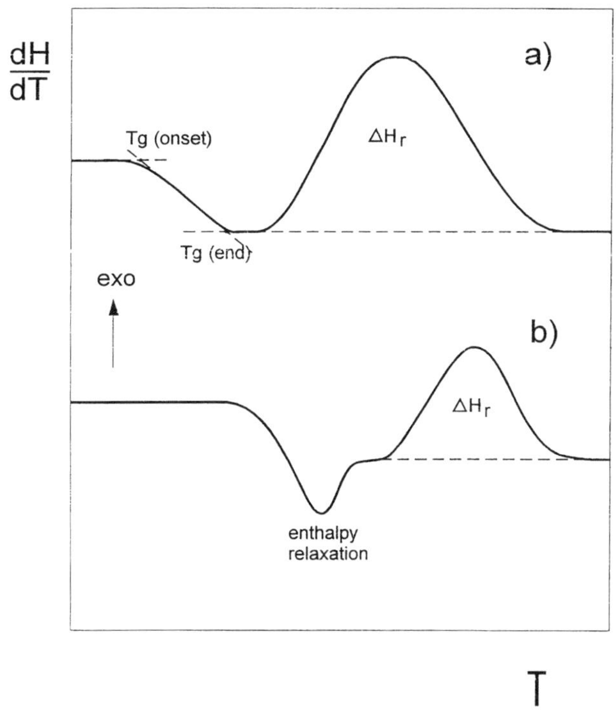

Figure 4.2 Typical DSC thermograms for partially reacted thermosetting polymers. (a) $T_g \leq T_{cure}$, (b) $T_g > T_{cure}$.

technique employed, the heating rate and the particular history of the specimen in the glassy state. Therefore, contrary to gelation, vitrification is a complex transition that takes place over a relatively broad temperature range.

Regarding the conversion value, it may be obtained from

$$x = (\Delta H_u - \Delta H_r)/\Delta H_T \quad (4.1)$$

where ΔH_u is the heat of reaction evolved by an unreacted sample (the initial mixture of monomers), ΔH_r is the residual heat of reaction obtained by rescanning a partially reacted sample, and ΔH_T is the total heat of reaction expected for the generic class of reaction under consideration (i.e. determined from model compounds). Other techniques, such as Fourier-transform infrared spectroscopy (FTIR) may also be employed to determine the conversion value.

The maximum conversion attained in network build-up is frequently close to 1, within the experimental error of the method used to follow the disappearance of functional groups. This means that, in equation (4.1), $\Delta H_u = \Delta H_T$. However, certain systems exhibit severe topological restrictions that prevent the complete reaction of functional groups. This is the case for vinyl–divinyl (or f-vinyl) systems and epoxidized novolacs cured with either novolacs [6] or aromatic diamines [7]. For the latter, a maximum conversion close to $x_{max} = 0.8$ has been reported [7].

Increasing reaction time at constant cure temperature leads to an increase in both T_g and x. At a particular reaction time, called the vitrification time t_{vit}, T_g equals T_{cure}. For $t > t_{vit}$ two different effects are associated with the further increase of T_g: (i) continuation of the polymerization in the glassy state at a rate that decreases significantly with conversion, and (ii) physical aging, leading to an endothermic relaxation during devitrification, as shown in Figure 4.2(b). The influence of the physical aging may be erased by first cooling the sample, then heating to a temperature just beyond the observed T_g, cooling rapidly to avoid further reaction and heating again to determine T_g in the absence of physical aging [5, 6, 8].

For network-forming polymers exhibiting a maximum T_g ($T_{g\infty}$), lower than the cure temperature, vitrification will not take place during polymerization. This is the case for materials characterized by low values of $T_{g\infty}$, i.e. showing an elastomeric behaviour at room temperature or, conversely, for thermosetting polymers cured at high temperatures ($T_{cure} > T_{g\infty}$).

Many network-forming polymers, such as epoxy–amine and epoxy–anhydride systems, polycyanates, polyimides, polyurethanes, vinyl–divinyl copolymers, vinyl ester–styrene systems, unsaturated polyesters, crosslinked natural rubber etc. [6–30], exhibit a one-to-one relationship between T_g and conversion. This means that experimental points of T_g versus x, obtained at different cure temperatures, may be superimposed on a master curve. This behaviour is expected for polymers where the network structure depends exclusively on conversion. However, it is also found for thermosetting polymers where the reaction path may vary with cure temperature, as is

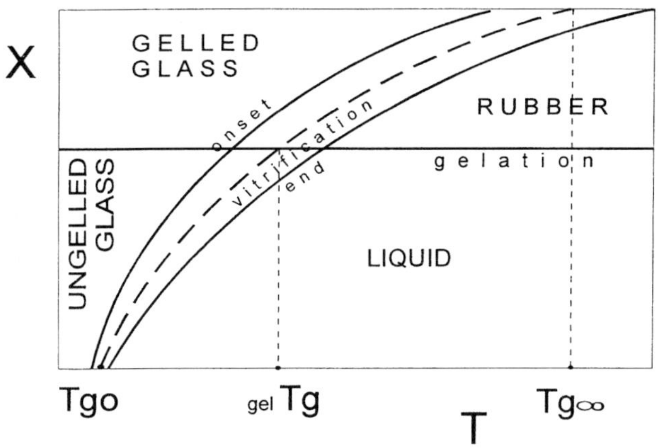

Figure 4.3 Conversion-versus-temperature transformation (CTT) diagram, showing gelation, vitrification and different states of the thermosetting polymer.

the case with most of the polymer networks just mentioned. The reason for the existence of a single T_g-versus-x relationship may be attributed to a low sensitivity of T_g to the resulting modifications of the network structure.

Figure 4.3 shows a representation of gelation and vitrification in conversion-versus-temperature coordinates (conversion-versus-temperature transformation (CTT) diagram [10]). Although gelation is a definite transition, taking place at a particular conversion, vitrification extends over a range of ΔT_g values ($T_{g,end}$ to $T_{g,onset}$) at a particular conversion level. This has important practical consequences as will be illustrated later. The width of the ΔT_g interval usually increases with conversion, reflecting the increase in the distribution of different structures in the polymer network (i.e. junctions and crosslinking units of different functionalities).

Unfortunately, vitrification has been presented traditionally as a definite transition, represented by a single curve, i.e. the dashed curve in Figure 4.3. Its curvature reflects the fact that particular structures, that have a significant incidence on the increase of T_g, such as high-functionality junction points, appear and increase their concentrations at high conversions. In fact, T_g constitutes the most sensitive indication of the advance of reactions at high conversions. The vitrification curve starts at T_{g0}, the glass transition temperature of the initial formulation, ${}_{gel}T_g$ represents the particular temperature at which gelation and vitrification take place simultaneously, and $T_{g\infty}$ is the glass transition temperature of the completely reacted polymer. When there are topological restrictions, $T_{g\infty}$ represents a virtual value (the experimental vitrification curve extends up to T_{gmax} at $x = x_{max}$).

Four different states of a network-forming polymer may be distinguished, depending on the particular location in the CTT diagram (Figure 4.3). At the right of the vitrification region, rubber ($x > x_{gel}$) or liquid ($x < x_{gel}$) states may be found. At the left of this region, gelled ($x > x_{gel}$) or ungelled ($x < x_{gel}$)

glasses are present. In the vitrification region properties vary continuously between two different states. For example, if a material initially present as a gelled glass is heated, a horizontal trajectory through the vitrification region will lead to a rubber. Contrariwise, if the starting monomers in the liquid state are polymerized at a constant temperature, the material will change from a liquid to a glass across the vitrification region. (Depending on the particular temperature, gelation may take place before entering the vitrification region, while traversing it or, eventually, after leaving this region.) It may be observed that, following a vertical trajectory, the polymerizing system will first find the 'end' of the vitrification region and will exit from the region by the 'onset' curve. In this sense, the behaviour is similar to cooling at constant conversion.

4.2.2.2 Glass-transition temperature-versus-conversion relationships

A rigorous treatment of the glass transition in highly crosslinked systems is a formidable task [6]. The development of theoretical T_g-versus-x relationships requires the combined effects of decreasing chain-end concentration, formation of junction points, departure from Gaussian behaviour and steric hindrance on the chain conformations at high junction-point densities to be taken into account. Several models have been described in the literature [6, 19, 28, 31]. A common characteristic is that a statistical description of network build-up has to be available to calculate the concentration of junction points of different functionalities as a function of conversion. To complicate matters, the effects of branch points (junctions) that are not elastically effective are not yet understood clearly. While some authors [6, 19] count only elastically effective junction points (i.e. those with three or more arms in the gel), others [28, 31] take into account all branch points in the system (i.e. those which have three or more reacted functionalities). Examination of a series of stoichiometrically imbalanced networks based on poly(dimethylsiloxane) (PDMS) revealed that elastically effective junctions rather than any branch points dominate the contributions of junction points to T_g [32]. In any case, all the models require the use of fitting parameters for application to particular systems.

A different approach for establishing a T_g-versus-x relationship has been proposed by Pascault and Williams [9]. The starting point is Couchman's analysis [33] for the compositional variation of T_g of a solution. Assuming random mixing, the following equation was obtained:

$$\sum \phi_i \int_{T_{gi}}^{T_g} (\Delta c_{pi}/T)\,\mathrm{d}T = 0 \qquad (4.2)$$

where T_g is the glass transition temperature of the solution, T_{gi} is the glass transition temperature of pure component i, $\Delta c_{pi} = c_{pi}(\mathrm{l}) - c_{pi}(\mathrm{g})$ is the change in the isobaric heat capacity of component i between liquid and glassy states, and ϕ_i is the concentration of component i in the solution (with units matching those of Δc_{pi}).

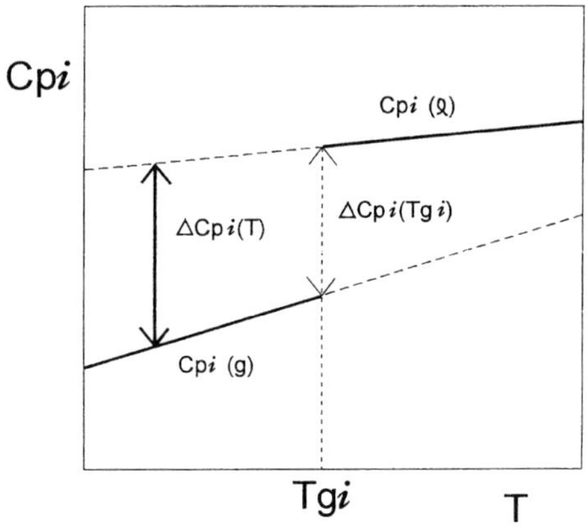

Figure 4.4 Isobaric specific heat capacity of a component i as a function of temperature.

The variation of c_{pi} with T, shown in Figure 4.4, illustrates the general tendency of Δc_{pi} to decrease with T [33]. To perform the integration of equation (4.2), it is necessary to introduce the $\Delta c_{pi}(T)$ relationship. The assumption that Δc_{pi} is inversely proportional to temperature leads to

$$\Delta c_{pi} = \Delta c_{pi}(T_{gi})T_{gi}/T \tag{4.3}$$

Introducing equation (4.3) into (4.2) and integrating, leads to

$$T_g = \frac{\sum \phi_i \Delta c_{pi}(T_{gi})T_{gi}}{\sum \phi_i \Delta c_{pi}(T_{gi})} \tag{4.4}$$

If the polymerizing system is modelled as a random solution of reacted functionalities with a concentration equal to the conversion ($\phi_i = x$, $T_{gi} = T_{g\infty}$, $\Delta c_{pi}(T_{g\infty}) = \Delta c_{p\infty}$), and unreacted functionalities with a concentration equal to $1 - x$ ($\phi_i = 1 - x$, $T_{gi} = T_{g0}$, $\Delta c_{pi}(T_{g0}) = \Delta c_{p0}$), equation (4.4) may be written as

$$T_g = \frac{x\Delta c_{p\infty}T_{g\infty} + (1-x)\Delta c_{p0}T_{g0}}{x\Delta c_{p\infty} + (1-x)\Delta c_{p0}} \tag{4.5}$$

Making

$$\Delta c_{p\infty}/\Delta c_{p0} = \lambda \tag{4.6}$$

and rearranging equation (4.5) leads to

$$\frac{T_g - T_{g0}}{T_{g\infty} - T_{g0}} = \frac{\lambda x}{1 - (1-\lambda)x} \tag{4.7}$$

Equation (4.7) has been applied with success to correlate experimental values of T_g-versus-x for many thermosetting polymers, using λ as an adjustable parameter. This equation has been popularized as the Di Benedetto equation, due to an incorrect interpretation [10] of an equation relating T_g with junction-point density and a virtual T_g of the linear polymer, proposed by Di Benedetto, as stated by Nielsen [34].

Reasonable predictions of the T_g-versus-x relationship were obtained for several thermosetting polymers, by introducing the values of λ arising from equation (4.6) into equation (4.7) [9, 27, 35, 36]. In this case, the full T_g-versus-x curve may be predicted by measuring T_g and Δc_p of the starting monomers and the fully reacted polymer.

When topological restrictions are present, it is not possible to obtain complete conversion. The polymerization may be conducted up to a particular value x_{max}, at which $T_g = T_{gmax}$. By defining a new coordinate $x' = x/x_{max}$, the system may be described as a random solution of reacted functionalities ($\phi_i = x'$, $T_{gi} = T_{gmax}$, $\Delta c_{pi}(T_{gmax}) = \Delta c_{pmax}$), and unreacted functionalities ($\phi_i = 1 - x'$, $T_{gi} = T_{g0}$, $\Delta c_{pi}(T_{g0}) = \Delta c_{p0}$). Substituting into equation (4.4) and rearranging gives

$$\frac{T_g - T_{g0}}{T_{gmax} - T_{g0}} = \frac{\lambda' x'}{1 - (1 - \lambda')x'} = \frac{\lambda' x}{x_{max} - (1 - \lambda')x} \tag{4.8}$$

with

$$\lambda' = \Delta c_{pmax}/\Delta c_{p0} \tag{4.9}$$

A virtual value of $T_{g\infty}$ may be obtained by making $x' = 1/x_{max}$. This gives

$$T_{g\infty} = T_{g0} + \lambda'(T_{gmax} - T_{g0})/(\lambda' + x_{max} - 1) \tag{4.10}$$

Figure 4.5 shows experimental T_g versus x obtained for an epoxidized novolac cured with an aromatic diamine, fitted by equation (4.8) [12]. In this case, $x_{max} = 0.80$ and $\lambda' = 0.15$ was used as a fitting parameter. For a system based on an epoxy cresol novolac cured with a cresol novolac, Hale *et al.* [6] reported experimental values of $\Delta c_{pmax}/\Delta c_{p0}$, at $x_{max} = 0.78$, scattered over the range 0.10–0.35, i.e. including the value predicted by the fitting procedure.

Regarding the experimental determination of T_{g0}, a problem may arise if crystallization of one of the monomers is possible. This may be avoided by melting and quickly quenching to a temperature below T_{g0} [35].

A different assumption used in the integration of equation (4.2) is to consider the Δc_{pi} independent of temperature, i.e. an assumption that is only reasonable for small values of $T_{g\infty} - T_{g0}$. This leads to

$$\ln T_g = \frac{\sum \phi_i \Delta c_{pi}(T_{gi}) \ln T_{gi}}{\sum \phi_i \Delta c_{pi}(T_{gi})} \tag{4.11}$$

Modelling the network-forming polymer as a random solution of reacted and unreacted functionalities, equation (4.11) becomes

$$\ln T_g = \frac{\lambda x \ln T_{g\infty} + (1 - x) \ln T_{g0}}{1 - (1 - \lambda)x} \tag{4.12}$$

where λ is given by equation (4.6).

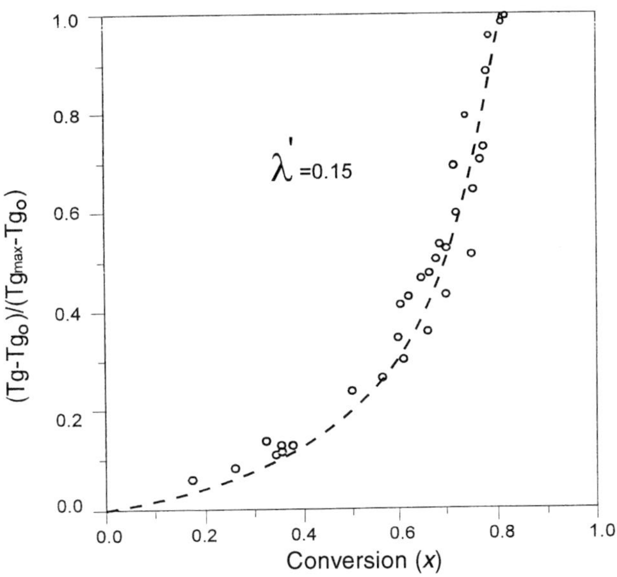

Figure 4.5 Glass transition temperature as a function of conversion for an epoxidized novolac cured with an aromatic diamine ($x_{max} = 0.80$)[12]. (Reprinted with permission.)

Equation (4.12) has been proposed as an alternative to equation (4.7) for adjusting experimental data of T_g versus x [22]. In particular, it has been stated that independent values of λ, obtained from the ratio $\Delta c_{p\infty}/\Delta c_{p0}$, fit better experimental results when used in equation (4.12) rather than in (4.7) [22]. However, the use of equation (4.7) with λ, obtained from the ratio of specific heats, was also found satisfactory for several experimental systems [9, 27, 35, 36].

A limiting case of equation (4.7) will now be considered. The general trend observed for network-forming polymers is a decrease in $\Delta c_p(T_g)$ with conversion or, what may be considered equivalent, a decrease in $\Delta c_p(T_g)$ with T_g. The following equation has been proposed to fit experimental data [36]:

$$\Delta c_p(T_g) = a' + b'/T_g \tag{4.13}$$

For the particular case in which the constant a' may be neglected, equation (4.13) leads to

$$\frac{\Delta c_{p\infty}}{\Delta c_{p0}} = \frac{T_{g0}}{T_{g\infty}} \tag{4.14}$$

Using equation (4.14) in equation (4.5) and rearranging results in

$$\frac{1}{T_g} = \frac{(1-x)}{T_{g0}} + \frac{x}{T_{g\infty}} \tag{4.15}$$

This simple rule-of-mixtures equation has not been found satisfactory for representing the behaviour of actual systems. A correction term is usually added, in the form of [14]

$$\frac{1}{T_g} = \frac{(1-x)}{T_{g0}} + \frac{x}{T_{g\infty}} + c'x(1-x) \tag{4.16}$$

where c' is a fitting parameter. Equation (4.16) has also been used as an alternative equation to interpret experimental data of T_g versus x [14, 17].

In conclusion, the following concepts must be taken into account when examining equations for representing T_g as a function of conversion:

1. Vitrification occurs over a temperature range, ΔT_g. The best representation of vitrification consists of two curves, T_g(onset) and T_g(end) as functions of conversion. When selecting a single curve it should be indicated which T_g value has been selected (i.e. onset, mid-point or end).
2. Equations based on the polymer network structure require a statistical description of the network build-up, assumptions related to the effect of f-functional branch points and crosslinks and, at least, one fitting parameter to account for the influence of several non-idealities on chain conformations, i.e. non-Gaussian and steric effects.
3. Equations derived from simple thermodynamic relationships lead to T_g-versus-x expressions bounded by initial and maximum T_g values, i.e. equations (4.7) or (4.8), (4.12) and (4.16). When the ratio $\Delta c_{p\infty}/\Delta c_{p0}$ is also available, equation (4.7) or equation (4.12) may be employed without the use of any fitting parameter. In this case, the whole T_g-versus-x curve may be generated from calorimetric information obtained for the initial and final materials.

4.2.3 Degradation

When a polymerization is carried out at relatively high temperatures or when the fully cured polymer is exposed to these conditions, degradation reactions may occur. They are usually accompanied by loss of mass, eventually leading to char formation [1, 17]. As covalent bonds are broken, a decrease in T_g is usually observed.

For example, an epoxy novolac cured with an aromatic diamine (4,4′-diaminodiphenylsulfone, DDS) showed a maximum glass transition temperature, $T_{gmax} = 215°C$ at a conversion $x_{max} = 0.80$, limited by topological restrictions [7]. When the polymerization was carried out at 270 °C, T_{gmax} was attained after some 50 min but, after 150 min, a decrease of 20 °C in T_g was observed. In this particular case, the polymer network was formed at a higher rate than it was decomposed, i.e. T_g first increased to T_{gmax} and then decreased due to degradation reactions.

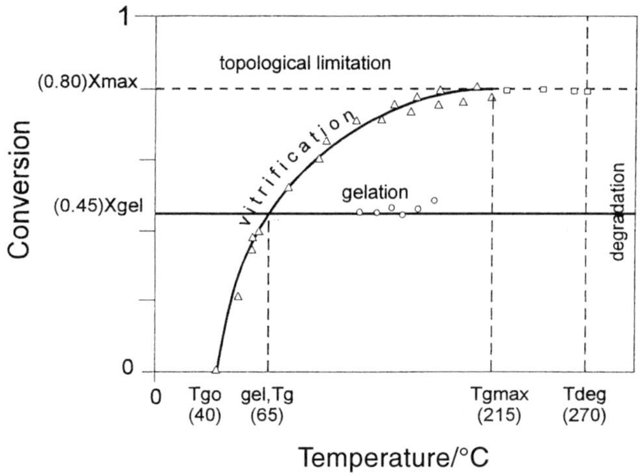

Figure 4.6 Conversion-versus-temperature transformation (CTT) diagram, showing gelation, vitrification, degradation and topological restrictions for an epoxy novolac–aromatic diamine system [7]. (Reprinted with permission.)

Figure 4.6 shows gelation, vitrification (mid-point) and degradation, for an epoxy novolac–aromatic diamine system in a conversion-versus-temperature transformation (CTT) diagram [7]. The degradation region was arbitrarily defined on the basis of a 2 h heating. The region would have been possibly shifted to lower temperatures if the selected degradation time had consisted of a period of days or months instead of hours.

For certain polymer networks showing high values of T_{gmax}, thermal degradation may compete with cure in such a way that vitrification and degradation regions overlap. In this case, the final structure of the polymer network will depend significantly on the selected cure cycle.

On some occasions, degradation of a polymer network is promoted to obtain a desired material. This is the case for carbon–carbon composites manufactured by controlled pyrolysis of a material composed of a thermosetting polymer and carbon fibres. Another example is the use of poly(vinylmethyl ether) (PVME)-modified epoxy/phenolic formulations to produce magnetic-disc coatings. First, a phase rich in PVME segregates during the cure of the thermosetting polymer. Then, the material is heated to 220 °C in air, producing the oxidative degradation of PVME and the generation of a continuous microporous structure. This enhances the retention of a lubricant used to increase the durability of the head–disc interface.

4.2.4 *Phase separation*

Several families of thermosetting polymers exhibit high values of strength, elastic modulus and glass-transition temperature. These properties result from the chemical nature of the starting monomers and the junction-point densities of the polymer networks. However, this kind of rigid structure also leads to a low resistance to crack initiation and propagation. The matrices may be toughened by generating a random dispersion of a second phase, either in the initial formulation or *in situ* during polymerization [37–39].

The domains of the second phase may be rich in a rubber or a thermoplastic polymer. Their dimensions must be in the micrometer range with a microstructure that may be homogeneous or heterogeneous (core-shell or secondary-phase dispersion).

When the phase separation is produced *in situ* during polymerization, the initial system consists of a homogeneous solution of the modifier in the monomers. At a particular extent of reactions, i.e. the cloud-point conversion, domains rich in the modifier are segregated from the thermosetting polymer. The increase in the average molar mass of the polymer and the consequent decrease in the contribution of the entropy of mixing to free energy, is the main driving force for the phase-separation process [40–43]. The process is usually initiated in the liquid state, well before gelation.

Figure 4.7 shows a CTT diagram of a castor-oil-modified epoxy [11]. Cloud-point conversions are located in the liquid region of the diagram and show an increasing trend with temperature (the full line represents the prediction of a thermodynamic model [11]). The increase in cloud-point conversion with temperature is indicative of a corresponding increase in

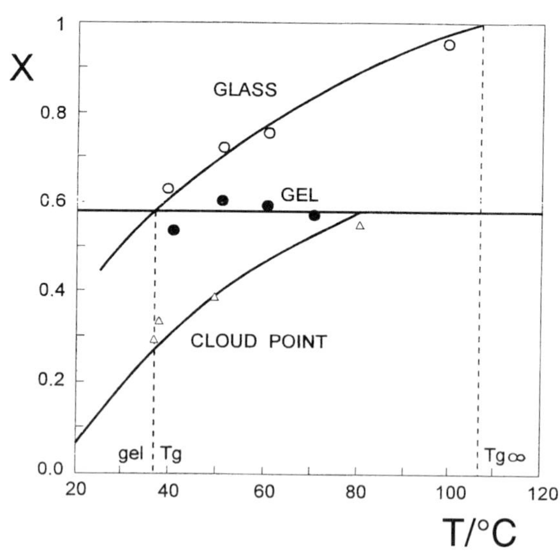

Figure 4.7 CTT diagram showing phase separation, gelation and vitrification (onset value), for a castor-oil-modified epoxy (15 wt% of castor oil in the initial formulation) [11]. (Reprinted with permission.)

the castor-oil miscibility (upper critical solution temperature behaviour (UCST)). Most modifiers employed in thermosetting polymers exhibit a similar behaviour. Exceptions are some thermoplastic modifiers, such as polyethersulfone (PES), that decrease in solubility as when temperature increases (lower critical solution temperature (LCST) behaviour). In this case, cloud-point conversions show a decreasing trend with temperature.

A different type of phase separation takes place in the network-forming process described in Figure 3.9, where microgel formation occurs from the beginning of the reaction. However, microgel formation is not a particular transition but constitutes the mechanism by which the network structure is built up. Macrogelation, i.e. the percolation of the microgel structure, is a characteristic transition associated with this process.

4.3 Use of conversion-versus-temperature transformation (CTT) diagrams

The various transitions occurring during network formation were conveniently represented in terms of CTT diagrams (i.e. Figures 4.3, 4.6 and 4.7). The aim of this section is to discuss the use of such diagrams in polymerizations taking place under different conditions, using isothermal, constant heating rate, adiabatic transformations and general nonisothermal transformations. Practical consequences of the analysis will be illustrated with particular examples.

4.3.1 Isothermal transformations

4.3.1.1 Vitrification

Isothermal transformations are typical of polymerizations taking place in laboratory devices used for characterization purposes. Figure 4.8 illustrates a particular isothermal trajectory intercepting the vitrification region. Both the onset and end of the vitrification transition have been represented in the CTT diagram.

A question that is often asked is how far can a trajectory enter the glassy state, i.e. how large can the difference be between glass-transition and cure temperatures. The answer to this question depends primarily on the particular way of defining and measuring T_g. For techniques such as differential scanning calorimetry (DSC), usual definitions of T_g are the onset or mid-point values of the experimental transition. For measurements based on the damping peak in dynamic–mechanical analysis (i.e. torsional braid analysis (TBA) [15]), reported values are close to the end part of the transition, as discussed by Pang and Gillham [17].

First, it will be assumed that T_g is measured with a dynamic–mechanical thermal analyser (DMTA) as the maximum of the damping peak ($\tan\delta$). In

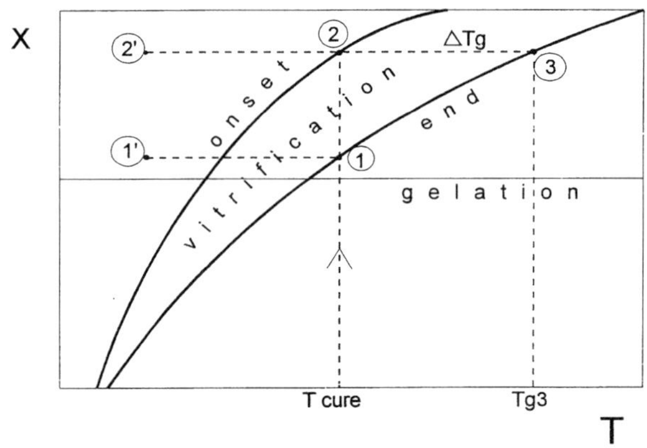

Figure 4.8 Particular isothermal trajectory in a CTT diagram.

this way, T_g is associated with the end part of the transition. The dashed trajectory shown in Figure 4.8 leads to vitrification at point 1 (i.e. cooling to 1′ and heating at a constant rate in the DMTA will give a maximum in $\tan\delta$ located at $T_g = T_{cure}$). Then, polymerization continues in the vitrification region, albeit at an ever-decelerating rate, until point 2 is reached on the onset curve of the vitrification region. Now, the experimental determination of T_g gives T_{g3} (i.e. the temperature of point 3 in Figure 4.8). Thus, a large ΔT_g over T_{cure} will be reported. However, this does not mean that polymerization has taken place in the glassy state. It is merely a consequence of the advance of reaction in the vitrification region and the arbitrary way in which T_g was defined. In this sense, if T_g had been defined at the onset value of the transition, no ΔT_g would have been reported up to point 2.

Figure 4.9 represents experimental points reported by Peng and Gillham [44] for the prolonged (14 400 min) isothermal cure of a trifunctional epoxy monomer (triglycidyl ether of tris(hydroxy phenyl)methane), with an aromatic diamine (trimethylene glycol di-*p*-aminobenzoate (TMAB)). Cure was performed at 80, 150 and 190 °C and T_g was measured using a TBA device. The corresponding values of ΔT_g were 39, 53 and 70 °C, respectively. However, as already stated, they mainly reflect cure in the vitrification region rather than in the glassy state. The fact that ΔT_g increases with T_{cure}, is explained by the increase in width of the vitrification region with conversion.

Kloosterboer and Lijten [45] reported a ΔT_g of 70 °C in the chain-crosslinking photopolymerization of 1,6-hexanediol diacrylate. This large difference between T_g and T_{cure} was attributed to (i) the presence of a significant amount of unreacted monomer up to high conversions and (ii) the persistence of a localized mobility of radical sites in the glassy state through tertiary-hydrogen abstraction. It was proposed that the continuation of the reaction was not limited by vitrification but by the freezing

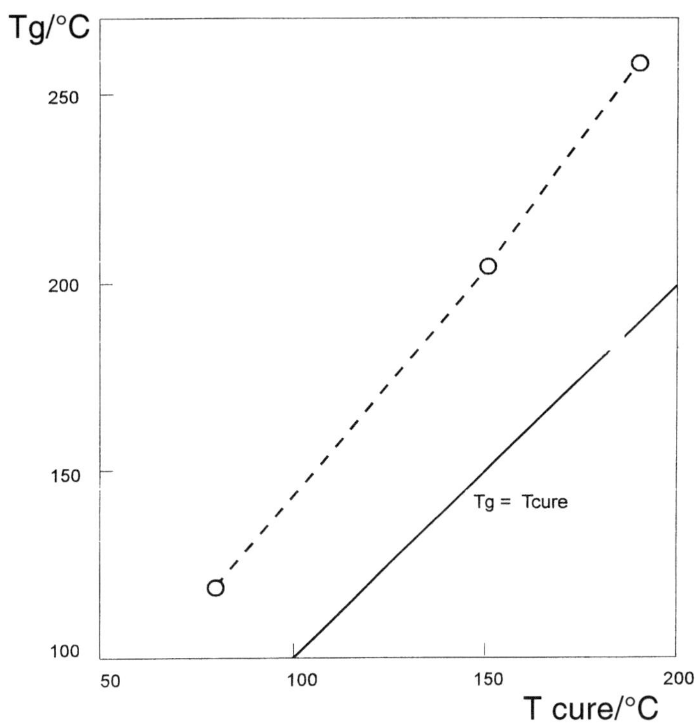

Figure 4.9 T_g (measured by TBA) as a function of T_{cure} for a 14 400 min cure of an epoxy–amine system.

of secondary relaxations. However, T_g was measured as a maximum in tan δ using a DMTA technique, i.e. reported T_g values are close to the end of the transition region. In fact, broad relaxation peaks were obtained, starting at temperatures close to T_{cure} (20 °C). Again, these results show that polymerization can proceed in the vitrification region but not necessarily in the glassy state. The larger the width of the vitrification region the higher the difference between T_g and T_{cure}.

On the basis of the preceding analysis, it is proposed that, in order to validate future reports of increases in T_g with T_{cure}, onset values of T_g should be measured experimentally.

Another type of experiment, on which it is interesting to comment, was performed by Stutz and Mertes [46]. Experiments in a TBA device were carried out up to the vitrification maximum and samples were quenched for analysis. The corresponding glass-transition temperatures were determined by DSC (mid-point). Values were 10–15 °C lower than the cure temperature. Hence, it was concluded that vitrification is indicated somewhat too early using TBA. However, this conclusion is not correct as can be proved simply by looking again at Figure 4.8. When a sample is quenched at point 1 to get 1′, and it is again heated to determine T_g (mid-point), the resulting value, located inside the vitrification region, will be lower than T_{cure}. This does not mean that vitrification was not attained during the experiment in the TBA device but simply reflects the fact that two different indications of vitrification were used. Obviously, T_g (end) values taken from DSC experiments would have been much closer to T_{cure}.

4.3.1.2 Effect of vitrification on polymerization rate

The polymerization rate r may be defined as a generic function of conversion and temperature:

$$r = f(x, T) \tag{4.17}$$

For the particular case where r can be expressed as a function of a single rate constant k, equation (4.17) may be written as

$$r = k(x, T)g(x) \tag{4.18}$$

where $g(x)$ reflects the dependence of the polymerization rate on the concentration of functional groups, and k may be expressed in terms of the Rabinowitch model [8, 14]:

$$\frac{1}{k(x, T)} = \frac{1}{k_T(T)} + \frac{1}{k_D(x, T)} \tag{4.19}$$

The overall rate constant k is written in terms of an Arrhenius rate constant k_T and a diffusion rate constant k_D (compare with equation (3.36)).

In the liquid or rubbery state (Figure 4.3), $k_D \gg k_T$, so the polymerization rate is chemically controlled. However, when the system enters the vitrification region, k_D becomes of the order of k_T and an abrupt decrease of the overall rate constant is observed. Figure 4.10 shows such a behaviour for the cure of a diepoxide based on the diglycidyl ether of bisphenol A (DGEBA) with an aromatic diamine (TMAB) [17]. The arrows indicate the conversions at which T_g (measured with TBA) equals T_{cure}, i.e. a point close to the end value of the vitrification region (like point 1 in Figure 4.8). Clearly, chemical control is observed before the beginning of vitrification, with a sharp drop of the polymerization rate inside the vitrification region.

4.3.1.3 TTT diagram

The time-versus-temperature transformation (TTT) diagram, developed by Gillham and co-workers [1, 8, 15–21, 44], is useful to describe the transitions occurring during network formation in the course of isothermal transformations. The diagram may be generated from the CTT diagram with knowledge of the polymerization kinetics (equation (4.17)).

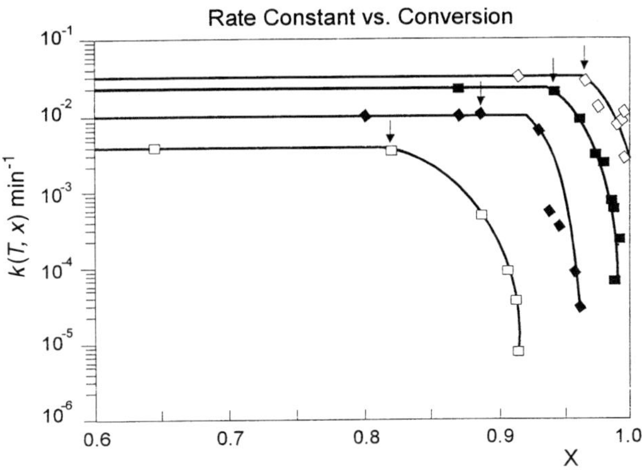

Figure 4.10 Overall rate constant as a function of conversion for the isothermal cure of an epoxy–amine system: (□) 100 °C, (◆) 120 °C, (■) 140 °C, (◇) 150 °C [17]. (Reprinted with permission.)

For an isothermal transformation occurring at a generic temperature $T = T_{\text{cure}}$, the balance of functional groups may be written as

$$\frac{dx}{dt} = r(x, T_{\text{cure}}) \tag{4.20}$$

so that times are unequivocally related to conversions through

$$t = \int_0^x dx/r(x, T_{\text{cure}}) \tag{4.21}$$

Therefore, times to gel (t_{gel}) and times to vitrify (t_{vit}) may be obtained by making $x = x_{\text{gel}}$ and $x = x_{\text{vit}}$ (taking any arbitrary point of vitrification, i.e. onset, mid-point or end):

$$t_{\text{gel}} = \int_0^{x_{\text{gel}}} dx/r(x, T_{\text{cure}}) \tag{4.22}$$

$$t_{\text{vit}} = \int_0^{x_{\text{vit}}(T_{\text{cure}})} dx/r(x, T_{\text{cure}}) \tag{4.23}$$

Figure 4.11 shows a typical TTT diagram. As x_{gel} is typically constant, $t_{\text{gel}}(T_{\text{cure}})$ depends exclusively on the effect of temperature on polymerization rate. Outside the vitrification region, an exponential decrease of t_{gel} with T_{cure} is usually observed. This is the case when a single overall rate constant may be used to represent the polymerization rate, as in equation (4.18). In the chemical-controlled regime,

$$r = k_{\text{T0}} \exp(-E/RT_{\text{cure}})g(x) \tag{4.24}$$

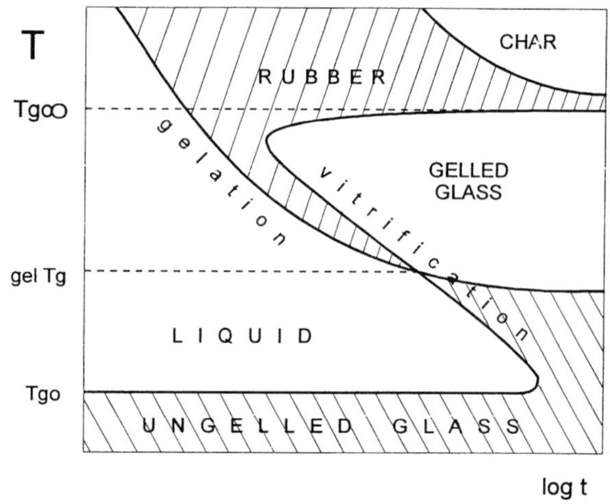

Figure 4.11 Time-versus-temperature transformation (TTT) diagram.

This leads to

$$t_{\mathrm{gel}} = [\exp(E/RT_{\mathrm{cure}})/k_{\mathrm{T0}}] \int_0^{x_{\mathrm{gel}}} \mathrm{d}x/g(x) \tag{4.25}$$

From the slope of a ln t_{gel}-versus-$1/T_{\mathrm{cure}}$ plot, the activation energy E may be obtained. In the vitrification region, the sharp drop in r brings a large increase in gelation times, as indicated in Figure 4.11.

Regarding vitrification, an S-shaped curve results. This arises from the presence of T_{cure} both in the rate equation and in the upper limit of the integral defined in equation (4.23). For $T_{\mathrm{cure}} = T_{\mathrm{g0}}$, $x_{\mathrm{vit}} = 0$, as the system is already in the glassy state; so $t_{\mathrm{vit}} = 0$. Increasing T_{cure}, starting from T_{g0}, requires a certain advance in conversion to cause vitrification. So, an increase in vitrification time is first observed. However, on further increasing T_{cure}, the exponential decrease of the reaction rate predominates, resulting in a decrease in vitrification time. The situation is reversed again at T_{cure} values close to $T_{\mathrm{g}\infty}$, where low values of r, arising from the small concentration of functional groups available for reaction, produce a new increase in t_{vit}.

The char region, generated as a consequence of degradation reactions, is also illustrated in Figure 4.11.

Figure 4.12 is a TTT diagram showing experimental values of times for phase separation, gelation and vitrification, for a castor-oil-modified epoxy (15 wt% castor oil in the initial formulation) [11]. Experimental points are the same as those represented in the CTT diagram of Figure 4.7. The link between both diagrams is the polymerization kinetics of the castor-oil-modified DGEBA–ethylenediamine (EDA) system, given by [11]:

$$r = k\,\mathrm{e}_0^2\{x + [\mathrm{OH}]_0/[\mathrm{e}]_0\}(1 - x)^2 \tag{4.26}$$

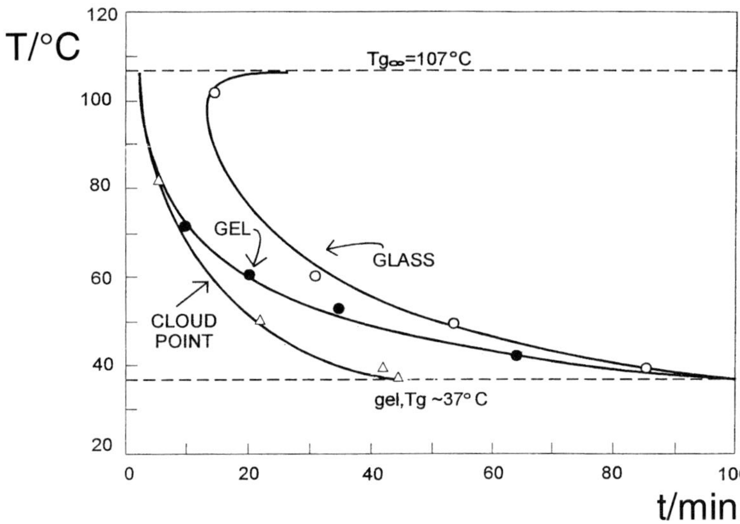

Figure 4.12 TTT diagram showing phase separation, gelation and vitrification (onset), for a castor-oil-modified epoxy (15 wt% of castor oil in the initial formulation) [11]. (Reprinted with permission.).

where $[e]_0$ and $[OH]_0$ are the initial concentration of epoxy and hydroxy groups, respectively, and

$$k/(l^2\,eq^{-2}\,s^{-1}) = 1.72 \cdot 10^5 \exp[-7060/(T/K)] \tag{4.27}$$

For isothermal transformations, the CTT and the TTT diagrams may be considered as equivalent ways of representing the transitions taking place during network formation. For the study of high-T_g materials where degradation may compete with polymerization, or may produce devitrification followed by revitrification processes [1], the TTT diagram may be regarded as more useful than the CTT diagram. It must be remembered, however, that values of times only have physical significance under conditions which are the same as those used to obtain them. Therefore, a TTT diagram generated for isothermal transformations cannot be used for arbitrary non-isothermal cure cycles. Unfortunately, this fact is sometimes forgotten [47]. In this regard, the CTT diagram is more versatile, as will be illustrated in the following sections.

4.3.2 *Transformations at a constant heating rate*

These transformations are also typical of laboratory devices used for characterization purposes. As a small mass of material is frequently used, temperature may be considered uniform in the specimen and increasing linearly with time:

$$dT/dt = q \tag{4.28}$$

Substitution in equation (4.20), written for any T, gives

$$dx/dT = r(x, T)/q \tag{4.29}$$

Numerical integration of equation (4.29) leads to conversion-versus-temperature trajectories in the CTT diagram [10]. In fact, the typical situation is the inverse, i.e. from experimental x versus T trajectories (dynamic runs on a DSC), the rate equation $r(x, T)$ is obtained. To perform these calculations, trajectories that do not enter the vitrification region must be selected in order to avoid the influence of k_D (equation (4.19)).

A particular case will be analysed. For a stoichiometric epoxy (DGEBA)–diamine (TMAB) system, Wisanrakkit and Gillham [18] reported the kinetics expression

$$r = k(x + 0.05548)(1 - x)^2 \tag{4.30}$$

with

$$\frac{1}{k} = \frac{1}{k_T(T)} + \frac{1}{k_D(x, T)} \tag{4.31}$$

$$k_T/\text{min}^{-1} = 7.68 \cdot 10^6 \exp[-7649.7/(T/\text{K})] \tag{4.32}$$

$$k_D/\text{min}^{-1} = 30.64 \exp\{42.61(T - T_g)/(51.6 + |T - T_g|)\} \tag{4.33}$$

T_g (determined by TBA) was expressed as a function of conversion, using equation (4.7) with $T_{g0} = 270\,\text{K}$, $T_{g\infty} = 453\,\text{K}$ and $\lambda = 0.31$.

Some trajectories, arising from numerical integration of equation (4.29) for constant heating rates, are shown in Figure 4.13. The starting temperature T_0 is 298 K (25 °C). For heating rates $q < 0.13\,°\text{C}\,\text{min}^{-1}$, the system enters the vitrification region in the course of polymerization. As the T_g determined by TBA is close but does not coincide exactly with the end of the vitrification range, it may be expected that vitrification begins to take place somewhere to the right of the T_g-versus-x curve in Figure 4.13. Analysis of experimental TBA relaxations, obtained at different heating rates [18], showed evidence of a vitrification peak for heating rates $q < 0.31\,°\text{C}\,\text{min}^{-1}$. This proves that, in fact, the T_g determined by TBA is very close to the end of the vitrification region.

The effect of going through the vitrification region during a polymerization at a constant heating rate will now be analysed. Figure 4.14 shows dynamic DSC thermograms for a formulation consisting of a diepoxide based on DGEBA, a diamine (triethylenetetramine (TETA)) and 10 wt% of an amine-terminated butadiene-acrylonitrile random copolymer (ATBN), as modifier [26]. At the lowest heating rate (0.5 °C min^{-1}), a second peak at high conversions is present in the thermogram. Conversion-versus-temperature trajectories, calculated from the various heating rates, are represented in the CTT diagram shown in Figure 4.15 [26]. The T_g-versus-x relationship is also represented. It was fitted by equation (4.7) with $T_{g0} = 244\,\text{K}$,

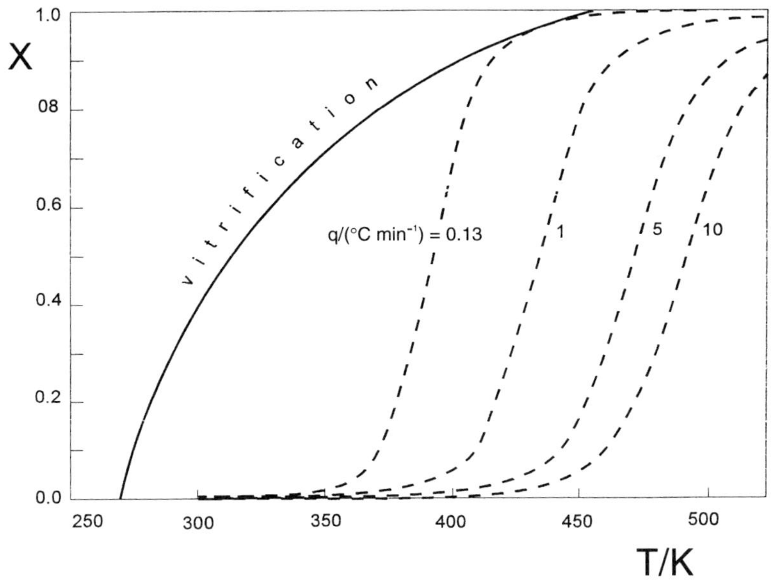

Figure 4.13 CTT diagram for a diepoxy (DGEBA)–diamine (TMAB) system showing trajectories at constant heating rates (q), starting at 298 K (25 °C).

$T_{g\infty} = 350\,\text{K}$ and $\lambda = 0.74$ (T_g was determined by DSC but no indication of the definition used was provided) [26].

From Figure 4.15 it is observed that, at high conversions, the trajectory at $0.5\,°\text{C}\,\text{min}^{-1}$ is possibly traversing the vitrification region (extending

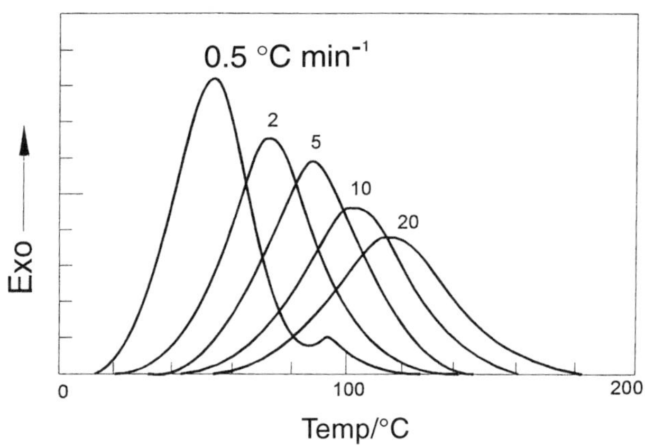

Figure 4.14 Dynamic DSC thermograms, for a diepoxy (DGEBA)–diamine (TETA) formulation modified by 10 wt% of rubber (ATBN), at various scanning rates [26]. (Reprinted with permission.)

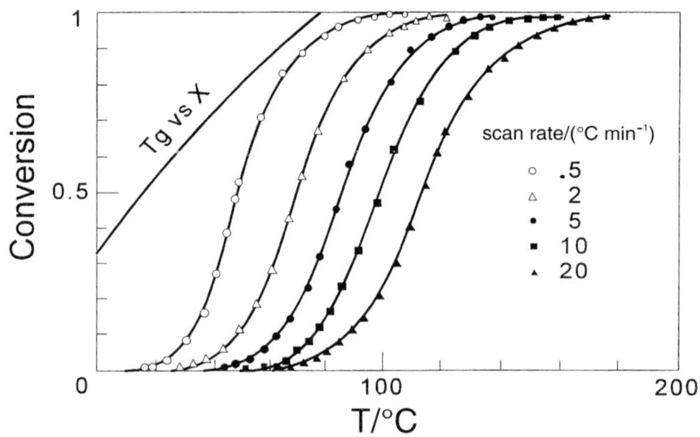

Figure 4.15 CTT diagram for the system of Figure 4.14, showing trajectories at constant heating rates [26]. (Reprinted with permission.)

somewhere to the right of the vitrification curve). This produces a sudden drop in the polymerization rate followed by an increase when the trajectory leaves the vitrification region (at very high conversions). This is the origin of the second peak in the DSC thermogram at $0.5\,^{\circ}\mathrm{C\,min^{-1}}$ in Figure 4.14. Arrhenius representations of kinetics rate constants were consistent for every heating rate, except for $0.5\,^{\circ}\mathrm{C\,min^{-1}}$ [26]. For this rate, the diffusional constant, k_{D}, has to be included in the rate equation to fit the experimental data.

4.3.3 Adiabatic transformations

Adiabatic transformations take place in the core regions of large parts, when the rate of diffusion of heat to or from the boundaries is much lower than the rate of generation of heat by the polymerization reaction [48, 49]. These transformations also occur when determining the kinetics of very fast polymerization reactions, using the adiabatic temperature-rise technique [48].

For an adiabatic system, the following hold:

1. Functional-group balance

$$\mathrm{d}x/\mathrm{d}t = r(x, T) \tag{4.34}$$

2. Thermal-energy balance

$$c_{\mathrm{p}}\,\mathrm{d}T/\mathrm{d}t = (-\Delta H)r(x, T) \tag{4.35}$$

where c_{p} and $(-\Delta H)$ are, respectively, the isobaric specific heat capacity and the heat of reaction per unit mass. If both c_{p} and $(-\Delta H)$ are assumed constant, using equation (4.34) in equation (4.35) and

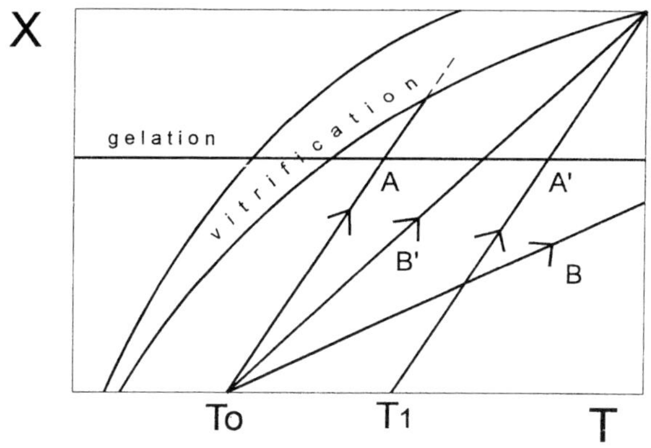

Figure 4.16 CTT diagram showing adiabatic trajectories.

integrating leads to

$$x = (T - T_0)/\Delta T_{ad} \tag{4.36}$$

where

$$\Delta T_{ad} = (-\Delta H)/c_p \tag{4.37}$$

is the adiabatic temperature rise produced by the polymerization reaction.

From equation (4.36), adiabatic trajectories in the CTT diagram are straight lines with slope ΔT_{ad}^{-1}. Some possible trajectories are represented in the CTT diagram shown in Figure 4.16. Trajectory A represents an adiabatic transformation starting at $T = T_0$, for a system characterized by a relatively low value of ΔT_{ad}, so that the vitrification region is attained during the course of polymerization. The trajectory will continue through the vitrification region and eventually into the glassy state. However, in an actual polymerization, when the reaction rate drops significantly (in the vitrification region), heat losses to the boundaries will cool the system preventing any further advance in conversion. Therefore, it may be stated that when an adiabatic trajectory enters the vitrification region (almost) no further reaction takes place. This may be a desirable situation in autoclave cure cycles, as will be analysed in the next section. However, if the adiabatic temperature-rise technique is being used to obtain the reaction kinetics, full cure must be attained in order to define the reference value, $(T_{max} - T_0)$, for the conversion scale. In this case, the initial temperature must be shifted to T_1, giving the trajectory A′ parallel to A.

When ΔT_{ad} has a high value, a trajectory such as B in Figure 4.16 results. This leads to degradation of the polymer network before full conversion is attained. The way to process these materials is to use an inert filler as reinforcement, so that the heat of reaction per unit total mass and the associated ΔT_{ad} decrease significantly. A trajectory like B′ (Figure 4.16) may be obtained, rendering the process feasible.

4.3.4 *General non-isothermal transformations*

These transformations are characteristic of industrial processes where network-forming polymers are subjected to particular cure cycles. In general, cure is not uniform throughout a specimen so that differential balances have to be stated in terms of partial derivatives. For example, the cure in a heated mould, where properties change along the thickness Z, is represented by the following equations [49]:

1. Functional-group balance

$$\partial x/\partial t = r(x, T) \qquad (4.38)$$

2. Thermal-energy balance

$$\rho c_p \, \partial T/\partial t = \partial/\partial z(\kappa \, \partial T/\partial z) + \rho(-\Delta H)r(x, T) \qquad (4.39)$$

   where ρ, c_p and κ are, respectively, density, isobaric specific heat capacity and thermal conductivity of the material. All of them may be functions of temperature and conversion. In order to complete the description, appropriate initial and boundary conditions have to be stated.

Reactive processing of formulations leading to polymer networks are described in Chapter 5. In this section, attention is focused on the way in which transformations occurring during network formation may affect the selection of cure cycles and the properties of the resulting materials. The CTT diagram will be used as a visual aid to comprehend a system's behaviour. Quantitative answers must always be obtained through numerical solutions of the differential balances.

4.3.4.1 *Selection of cure cycles in autoclave processing*

For graphite–epoxy composites based on tetraglycidyl 4,4′-diamino-diphenylmethane (TGDDM), crosslinked with 4,4′-diaminodiphenylsulfone (DDS), a typical autoclave cure cycle is shown in Figure 4.17. At $t = 45$ min, when the wall temperature has reached 135 °C, the autoclave pressure is applied. This step provokes the elimination of resin which is absorbed by bleeder plies, and the increase in the fibre volume fraction from its initial value in the preimpregnated tissue (close to 50%) to its final value in the composite (close to 65%). The next temperature increase to 177 °C leads to the cure of the thermosetting polymer.

Figure 4.18 represents a CTT diagram for the TGDDM-DDS system. The location of the vitrification curve was estimated by applying equation (4.15) [50]. T_{g0} is 258 K (−15°C) whereas the initial temperature of the cure cycle is

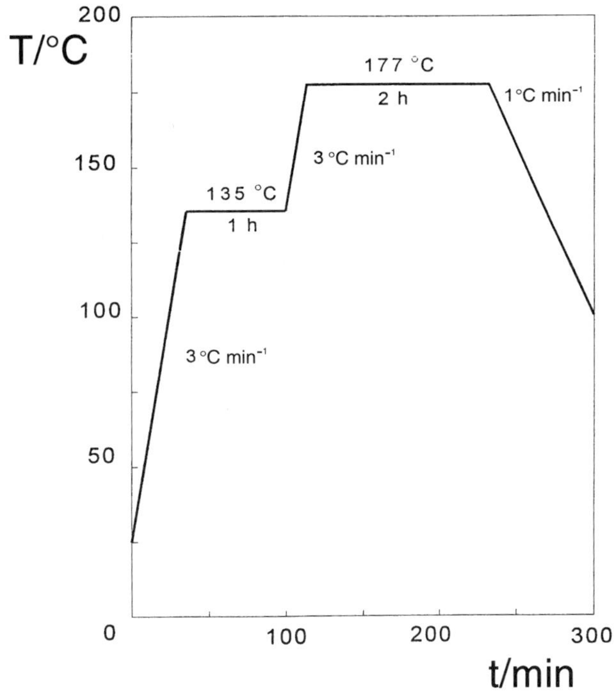

Figure 4.17 Typical autoclave cure cycle for a graphite–epoxy composite (TGDDM-DDS).

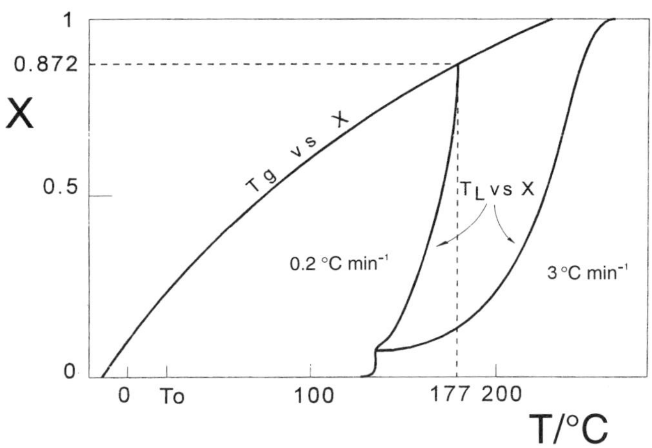

Figure 4.18 CTT diagram showing the vitrification curve for the graphite–epoxy composite (TGDDM-DDS) and trajectories at the boundary in contact with bleeder plies (T_L), for different cure cycles [50]. (Reprinted by permission.)

$T_0 = 298\,K$ (25 °C). The fact that $T_0 > T_{g0}$ is a necessary condition for the preparation of the laminate, an operation which requires some tack as well as drapability at room temperature.

Trajectories in the CTT diagram are particular for definite values of the coordinate Z. Only for very thin parts is the cure uniform throughout a specimen. As the maximum temperature is 177 °C, the maximum conversion is $x_{max} = 0.872$, read from the vitrification curve (neglecting any possible advance of reaction in the glassy state during the cure cycle). At this value of conversion, a good compromise is obtained between mechanical and thermal properties of the composite material [50].

For large parts, the layer in contact with bleeder plies exhibits the highest temperature in the part (T_L). For 32-ply laminates, T_L is 5–10 °C higher than the wall temperature, but for 128-ply laminates T_L may exceed the wall one by as much as 60–70 °C [51]. If it is desired to keep $x_{max} = 0.872$ at every point, it is necessary to modify the cure cycle.

Figure 4.18 shows two T_L-versus-x trajectories, for a 2 cm-thick part cured using two different cycles. These trajectories were obtained by numerically solving the differential mass and energy balances [50]. The cycle was modified by changing the rate at which temperature was increased from 135 °C to 177 °C in the second step of the cure cycle. Using the standard cure cycle shown in Figure 4.17 leads to a temperature increase of some 90 °C over the wall temperature. The half of the part closest to the bleeder plies attains full conversion [50]. However if, in the second step, temperature is increased at a slower rate, i.e. 0.2 °C min^{-1}, the trajectory of T_L ends at the vitrification curve, exactly at the same point as the trajectory at the wall ($T = 177$ °C, $x = 0.872$). This leads to a uniform conversion profile in the part, at the value required to optimize physical and mechanical properties at room temperature. The price which is paid is an increase in the cure time with respect to the standard cure cycle, by a factor close to 2. Such an increase may have sense for the production of short series of large parts.

Other examples dealing with the influence of vitrification on processing have also been reported [25, 52].

4.3.4.2 *Rapid thermoset processing (RTP)*

Breitigam *et al.* [47] proposed an alternative approach for the processing of thermoset composites. Their idea is illustrated in Figure 4.19 in terms of a CTT diagram. The first step of the cure is carried out in an autoclave until the laminate is consolidated (void free) and the conversion is advanced in the pre-gel region. At this time the specimen is cooled to reach the glassy state and taken out from the autoclave (trajectory A). The second step is carried out in an unattended pre-programmed oven, such that the cure temperature remains equal to or lower than T_g (trajectory B). While the part remains in the glassy state, no distortion may be expected in this free-standing procedure. Claimed advantages of this process are (i) lower

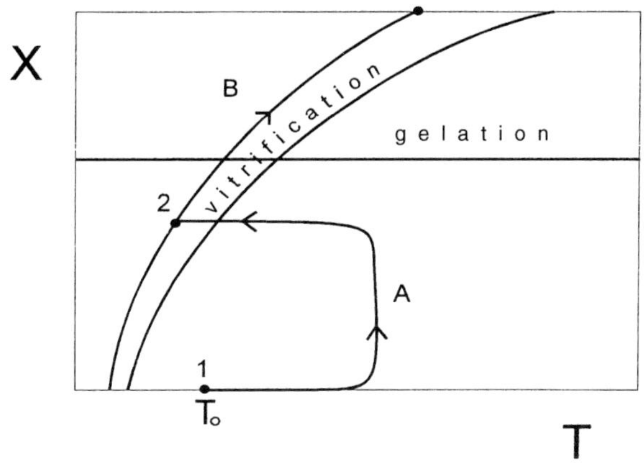

Figure 4.19 CTT diagram showing possible trajectories for rapid, thermoset processing (RTP): A, trajectory in the autoclave; B, trajectory in an unattended pre-programmed oven.

processing temperatures with respect to the standard cure cycle, (ii) shorter residence times in the autoclave (for this reason the new process is called rapid thermoset processing (RTP)), and (iii) reworkability of the part after the first step (thermoplastic-like behaviour). However, clear disadvantages are the presumably long time to complete the cure in the glassy state and the necessity of strict control of the oven temperature to keep T_{cure} close to T_g (onset). If $T_{cure} < T_g$ (onset), the polymerization rate will be extremely slow. If T_{cure} is close to T_g (end), the reduction in the elastic modulus may cause distorsion of the part.

Experimental evidence of the process's feasibility was provided using a 15-ply laminate, fabricated from a satin-weave glass-cloth impregnated with an epoxy–amine reaction system [47]. This laminate was processed in an autoclave, increasing temperature from 25 °C to 121 °C in 1 h, followed by a 0.5 h hold. Step 2 was carried out in a dynamic–mechanical spectrometer. Temperature was increased at a rate of 3 °C min^{-1} up to 140 °C, at which the beginning of devitrification was observed (reduction in real modulus). Then, the heating rate was decreased to 0.75 °C min^{-1} such that the shear modulus, that was being continuously monitored, did not drop. Complete cure could be attained under these conditions, as revealed by the fact that T_g reached the value of $T_{g\infty}$ (243 °C as measured by a maximum in tan δ), after a 3 h period in step 2.

The length of the cure cycle depends significantly on the final conversion attained in the autoclave. To illustrate this fact, the DGEBA-TMAB system, analysed in Section 4.3.2, will be considered again. Both the kinetics and the T_g-versus-x dependences are known. Hence, times to attain complete conversion, following $T_{cure} = T_g$ (trajectory B in Figure 4.19), may be calculated numerically starting from an arbitrary conversion value x_0, resulting from the first stage in the autoclave. For $x_0 = 0$, i.e. no reaction in the autoclave, the time necessary to reach $x = 0.99$ is 8.2 months. For $x_0 = 0.4$, this reduces to 8 days and for $x_0 = 0.68$ (beyond the gel point) 36 h. The high sensitivity of T_{cure} on x_0 arises from the shape of the T_g-versus-x curve. Hence, the effect is general for any thermosetting polymer. Times are still very much larger if the cure is performed below T_g, i.e. for $T_{cure} = T_g - 10$ starting at $x_0 = 0.4$, gives 19 days to reach $x = 0.99$.

Therefore, this novel process requires previous numerical calculations, for the particular system under consideration, to establish its economic feasibility. In any case, the name 'rapid thermoset processing' is misleading.

4.3.4.3 Phase separation profiles in a cured part

When a thermosetting polymer is cured in a heated mould, conversion-versus-temperature trajectories vary with position in the part, in a way that depends on the selected cure cycle and the mould geometry.

The cure of a modified thermosetting polymer in a heated mould will be considered. The CTT diagram, shown in Figure 4.20, indicates two possible T-versus-x trajectories during polymerization. Trajectory A (lower temperatures) goes through the phase-separation region, whereas trajectory B (higher temperatures) does not. This means that a two-phase morphology will be generated along A but the material will remain homogeneous along B. When curing in a heated mould, trajectory A may represent the material located near the walls whereas B may be assigned to the material at the core. This means that by an appropriate selection of the modified thermosetting polymer, cure cycle and mould geometry, 'sandwich structures' consisting of a homogeneous core surrounded by phase-separated boundaries, may be generated. This is the situation for modifiers that increase their

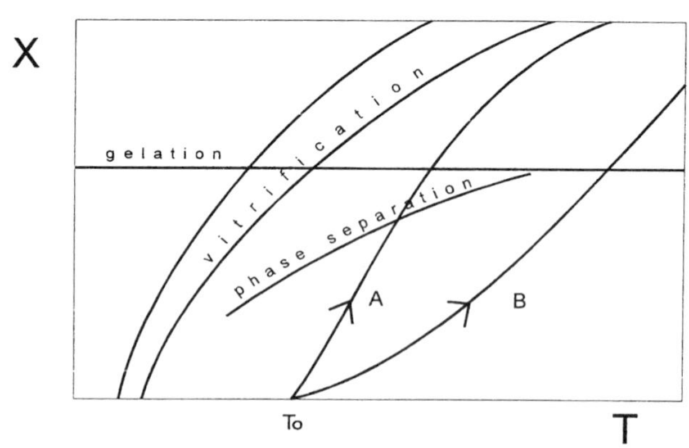

Figure 4.20 CTT diagram for a modified thermosetting polymer showing trajectories (A) going through or (B) avoiding the phase-separation region.

miscibility with temperature (UCST behaviour). The inverse situation may be observed if the modifier's miscibility decreases with increase in temperature (LCST behaviour).

An experimental verification of the possibility of producing sandwich structures was reported recently [53]. A castor-oil-modified epoxy-amine formulation was cured in a heated mould of 6.5 mm thickness. The wall temperature was increased at a rate of $5\,°C\,min^{-1}$, from room temperature to 125 °C. This cure schedule produced trajectories that are qualitatively equivalent to those represented in Figure 4.20. The maximum temperature for the material located at the wall was close to 125 °C, whereas the maximum temperature at the core was close to 175 °C. Visual inspection of the cured part revealed a sandwich structure with 1 mm opaque boundaries surrounding a transparent core. Scanning electron microscopy (SEM) proved the presence of a random dispersion of spherical domains, with sizes in the range 0.4–0.5 μm, in the material located near the wall. No segregated domains were observed in the transparent core. Distinct mechanical properties were determined for the sandwich-materials, when compared to the uniform structures obtained by reducing the rate of wall-temperature increase, thus allowing phase separation to take place throughout the part [53].

In conclusion, examples analysed along this section illustrate the practical significance in the selection of cure cycles of taking into account the various transitions occurring during network formation. In this regard, CTT diagrams constitute a convenient framework for analysing different possibilities of carrying out the polymerization processes.

Acknowledgements

The author expresses his gratitude to all the co-workers mentioned in the list of references. Helpful discussions with Prof. J.P. Pascault (INSA de Lyon, France), Prof. J.K. Gillham (Princeton University, USA) and Dr R.B. Prime (IBM, USA) are acknowledged.

Glossary of symbols and abbreviations

c_p	isobaric specific heat capacity
c_{pi}	isobaric specific heat capacity of component i
Δc_{pi}	change in c_{pi} between the liquid and the glassy states
Δc_{p0}	change in c_p between the liquid and the glassy states at zero conversion
$\Delta c_{p\infty}$	change in c_p between the liquid and the glassy states at complete conversion

$g(x)$	function describing the dependence of polymerization rate on conversion (x)
k	rate constant
k_T	Arrhenius rate constant
k_D	diffusion rate constant
q	heating rate
r	polymerization rate
t_g	gel time
t_{vit}	vitrification time
t_{cure}	cure time
x	conversion
x_{gel}	conversion at gel
x_{max}	maximum attainable conversion
x_{vit}	conversion at vitrification
$\Delta H_u, \Delta H_r, \Delta H_T$	heats, determined by DSC, characterizing a curing reaction
$-\Delta H$	heat of reaction per unit mass
T_0	temperature at zero conversion
T_{cure}	cure temperature
T_{deg}	degradation temperature
T_g	glass-transition temperature
T_{g0}	T_g at zero conversion
T_{gmax}	T_g at conversion x_{max}
$T_{g\infty}$	T_g at complete conversion
$T_{g,onset}$	T_g at the onset of vitrification
$T_{g,end}$	T_g at the end of vitrification
${}_{gel}T_g$	temperature of coincidence of gelation and the glass-transition temperature
T_L	highest temperature in a part during a non-isothermal transformation
ΔT_{ad}	temperature rise during an adiabatic polymerization process
Z	thickness dimension in a mould
ϕ_i	concentration of component i
CTT	conversion-versus-temperature transformation (diagram)
DGEBA	diglycidyl ether of bisphenol A
DMTA	dynamic–mechanical analyser
LCST	lower, critical solution temperature
PDMS	poly(dimethylsiloxane)
RTP	rapid thermoset-processing
TBA	torsional braid-analysis
TMAB	trimethylene glycol di-*p*-aminobenzoate
TTT	time-versus-temperature transformation diagram
UCST	upper, critical solution temperature

References

1. Chan, L.C., Naé, H.N. and Gillham, J.K. (1984) *J. Appl. Polymer Sci.*, **29**, 3307.
2. Riccardi, C.C. and Williams, R.J.J. (1986) *J. Appl. Polymer Sci.*, **32**, 3445.
3. Zukas, W.X. (1994) *J. Appl. Polymer Sci.*, **53**, 429.
4. Riccardi, C.C. and Williams, R.J.J. (1986) *Polymer*, **27**, 913.
5. Choy, I.Ch. and Plazek, D.J. (1986) *J. Polymer Sci. B*, **24**, 1303.
6. Hale, A., Macosko, C.W. and Bair, H.E. (1991) *Macromolecules*, **24**, 2610.
7. Oyanguren, P.A. and Williams, R.J.J. (1993) *J. Appl. Polymer Sci.*, **47**, 1361.
8. Wisanrakkit, G. and Gillham, J.K. (1990) *J. Appl. Polymer Sci.*, **41**, 2885.
9. Pascault, J.P. and Williams, R.J.J. (1990) *J. Polymer Sci. B*, **28**, 85.
10. Adabbo, H.E. and Williams, R.J.J. (1982) *J. Appl. Polymer Sci.*, **27**, 1327.
11. Ruseckaite, R.A., Hu, L., Riccardi, C.C. and Williams, R.J.J. (1993) *Polymer International*, **30**, 287.
12. Oyanguren, P.A. and Williams, R.J.J. (1993) *J. Appl. Polymer Sci.*, **47**, 1373.
13. Lucas, J.C., Borrajo, J. and Williams, R.J.J. (1993) *Polymer*, **34**, 3216.
14. Havlíček, I. and Dušek, K. (1987) in *Crosslinked Epoxies* (eds B. Sedláček and J. Kahovec), de Gruyter, Berlin, pp. 417–24.
15. Enns, J.B. and Gillham, J.K. (1983) *J. Appl. Polymer Sci.*, **28**, 2567.
16. Wisanrakkit, G., Gillham, J.K. and Enns, J.B. (1990) *J. Appl. Polymer Sci.*, **41**, 1895.
17. Pang, K.P. and Gillham, J.K. (1990) *J. Appl. Polymer Sci.*, **39**, 909.
18. Wisanrakkit, G. and Gillham, J.K. (1991) *J. Appl. Polymer Sci.*, **42**, 2453.
19. Wang, X. and Gillham, J.K. (1992) *J. Appl. Polymer Sci.*, **45**, 2127.
20. Simon, S.L. and Gillham, J.K. (1992) *J. Appl. Polymer Sci.*, **46**, 1245.
21. Simon, S.L. and Gillham, J.K. (1993) *J. Appl. Polymer Sci.*, **47**, 461.
22. Venditti, R.A. and Gillham, J.K. (1993) *Amer. Chem. Soc. Polymeric Materials Science and Engineering Preprints*, **69**, 434.
23. Mijović, J. and Wijaya, J. (1990) *Macromolecules*, **23**, 3671.
24. Hong, M.S. and Chung, I.J. (1991) *Polymer J.*, **23**, 747.
25. Kim, D.H. and Kim, S.Ch. (1989) *Polymer Eng. Sci.*, **29**, 456.
26. Kim, D.S. and Kim, S.Ch. (1994) *Polymer Eng. Sci.*, **34**, 625.
27. Jordan, C., Galy, J. and Pascault, J.P. (1992) *J. Appl. Polymer Sci.*, **46**, 859.
28. Stutz, H., Illers, K.H. and Mertes, J. (1990) *J. Polymer Sci. B*, **28**, 1483.
29. Montserrat, S. (1992) *J. Appl. Polymer Sci.*, **44**, 545.
30. Montserrat, S. (1993) *J. Therm. Anal.*, **40**, 553.
31. Di Benedetto, A.T. (1987) *J. Polymer Sci. B*, **25**, 1949.
32. Shefer, A. and Gottlieb, M. (1992) *Macromolecules*, **25**, 4036.
33. Couchman, P.R. (1987) *Macromolecules*, **20**, 1712.
34. Nielsen, L.E. (1969) *J. Macromol. Sci. – Reviews in Macromolecular Chemistry*, **C3**, 69.
35. Georjon, O., Galy, J. and Pascault, J.P. (1993) *J. Appl. Polymer Sci.*, **49**, 1441.
36. Montserrat, S. (1995) *Polymer*, **36**, 435.
37. Riew, C.K. and Gillham, J.K. (eds) (1984) *Rubber-Modified Thermoset Resins*, American Chemical Society, Washington, DC..
38. Riew, C.K. (ed) (1989) *Rubber-Toughened Plastics*, American Chemical Society, Washington, DC..
39. Riew, C.K. and Kinloch, A.J. (eds) (1993) *Toughened Plastics I: Science and Engineering*, American Chemical Society, Washington, DC..
40. Williams, R.J.J., Borrajo, J., Adabbo, H.E. and Rojas, A.J. (1984) in Ref. 37, pp. 195–213.
41. Vázquez, A., Rojas, A.J., Adabbo, H.E., Borrajo, J. and Williams, R.J.J. (1987) *Polymer*, **28**, 1156.
42. Moschiar, S.M., Riccardi, C.C., Williams, R.J.J., Verchère, D., Sautereau, H. and Pascault, J.P. (1991) *J. Appl. Polymer Sci.*, **42**, 717.
43. Ruseckaite, R.A. and Williams, R.J.J. (1993) *Polymer International*, **30**, 11.
44. Peng, X. and Gillham, J.K. (1985) *J. Appl. Polymer Sci.*, **30**, 4685.
45. Kloosterboer, J.G. and Lijten, G.F.C.M. (1990) *Polymer*, **31**, 95.
46. Stutz, H. and Mertes, J. (1989) *J. Appl. Polymer Sci.*, **38**, 781.
47. Breitigam, W.V., Bauer, R.S. and May, C. (1993) *Polymer*, **34**, 767.

48. Williams, R.J.J. (1985) in *Developments in Plastics Technology-2* (eds A. Whelan and J. Craft), Elsevier Applied Science, London, p. 339.
49. Williams, R.J.J., Rojas, A.J., Marciano, J.H., Ruzzo, M.M. and Hack, H.G. (1985) *Polymer-Plastics Tech. Eng.*, **24**, 243.
50. Williams, R.J.J., Benavente, M.A., Ruseckaite, R.A., Churio, M.S. and Hack, H.G. (1990) *Polymer Eng. Sci.*, **30**, 1140.
51. Mijović, J. and Wang, H.T. (1988) *SAMPE Journal*, March/April, 42.
52. Osinski, B.A. (1992) *Amer. Chem. Soc. Polymer Preprints*, **33**, 380.
53. Fang, D.P., Frontini, P.M., Riccardi, C.C. and Williams, R.J.J. (1995) *Polymer Eng. Sci.*, **35**, 1359.

5 Reactive processing of polymer networks: Structure–property relationships in copolymer networks formed by reaction injection moulding

J.L. STANFORD

5.1 Introduction

Reactive processing of polymers and composites involves simultaneous polymerization, development and modification of polymer structure and the shaping of materials to produce polymer artefacts with final properties suitable for particular applications. In engineering terms, reactive processing is concerned primarily with the design and control of the equipment, and of the heat transfer and rheological processes required for particular polymer-forming systems. Equally important, in scientific terms, is the understanding of the molecular processes involved in the transformation from reactive liquid state (either as low-viscosity monomers or as high-viscosity polymer melts) to solid polymer material that occurs during reactive processes, such as reaction injection moulding (RIM) and reactive extrusion (REX). The transformation processes defining the development of polymer structure, require *in situ* and *a posteriori* measurements and analyses to be made to provide the fundamental understanding upon which appropriate process models can be developed for the control and improvement of reactive processing and materials properties.

This chapter deals with the relationships between the chemistry, molecular and morphological structure and properties of reactively processed polymers. Particular emphasis is placed on polymer structure development and properties in RIM systems in terms of polymerization, gelation and network formation, phase separation and vitrification processes. The basic theories underpinning these processes in single-phase and multiphase (linear and non-linear) systems are presented. Often such processes occur competitively, either simultaneously or consecutively, in terms of the kinetics of chemical reactions and of the thermodynamics that define structure development. As a consequence, reactively processed materials are characterized by complex morphologies that largely determine their final physical properties, and *a posteriori* studies on structure–property relationships in such materials are extensive. Also important are *in situ* studies on reacting systems, involving the use of and the results obtained from time-resolved analytical techniques. Examples of both types of study are reported on systems that produce a wide range of multiphase polymers classified, in general terms, as linear thermoplastics, block copolymers and network thermosets.

Sections 5.1.1 and 5.1.2 give a brief overview of some of the more common reactive polymer processes, such as reactive extrusion, casting, coating and moulding. Later sections present, in more detail, a review of the fundamentals of RIM in terms of the process (Section 5.2), molar-mass development and the thermodynamic principles of phase separation (Section 5.3). Sections 5.4 and 5.5 give accounts of the kinetics of copolymerization and phase separation, and of the structure–property relationships typical of RIM copolymers using, as specific examples and in keeping with the theme of this book, non-linear RIM systems that produce ultimately phase-separated network materials.

5.1.1 *Reactive extrusion*

Reactive extrusion (REX) [1, 2] involves the use of extruders as continuous reactors to combine polymer processing and chemical reactions such as polymerization, crosslinking, polymer modification and functionalization, compatibilization of polymer blends and controlled polymer degradation. REX provides the ability to pump and mix highly viscous materials and, as a result of the laminar flow patterns generated during the extrusion process, facilitates the staging of multiple process steps in a single machine. Such steps include feeding, melting, metering, mixing, reacting, side-stream addition, venting and, where appropriate, shaping. In many respects, the extruder with dimensions characterized by high length-to-diameter ratios ($15 < L/D < 50$) may be considered to be a tubular reactor in which chemical reactions are achieved in continuous processes with short residence times ranging, typically, from a few seconds to about 20 min. The ability to process reactive systems with a wide range of viscosities (10–10 000 Pa s) gives REX an advantage over other conventional reactors, particularly for high viscosities. This advantage is clearly exemplified by equation (5.1) which expresses in simple form the basic drag flow mechanism of an extruder:

$$Q = \alpha N - \frac{\beta(\Delta P)}{\mu} \tag{5.1}$$

where Q is the volumetric throughput (flow rate), N is the screw speed, ΔP is the pressure drop, μ is the viscosity, and α and β are extruder design parameters. Essentially, the negative term on the right-hand-side, containing the reciprocal viscosity factor, demonstrates clearly why extruders are so efficient at processing high-viscosity polymer systems. Additionally, viscosity increases from 50 to 1000 Pa s over the extruder length, not uncommon during polymerizations, are easily accommodated in REX. Another obvious advantage of REX is the ability to carry out reactions in bulk thereby obviating the need for solvents and dramatically reducing costs of raw materials, solvent recovery, product separation and drying equipment, and providing improved environmental production through the elimination of emissions otherwise associated with solvents. Bulk reactions also give higher overall reaction rates compared with solution reactions, due mainly to the higher concentrations of reactive species in the melt and to the reaction temperatures, often 50–150 °C higher than in solution, achievable in an extruder. The main disadvantages of REX are the difficulties in handling reactive systems that produce large heats of reaction, due to the limited heat transfer capacity of extruders. For example, the bulk polymerization of 1 kg of styrene produces ~700 kJ of heat which is too much for most conventional extruders. Also, the relatively short residence times typical of extruders make them uneconomical as reactors for intrinsically slow reactive systems requiring residence times greater than ~30 min.

Both single-screw and twin-screw extruders are used for REX processes. In single-screw extrusion, material transport is by drag-induced flow, namely frictional drag in solids-conveying zones and viscous drag in melt-conveying zones, and such flow is often sufficient to give efficient mixing for homogeneous reactions involving miscible reactants. For heterogeneous reactions such as those involving immiscible reactants or reactions across polymer phase boundaries, more intensive mixing is required and twin-screw extruders are more appropriate. Compared with single-crew extrusion, the velocity, pressure and temperature profiles in twin-screw extrusion are more complex, and material transport occurs by a positive displacement mechanism which depends on the extent to which the two screws are intermeshed (i.e. the proximity between the flight of one screw and the opposing channel of the second screw). Generally, counter-rotating, closely intermeshing twin-screw extruders are considered to provide the most positive displacement conditions during REX.

The general types of reactive extrusion process are summarized in Table 5.1. One of the principal factors underpinning such REX processes, from a chemical standpoint, is the kinetics of the desired reaction. Designing a chemical reaction to take place efficiently within the residence time of available extrusion equipment is clearly vital to maximize the major advantage of REX processes for the rapid production of linear and network polymers from monomers or prepolymers (or both), or the production of chemically modified polymers, or compatibilized polymer blends, some examples of which are given in Table 5.1. The general requirements in designing new chemistries for successful REX processes include activation energies for reactive systems which must be low enough to ensure chemical reaction occurs under acceptable extrusion conditions. Intrinsically slow chemical reactions or those involving low-melting polymers may require catalysis in order to obviate the need to raise extrusion temperatures as high as the point at which degradation commences. In contrast, polymers with melting points higher than the temperatures required to initiate reaction require

Table 5.1 Classification, description and examples of the various chemical reactions carried out using reactive extrusion

Reaction type	Description	Example [1, 2]
Bulk polymerization	formation of high molar mass polymer from either monomer, prepolymer or monomer–prepolymer mixtures	(i) chain polymerization (free radical) of styrene/acrylonitrile to SAN copolymer (ii) polyaddition of hydroxy-terminated polyester, butane diol and MDI to copoly(ester–urethane)
Grafting	formation of either grafted polymer or copolymer from reactions of polymer and monomer	(i) free-radical grafting of maleic anhydride onto polypropylene (ii) ester-exchange grafting of ethylene–ethyl acrylate copolymer with acetooxyethyltrimethoxy silane using tetraalkoxy titanate catalyst
Interchain copolymerization	formation of copolymers (random, graft or block) from reactions of two or more polymers via ionic or covalent bonding	blend compatibilization of nylon-6 and maleic anhydride-grafted polypropylene
Coupling and crosslinking	development of polymer molar mass and melt viscosity by chain extension, branching and crosslinking reactions with polyfunctional coupling reactants and crosslinking agents	(i) coupling of polyesters (PET or PBT) with polyisocyanates (MDI) (ii) crosslinking maleic anhydride-grafted polypropylene with poly(oxyethylene) diol
Controlled degradation	molar mass reduction of polymers for rheology control and controlled degradation of polymer to monomer feedstock	(i) hydrolysis of nylon-6.6 (ii) degradation of polypropylene with *t*-butyl hydroperoxide (iii) hydrolysis (water, sulfuric acid, steam) of paper pulp to glucose monomer

plasticization (preferably with a reactive diluent) or lubrication to lower the processing temperature for controlled REX. In addition, for the successful commercialization of REX processes, efficient and complete chemical reaction must occur within the constraints of the extruder, any by-products and noxious fumes must be removed totally and the chemical and structural changes in the materials produced by REX must be stable to the conditions used in subsequent manufacturing processes.

By far the most commonly reported examples of REX involve grafting reactions of either polyalkenes (either saturated or unsaturated) or of polyphenylene ethers with unsaturated monomers, particularly maleic anhydride or vinyl silanes. Also important are reactions enabling polymer modification, crosslinking or *in situ* copolymer formation in polymer blends, and such reactions involve, typically, amine or carboxylic groups (attached to polymers as end groups or side groups) with electrophilic reactants that include anhydrides, oxazolines, orthoesters, epoxides, isocyanates and carbodiimides.

Comprehensive accounts of the engineering aspects of REX, of the wide range of chemical systems used and the polymers produced in REX processes, and detailed and systematic state-of-the art descriptions of REX are available [1, 2].

5.1.2 *Reactive casting, coating and moulding*

In this section, a brief overview is presented of some of the other, more common reactive polymer processes. More detailed descriptions are given elsewhere [3, 4]. The methods used in reactive processing come under the general headings of casting, coating and moulding, and involve the use of viscous (as opposed to viscoelastic) materials which include liquid monomers and prepolymers, polymer solutions and dispersions and (low-molar-mass) polymer powders. The reactive processing of such materials may be characterized by the flow (usually under gravity and requiring little pressure) of liquids either within a mould cavity or on a substrate surface.

The simplest reactive polymer process is perhaps monomer casting, and examples include monomers such as methyl methacrylate, styrene, ε-caprolactam and *N*-vinyl carbazole. The process requires inexpensive moulds and is used for small-volume production of shaped articles. The polymerizations are exothermic and require good temperature control and long reaction times. Structure development is via molar-mass development during the process and may, in some cases, involve polymer crystallization.

Coating is a surface-modification method that may be divided into planar and contour techniques. The former include spread-, roller- and spray-coating of flat substrates, and the latter involve the coating of objects having complex shapes using liquid-dipping and fluidized-bed equipment.

Reactive-moulding processes include compression moulding (CM), resin-transfer moulding (RTM) and reaction injection moulding (RIM). Compression moulding is used almost exclusively for temperature-activated thermosetting polymers, usually requiring temperatures in the range 140–200 °C and mould pressures in the range 35–700 bar. Structure development involves network-forming processes occurring in ~1 min as, for example, in the production of phenolic mouldings, and vulcanization processes taking >60 min for the manufacture of very thick rubber components.

Resin-transfer moulding is a process intermediate between compression and injection moulding and is used primarily for the manufacture of composite-materials products based on unsaturated polyester and epoxy resins as matrices. The low-viscosity thermosetting resin, containing catalyst, is pumped under moderate pressure (~5 bar) into a mould cavity containing preplaced reinforcing fibre mats. Gel- or in-mould coating prior to mat placement is frequently employed. Mould filling and fibre impregnation are facilitated by evacuation of the cavity. Polymerization is temperature activated (140–200 °C) and structure development occurs by polymerization, gelation and network formation in cycle times of up to 20 min. Additionally, phase-separation processes are involved when a reactive liquid rubber or a thermoplastic is included in the resin formulations to improve the toughness of the otherwise brittle polyester and epoxy resins.

5.2 Reaction injection moulding (RIM)

5.2.1 The RIM process

Reaction injection moulding (RIM) is a relatively new process used for the rapid and automated production of large, thin (3–6 mm) and complex-shaped parts [5–8]. In the RIM process, shown schematically in Figure 5.1, two or more low-viscosity reactant liquids (monomers or prepolymers, or both) are accurately metered according to chemical stoichiometry and impingement mixed at high pressure (200–300 bar), giving turbulent flow conditions, in a hydraulically operated mixhead. Mixing initiates polymerization and the reactant mixture then flows with low pressure (<10 bar) under laminar conditions into a mould cavity. Reactant throughputs can be up to 100 kg s^{-1}, and mould-fill times of ~1 s and cycle times of <60 s are typical. Conversion of liquid reactants into moulded material involves a

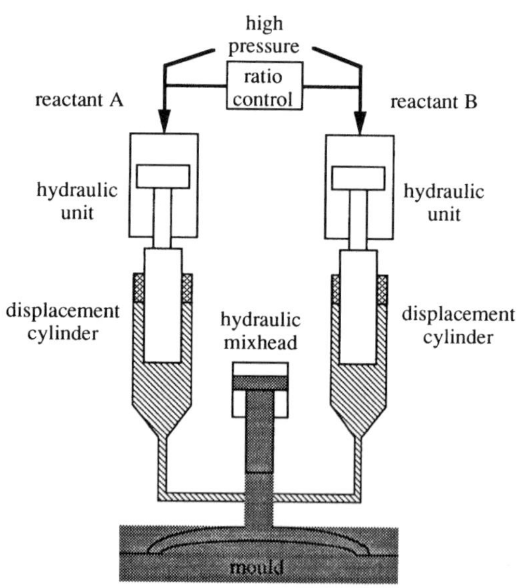

Figure 5.1 Schematic representation of the equipment used typically in the RIM process. Lance displacement cylinders, driven by hydraulic units, are used to dispense metered amounts of liquid reactants (A and B), under high pressure, to the hydraulically operated mixhead. The reactants are impingement mixed, which initiates reaction, and the polymerizing mixture flows under low pressure into the mould.

solidification process and, compared to the other reactive process methods described in the previous section, structure development in RIM copolymer-forming systems [9] is much more complex, involving polymerization (often with network formation and gelation) processes and competitive phase-separation and vitrification processes.

The principal advantage of RIM is the ability to process low-viscosity liquid reactants, typically 0.1–1.0 Pa s, using low temperatures and pressures, particularly during the mould-filling stage. This means that small-scale hydraulic equipment and lighter-weight and lower-cost moulds can be used, facilitating short production runs and prototype applications. Low-viscosity reactant systems also facilitate composite materials production, so-called structural RIM (SRIM) composites [10], in which continuous-fibre reinforcement mats are pre-placed in mould cavities prior to injection. Capital investment and operational costs in RIM are therefore much less than those for conventional injection moulding.

Mix-activated copolyurethanes (PU)s, copoly(urethane–urea)s (PUU)s and copolyureas (PUr)s are the most common systems used in RIM, and constitute over 95% of RIM materials produced. Several bulk polymerization chemistries [7] have been adapted and developed for the RIM process, including unsaturated ester–urethanes (acrylamates), polyepoxides, unsaturated polyesters and phenolics, for use mainly as thermosetting matrices for SRIM composites. Other, more commercially important RIM systems use copolyamides (based on ε-caprolactam) and polydicyclopentadiene.

Polyamides were developed for the RIM process from the anionic polymerization of ε-caprolactam used in the production of cast nylon-6. Polymerization, involving a two-stage propagation mechanism, occurs by ring-opening ε-caprolactam using acyl lactam initiators and metal catalysts (usually sodium caprolactam or magnesium bromide caprolactam). In RIM formulations used to produce rubber-toughened nylon-6 [11], the initiators are polyether prepolymers which have been pre-reacted to convert hydroxy functional groups to acyl lactam groups, using either terephthaloyl biscaprolactam or hexamethylene diisocyanate and ε-caprolactam. Ideally, polymerization should form ABA triblock copolymer but, during RIM, side reactions such as hydrogen abstraction and amide interchange occur, which produce chain branching and crosslinking, resulting in nylon-6 with an alternating block copolymer structure. Processing of nylon-6 formulations by RIM does not require accurate stoichiometric control of reactant streams (since formation of nylon-6 occurs by chain polymerization), although high-temperature RIM equipment must be used to provide relatively high processing temperatures. The melting temperature (67 °C) of ε-caprolactam requires reactant temperatures of 90 °C and mould temperatures of 135 °C, which are necessary to optimize polymerization and crystallization rates [12] so that acceptable demould times (~90 s) are achieved. The RIM materials produced are segmented block copolymers with phase-separated morphologies comprising soft rubbery blocks dispersed as lamellar domains (~3 μm) in rigid crystalline nylon-6. The materials have high moduli (~1.5 GPa) and good impact and high-temperature properties ($T_g = 75\,°C$, $T_m = 220\,°C$) although, because of their hygroscopic nature, moulded components have poor dimensional stability and exhibit mechanical-property variation with humidity changes.

Polydicyclopentadiene (PDCP), a crosslinked alkenic polymer introduced in 1983 by Hercules, was the first polymer to be developed specifically for RIM [13,14]. PDCP systems form rapidly at low temperatures (~30 °C) by metathesis polymerization involving a coordination catalyst formed from a tungsten or molybdenum compound and an alkylaluminium chloride. Polymerization of the dicyclopentadiene monomer involves opening the norbornene ring to produce linear chains and reaction of the cyclopentene ring to give crosslinking, with chemical gelation occurring at very low monomer conversion (<1%). Processing of PDCP formulations by RIM is readily accomplished using low mould temperatures (50–80 °C), although the need to exclude oxygen and water (which destroy the initiator) and the narrow processing temperature range (28–40 °C), defined by the freezing point and flash point of the monomer, are limitations. PDCP produced by RIM is a single-phase material with a high modulus (1–2 GPa), good impact properties and, due to its alkenic nature, excellent solvent resistance. The relatively low T_g (110–135 °C) of PDCP, however, gives the material a low heat-distortion temperature.

5.2.2 *Cure-time–transformation diagram*

In the RIM process, polymerization occurs simultaneously with shaping of the moulded component which, while hot, is removed from the mould cavity. Solidification in RIM is due to polymerization in the mould cavity, rather than by heat transfer from polymer melt to mould walls as in the injection moulding of thermoplastic (preformed) polymers. The transformation from liquid reactants to solid polymer must result from molar-mass increases, network formation or phase separation (or both), i.e. molecular processes that are accompanied by the development of rheological and mechanical properties. These processes are summarized by the cure-transformation–time diagram shown in Figure 5.2 (compare Chapter 4).

The polymerization kinetics are given in terms of the fractional conversion of chemical functional groups, on a scale from 0 to 1 (that is, from the instant that impingement mixing occurs to the point of complete chemical reaction), versus time, which varies typically between 30 s and 60 s for most RIM systems. As conversion increases and molar mass develops during RIM, the viscosity of the reactant mixture rises exponentially, but the rise must not be so rapid as to prevent complete filling of large moulds. The point representing maximum processing viscosity, allowing ready mould filling

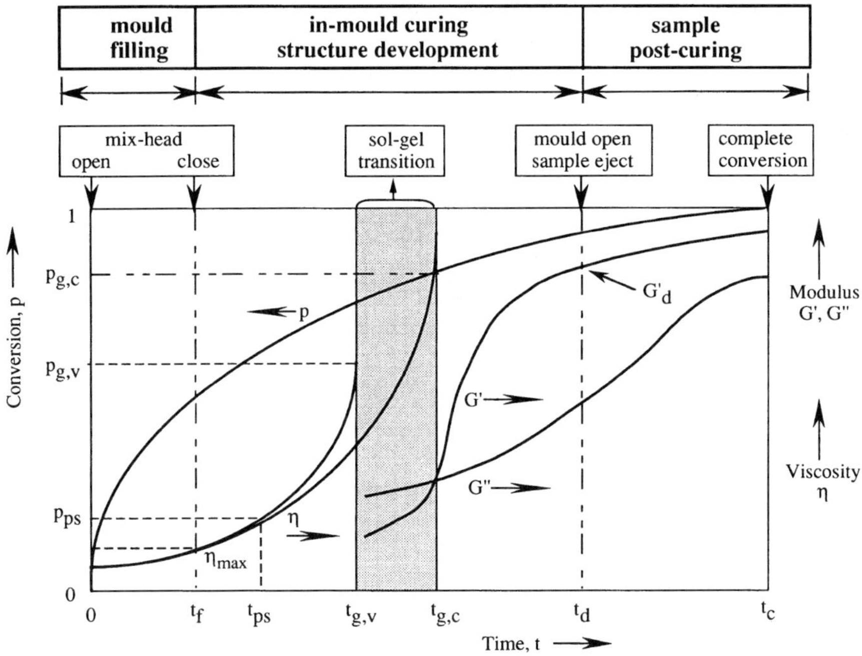

Figure 5.2 Schematic conversion-time–transformation (CTT) diagram representing the increases in viscosity (with and without phase separation) and modulus as functions of time during the RIM process. The points (p_{ps}, t_{ps}) and ($p_{g,v}$, $t_{g,v}$) represent, respectively, the onset of phase separation and vitrification (physical gel point); ($p_{g,c}$, $t_{g,c}$) represents the chemical gel point. Other symbols are defined in the text.

under controlled laminar flow conditions, is represented by the mould-filling time t_f which is of the order of 1–2 s. Within the filled mould for a completely miscible, non-linear polymerization leading to the formation of a single-phase network material, the viscosity becomes infinite at the chemical gel point, and the chemical conversion $p_{g,c}$ at gel, according to the Flory–Stockmayer expression [15], is determined to a first approximation by the weight-average functionalities of the reactants (Section 2.2). In the sol–gel transition region, the shear modulus G begins to develop: prior to gelation the loss modulus G'' exceeds the elastic modulus G' and, in the post-gel region, the reverse is the case and $G' > G''$. The gel time $t_{g,c}$, corresponding to $p_{g,c}$, is given [16] approximately by the crossover of G' and G'' (Chapter 8). Stiffness increases rapidly in the post-gel region and when the value of G' is large enough ($G' = G'_d \approx 10^6$–10^7 Pa), the mould can be opened and the moulded component can be ejected in demould times (t_d) typically ~30 s.

In the majority of RIM systems, polymerization is accompanied by phase separation which may occur quite early in the process [9], as indicated by the point p_{ps}, t_{ps} in Figure 5.2. In this case, viscosity and modulus rise much more rapidly during the RIM process and, as molecular structure develops, the glass-transition temperature of the phase-separating mixture in the mould increases rapidly and soon becomes equal to the processing temperature. Vitrification of the reaction system then occurs, indicated by the critical point $p_{g,v}$, $t_{g,v}$, and this solidification process is often referred to as physical gelation. Phase separation and physical gelation are phenomena that can occur in both linear and non-linear RIM systems and, once they occur, chemical reaction becomes diffusion controlled and its rate decreases to such an extent that, within the timescale of the RIM process, polymerization is arrested and the morphological structure becomes 'frozen'. Hence, upon demoulding, RIM components often require postcuring.

5.2.3 *Polyurethane, poly(urethane–urea) and polyurea systems*

As stated in Section 5.2.1, polyurethane (PU), poly(urethane–urea) (PUU) and polyurea (PUr) systems are the most common materials formed via RIM, and Figure 5.3 shows the chemical structures of some of the monomer

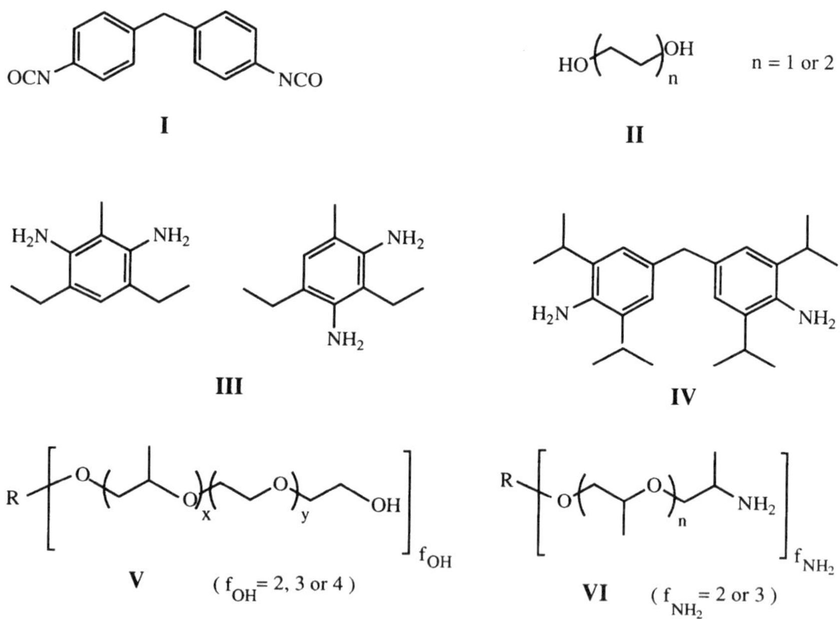

Figure 5.3 Idealized chemical structures of typical reactants used in the RIM process to form urethane, urethane–urea and urea segmented copolymers (PU, PUU and PUr): **(I)** 4,4′-diphenylmethane diisocyanate (MDI); **(II)** ethylene glycol (EG; $n = 1$) or 1,4-butane diol (BD; $n = 2$); **(III)** 2,4 and 2,6 isomers of 3,5-diethyltoluene diamine (DETDA); **(IV)** methylene-bis(2,6-diisopropylamine) (MDIPA); **(V)** hydroxy-functionalized poly(oxypropylene–oxyethylene) (POE-tipped POP polyols), diol ($f_{OH} = 2$), triol ($f_{OH} = 3$) or tetrol ($f_{OH} = 4$); **(VI)** amine-functionalized polyoxypropylene (POP polyamines), diamine ($f_{NH_2} = 2$) or triamine ($f_{NH_2} = 3$).

and prepolymer reactants used typically to form these RIM copolymers. The reactant common to the three copolymers is 4,4′-diphenylmethane diisocyanate (MDI (**I**)). However, because of its low melting point (42 °C), pure MDI is seldom used in RIM and its melting point is depressed by chemical modification to give a low-viscosity liquefied-MDI variant that is much easier to process. For example, a uretonimine-modified MDI (an equilibrium mixture of pure MDI, dimerized-MDI carbodiimide and trifunctional four-membered cyclic MDI uretonimine) is often used in RIM. More detailed information on the chemistry of isocyanates is available in several extensive reviews [17–20].

PU copolymers [7, 21], in which the linking groups are solely urethane groups, are formed from modified MDI and blends of aliphatic diol (**II**) and a hydroxy-functionalized polyether (**V**), using a catalyst mixture comprising organotin and tertiary amine compounds. The PU copolymer produced then comprises, at room temperature, a mixture of flexible polyether soft segments and rigid polyurethane hard segments. In the case of PUU copolymers [22], the aliphatic diol is replaced with a hindered aromatic diamine (**III**) or (**IV**), giving materials with rigid aromatic polyurea hard segments. Since, compared with OH groups of the polyether, the aromatic NH_2 groups are much more reactive toward the NCO groups, the slower urethane-forming reactions in RIM PUU formulations must be preferentially catalysed using solely organotin catalysts which have little effect on isocyanate–amine reactions. The PUr copolymers [8, 23], comprising solely urea linking groups, are formed from modified MDI and blends of a hindered aromatic diamine (**III**) or (**IV**) and an amine-functionalized polyether (**VI**). The intrinsically fast amine–isocyanate reaction enables very short moulding-cycle times to be achieved without the use of catalysts, in contrast to PU and PUU copolymers. Indeed, the rates of amine–isocyanate reactions are such that the formation of PUr copolymers, in bulk, can only be achieved via RIM. Furthermore, catalyst residues in PU and PUU copolymers have been shown to promote degradation at elevated temperatures, so that RIM PUr copolymers exhibit much superior high-temperature properties (Section 5.5).

The different types of RIM materials may be considered as either linear copolymers containing difunctional polyether soft segments (from prepolymers with f_{OH} or $f_{NH_2} = 2$) or polyfunctional network copolymers containing polyfunctional polyether soft segments (from prepolymers with f_{OH} or $f_{NH_2} > 2$). The competitive–consecutive polymerization reactions produce distributions of growing polymer chains of both soft- and hard-segment structures and, prior to phase separation, the copolymerizing system in the RIM mould-cavity comprises the complex mixture of molecular species, depicted in Figure 5.4. When phase separation occurs, followed by vitrification, copolymerization reactions are arrested and the solidifying mixture produced, as Figure 5.4 attempts to describe, resembles more a

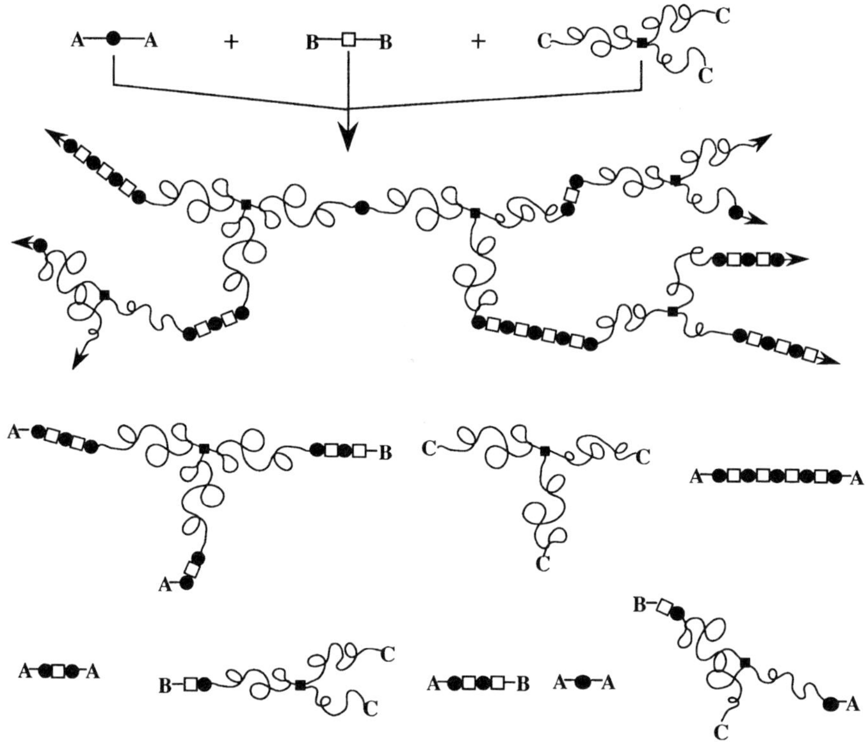

Figure 5.4 Schematic representation of the mixture of molecular species formed under the non-equilibrium conditions typical of the RIM process. The example shows a trifunctional network-forming system comprising three reactants, $RA_2 + R'B_2 + R''C_3$. Functional groups B and C are mutually unreactive but have different intrinsic reactivities towards A-functional groups. The sequences of closed circle and open square symbols represent hard-segment species formed from reactions between $RA_2 + R'B_2$, and the curled lines represent soft-segment chains. The nonequilibrium mixture contains finite (functionalized) species that include unreacted monomers and prepolymers, isolated short hard-segment species comprising di-, tri- and multiblock sequences, and 'infinite' trifunctional species, representing the multiblock RIM copolymer, with continuing chains indicated by the arrows.

non-equilibrium blend of copolymer, homopolymer and monomer structures. As discussed in Section 5.2.1, both linear and non-linear types of copolymer undergo phase separation and exhibit gelation phenomena as part of structure development during the RIM process. This competition between copolymerization kinetics and the thermodynamics of phase separation are described in more detail in the following sections.

In PUU and PUr systems, the change of chemical functionality from OH to NH_2, respectively, on the soft-segment prepolymer reactant, and the consequent differences produced in copolymerization kinetics, phase-separation dynamics and resulting materials morphology, have profound effects on RIM processing. For example, process demould times (t_d) are significantly affected since, as discussed with reference to the cure-transformation–time diagram (Figure 5.2), the magnitude of t_d is determined by the time at which the mechanical strength (the so-called 'green strength') of the RIM material has developed sufficiently to allow removal of the moulded component without causing visible or mechanical damage [24]. Generally, materials formed using relatively higher process temperatures or longer mould times, or both, are usually capable of withstanding more severe deformation during demoulding. In industrial terms, however, mould temperatures of 100–200 °C and demould times of 50–60 s are regarded as upper limits for high-speed RIM processing [7]. The combined effects of mould temperatures and demould times on the efficiency of RIM processing are conveniently represented by moulding area diagrams (MADs), such as those shown in Figure 5.5, which compares PUU and PUr systems, containing equal amounts (50% w/w) of polyurea hard segments, moulded as RIM plaques [24]. In Figure 5.5, to aid discussion, curves have been drawn

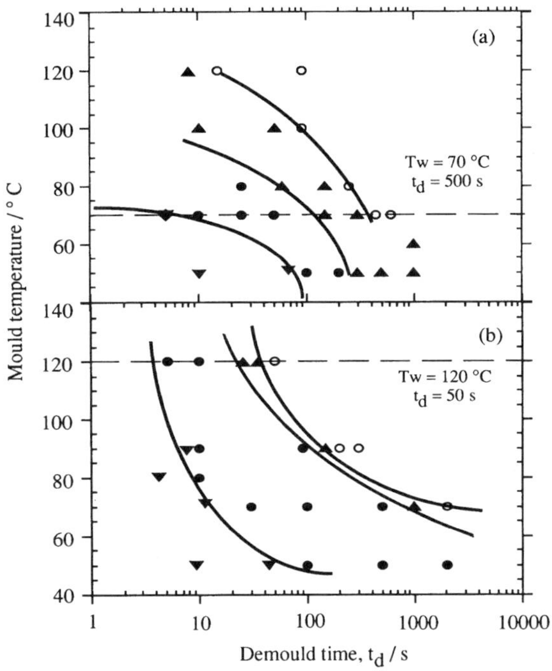

Figure 5.5 Moulding area diagrams (MADs) for trifunctional RIM-copolymer networks containing 50% w/w of hard segments. Network materials are designated as (▼) very brittle, (●) brittle, (▲) tough and (○) very tough, as defined in the text. The dashed lines indicate typical mould temperatures T_w used for PUU and PUr materials with the corresponding demould times (t_d) needed to produce very tough materials also shown. (a) PUU-DETDA formed from POE-POP triol ($M_n \approx 6000\,g\,mol^{-1}$), MDI and DETDA; (b) PUr-DETDA formed from POP triamine ($M_n \approx 5700\,g\,mol^{-1}$), MDI and DETDA.

to separate regions relating to different classifications of the green strength of PUU and PUr mouldings. The areas below the bottom curves define the very brittle regions where the plaques always fractured on opening the mould because of the material's inability to sustain even moderate stresses, arising from shrinkage due to polymerization and thermal contraction. The middle curves define the transition regions from brittle-to-tough behaviour, i.e. regions where it was possible to demould complete plaques but, when immediately subjected to the 90° bend test [24], the plaques either fractured completely or suffered visible damage. Above the top curves are the regions defining very tough behaviour, where the RIM plaques showed no visible damage following deformation in the 90° bend test. The changes in shape of the MADs, comparing PUU and PUr materials, result directly from the use of hydroxy- and amine-functionalized prepolymers in the respective RIM formulations. The most significant difference is shown by the much reduced region bounded below by the brittle-to-tough transition (i.e. the middle curve) for the PUr material. Generally, the data in the MADs confirm the requirement for different conditions when processing PUU and PUr systems [25]. Higher mould temperatures are required to demould PUr materials without causing damage, as may be shown by comparing the data at 60 s demould time in Figure 5.5.

The PUU material changes behaviour from brittle to tough at mould temperature $T_w \approx 80\,°C$ and from tough to very tough at $T_w \approx 105\,°C$, whereas corresponding changes in the PUr material occur at higher temperatures, namely $T_w \approx 100\,°C$ and $T_w \approx 120\,°C$, respectively. Typically in RIM processing, mould temperatures of 70 °C and 120 °C, respectively, are used to produce PUU and PUr materials which, as the dashed lines in Figure 5.5 indicate, require demould times of 500 s and 50 s, respectively, to give very tough mouldings capable of being demoulded without damage. These and other studies [24] using MADs to optimize processing of related PUU and PUr materials show that, as expected, a more rapid development of demould toughness (or green strength) results from the direct relationship of soft-segment prepolymer functionality with copolymer molar-mass development, phase separation and materials morphology. Structure development is strongly affected by both the chemical nature (OH or NH_2) and average functionality (f_{OH} or f_{NH_2}) of the prepolymer. For example, during the processing of PUU materials using $T_w = 70\,°C$, very tough mouldings were produced in demould times of ~2000, 500 and 10 s upon increasing the nominal functionality from $f_n = 2$ to 3 to 4 of the polyether polyol (see Figure 5.3).

Although the focus of this chapter is on non-linear RIM copolymers, linear systems have been extensively studied and a brief description here is relevant. Much of the work has been carried out on linear segmented copolyurethanes formed by RIM and has been reviewed by Macosko [7]. For example, structure–property relationships were determined for copolyurethanes comprising a poly(oxyethylene) (POE)-tipped poly(oxypropylene)

(POP) diol ($M_n \approx 2000\,\mathrm{g\,mol^{-1}}$) as the soft segment, and hard segments formed from 4,4′-diphenylmethane diisocyanate (MDI) and 1,4-butane diol (BDO). Factors that determined the final properties of the copolyurethane included the hard-segment content, structure and degree of crystallinity, the copolymer molar mass and the degree of microphase separation. The effects of processing variables on the segmental structure, molar mass and degree of phase separation of a linear polyether-based copolyurethane elastomer have also been investigated [26–28]. In these studies, copolyurethanes prepared in the absence of catalyst exhibited macrophase separation and were of low molar mass. ('Macrophase' and 'microphase separation' mean the formation of macrodomains and microdomains via phase separation.) Consequently, the materials were brittle and possessed a highly heterogeneous morphology with hard-segment-rich globules containing spherulitic texture. In contrast, the use of catalysts drove the copolyurethane-forming system into the high-temperature single-phase region of the phase diagram. The resulting materials were microphase separated and of higher molar mass. Cooling after copolymerization produced materials with a fine interconnected microstructure. Monitoring structure development in slow reacting systems at low temperatures using FTIR spectroscopy, indicated that phase separation occurred at a characteristic number-average hard-segment sequence length ($N_n = 1.2$). This observation was in agreement with results obtained from earlier studies [29] using light transmission and differential scanning calorimetry. In a later study [28], the molar mass of the polyether and the structure of the chain extender were varied. The polyether molar mass was varied from 2000 to $4000\,\mathrm{g\,mol^{-1}}$ and the butane-1,4-diol (BDO) was replaced by ethane-1,2-diol (EDO). Employing a higher-molar-mass polyether at the same hard-segment content resulted in materials that exhibited a higher degree of microphase separation. The EDO-based materials gave hard segments with higher melting temperatures compared with the BDO-based materials.

The dynamics of microphase separation in fast, isothermal, bulk copolyurethane-forming systems were studied at different temperatures and concentrations of dibutyltin dilaurate (DBTDL) catalyst [30]. The copolyurethanes comprised 50% by weight hard segment (MDI and BDO), and the soft segment was a POE-tipped POP diol ($M_n \approx 2000\,\mathrm{g\,mol^{-1}}$). FTIR spectroscopy and dynamic rheometry were used to investigate structure development during the copolymerizations. Reaction kinetics, analysed by calculating the isocyanate conversion from the decay in the isocyanate absorption band, showed the copolymerization to follow second-order kinetics in the initial stages, but to become diffusion controlled at higher conversions. The phase-separation dynamics were monitored using the carbonyl absorption bands to determine the extent of interurethane hydrogen bonding. The point at which the concentration of free carbonyl passed through a maximum and that of the hydrogen-bonded carbonyl showed a rapid increase was designated as the onset of phase separation. The data indicated that phase separation occurred early in the reaction (<50% conversion of isocyanate). At low temperatures (≤100 °C), the onset of phase separation occurred at an average sequence length of 1.0 to 1.3, in good agreement with results from the earlier studies by Camargo *et al.* [26, 28]. The trend with temperature (at constant catalyst concentration) in the onset of phase separation, as detected by the increase in the dynamic-mechanical modulus, was consistent with the FTIR data. However, the two techniques probe different types of structural development so that the estimates of the time for characteristic phase separation were not in agreement.

Similar studies by Blake *et al.* [31] using adiabatic rheometry to investigate a linear, phase-separating RIM system, showed the sharp rise in viscosity to be coincident with physical gelation resulting from vitrification of the hard-segment-rich phase. As the hard-segment content, catalyst concentration and initial reactant temperature were increased, the gel time decreased and the conversion at gelation increased. Isothermal rheometry on similar but highly catalysed copolyurethane-forming systems was carried out [32] using laboratory-scale RIM equipment interfaced to a specially designed rheometer. Systems at four different hard segment contents were studied at various temperatures and catalyst concentrations. The results showed that the increase in viscosity during the reaction occurred more rapidly than predicted solely from molar-mass increases. Viscosity increases were shown to be due to phase separation of the hard-segment blocks, which subsequently associated via hydrogen bonds to form domains rich in hard segment that acted as physical (labile) crosslinks. As the glass transition temperature of the hard-segment-rich phase was approached and exceeded the copolymerization temperature ($T_g \geq T_{reaction}$), vitrification occurred, freezing-in the incipient morphology.

Synchrotron small-angle X-ray scattering (SAXS) has been used [33] to study microphase separation during the bulk polymerization of a copolyurethane comprising 48% by weight hard segment (MDI and BDO) and a POE-PPO diol soft segment. A constant concentration (0.02% by weight) of DBTDL catalyst was used. Small-scale micro-RIM equipment was used to meter, mix and inject stoichiometric amounts of the reactants into a thermostatted cell, positioned on an optical bench on the SAXS beam line. The results showed that the rate of microphase separation was faster, the higher the polymerization temperature, but that the degree of microphase separation in the finally formed material was lower. Although the scattered intensity at fixed angle was observed to increase with the square of time, and the length-scale characterizing the morphology of the material to increase with temperature, it was not possible to distinguish with certainty whether the ultimate morphology was due to crystallization of the hard segment or microphase separation of the non-crystalline regions. Despite further analysis [9] of the FTIR and SAXS data [30, 33] in terms of the Avrami equation, it was still not possible to resolve which of these processes is dominant, although structural growth was shown to occur, either by microphase separation involving nucleation and growth or by crystallization involving either homogeneous nucleation of discs or heterogeneous nucleation of rods. The use of wide angle X-ray diffraction (WAXD) to resolve this question has proved inconclusive. Other attempts to model phase separation in linear segmented copolyurethanes have been made by Speckhard *et al.* [34, 35] using Monte-Carlo simulations. The effects of molecular parameters such as extent of reaction, the structure and relative reactivity of monomer on copolyurethane composition, hard-segment-length distribution and molar-mass distribution were studied. Additionally, the effects of premature phase separation on the composition, molar mass and hard-segment sequence-length were predicted using a partitioning model which separates the reactants into two separate phases prior to reaction. Although the initial reaction conditions used in the model are not a true representation of reaction-induced phase separation, typical in many multiphase polyurethanes, the model does give some insight into the numerical values of important parameters, their distributions and their influences on the structuring processes.

5.3 Modelling structure development in RIM systems

In general, RIM materials may be described as segmented block copolymers of the $(AB)_n$ type which, at room temperature, comprise alternating blocks of hard segments (either crystalline or amorphous) and soft rubbery segments. During copolymer formation, the developing segments undergo microphase separation producing RIM materials with a characteristic two-phase morphology. The chemical structure of the copolymer (determined by those of the initial reactants), the reaction kinetics of copolymerization and the phase-separation process are the main factors that determine the structure and properties of the RIM materials. However, it is the nature and degree of microphase separation, occurring in competition with molar-mass development during copolymerization, that is the dominating factor. If the copolymerization reaction kinetics are much faster than the rate of phase separation, then high-molar-mass copolymer will form before phase development can begin, and the rate of phase separation is decreased. If, however, the reverse is the case and the rate of phase separation is faster, then hard-block sequences will precipitate via macrophase separation – a process that removes reactant functional groups from the reactant mixture. Thus, because of these dynamic factors, characteristic of the fast RIM process, the two-phase morphologies that develop in RIM copolymers are rarely in equilibrium, and are often easily modified by annealing [23, 36–38]. In this section, the two competitive processes, namely molar-mass development via copolymerization kinetics and structure development via phase separation, are dealt with in terms of the simple application of polymerization statistics and the thermodynamics of idealized block copolymers.

5.3.1 *Molar-mass development: statistical approach*

The development of structure during RIM copolymerizations has been modelled [24], in terms of the growth of weight-average copolymer molar mass (M_w) and hard-segment sequence length (N_w) as functions of conversion (p). The modelling uses recursive calculations [39] based on classical Flory–Stockmayer polymerization statistics. This is a simple method for calculating the average molar masses of polymers without first determining their distributions, and has been applied to linear [40] and non-linear [24, 41] RIM copolymerizations. The method utilizes the recursive nature of a copolycondensation or copolyaddition, modelled as a first-order Markov chain process [39]. Thus, to determine M_w in a typical three-component RIM system, a unit of each reactant is selected at random and the expected weight of the molecule of which it is a part is calculated – M_w is then obtained by averaging the weights of the three selected polymer molecules.

To demonstrate the recursive method, calculations have been made on a non-linear three-component system, $RA_2 + R'B_2 + R''C_f$, as a model for copoly(urethane–urea) (PUUr) and copolyurea (PUr) RIM systems [24]. Formulations used in these RIM copolymers are based, typically, on MDI and diethyltoluene diamine (DETDA) as the hard segment, and either a hydroxy- or an amino-functionalized polyether prepolymer, with $M_n \approx 2000\,\mathrm{g\,mol^{-1}}$ per functional group and $f = 2$, 3 or 4, as the soft segment. The various formulations data, for copolymers comprising 50% by weight of polyurea hard segment, were coupled with a set of kinetics differential equations to introduce both time dependence and unequal reactivities for different pairs of reactants. Assumptions used in the modelling required that homogeneous reaction conditions existed and that all groups reacted independently of each other. Additionally, the ratio K of the rate constants for the two basic reactions, isocyanate–polyether ($RA_2 + R''C_f$) and isocyanate–DETDA ($RA_2 + R'B_2$), was assumed to be independent of temperature and conversion. As in the case for linear systems [40], overall second-order kinetics were assumed, and the values of K used for the non-linear systems corresponded approximately to the degree of unequal reactivity expected in the formation of RIM copolymers, that is $0.1 \leq K \leq 1.0$ for PUU and $100 \leq K \leq 1000$ for PUr. The kinetics equations were solved numerically to determine the conversion of each reactant as a function of time which, when coupled with the recursive-method calculations, allowed the calculation of M_w and N_w as functions of isocyanate conversion (p).

Representative plots of M_w (normalized to M_{w0}, the molar mass of the initial reactant mixture) and N_w versus p are shown in Figure 5.6 for

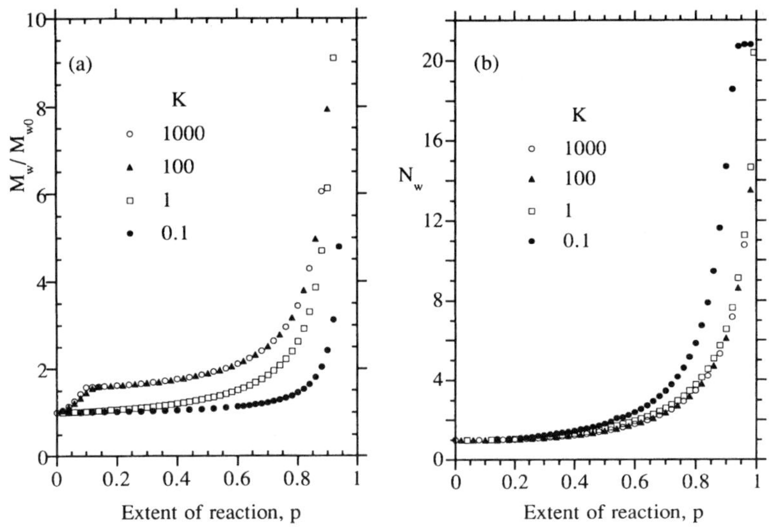

Figure 5.6 Distribution curves in terms of (a) relative molar mass M_w/M_{w0}, and (b) hard segment sequence length (N_w) versus extent of reaction (p) for a model three-component reactant system $RA_2 + R'B_2 + R''C_3$, forming a trifunctional RIM copolymer network containing 50% by weight of $(RA_2 + R'B_2)$ hard segments. Functional groups B and C are not mutually reactive but have different intrinsic reactivities toward A-functional groups. K is the ratio of the rate constants for the reactions $RA_2 + R'B_2$ and $RA_2 + R''C_3$.

trifunctional ($f = 3$) RIM systems [24]. When the reactivity of the soft-segment prepolymer is high ($K = 1000$ or 100), the prepolymer is consumed during the early stages of the reaction (that is, $0 \leq p \leq 0.1$), resulting in a sharp increase in M_w, as the prepolymer has undergone oligomerization and becomes capped with MDI. At lower values of K, the sharp increase in M_w is not observed, only a gradual increase in M_w/M_{w0} with p occurs and the rate of molar mass development increases with the value of K. For the non-linear RIM systems with $f = 3$, the ideal conversion of isocyanate groups at gelation, calculated according to the Flory–Stockmayer equation [15], is 0.96. The dependences of N_w on p for various values of K (Figure 5.6(b)) show the limiting value of N_w to be 20.8, which is dictated by the initial ratios of the three components. When the relative reactivity of the soft-segment prepolymer is very low ($K = 0.1$), this limiting value is reached at $p = 0.96$, whereas for higher values of K it is achieved only in the limit of complete reaction. The model therefore proposes that PUU copolymers develop longer hard-segment sequence lengths, compared to equivalent PUr copolymers, and PUUs should therefore exhibit superior properties such as stiffness and hard-segment glass-transition temperature T_g^H. However, experimental data, obtained using DMTA [22, 37, 42], show the reverse, with PUr materials having higher values of T_g^H than equivalent PUU. The differences between experiment and theory are readily explained by additional critical phenomena, such as phase separation and vitrification which inevitably occur in the experimental RIM systems. Phase separation creates heterogeneous polymerization conditions that slow the rates of reaction and the onset of vitrification effectively quenches copolymerization at relatively low conversions. Thus, chemical gelation in these non-linear RIM systems is often prevented. Despite these differences, the use of polymerization statistics to predict molar-mass development provides a uesful basis for modelling RIM systems.

5.3.2 Phase separation: thermodynamic principles

Qualitatively, there are parallels between the thermodynamics of phase separation of polymer blends and the thermodynamics of microphase separation of block copolymers. Only in the special case of complete reaction, may RIM materials be considered as $(AB)_n$ block copolymers [43, 44]. The dynamic phase-separation processes that occur during RIM are more likely to produce materials with the characteristics of a complex blend of homopolymers and different types of block copolymer, as depicted schematically in Figure 5.4. For a polymer blend, liquid–liquid phase separation leads to two distinct phases whereas, in the case of a pure block copolymer, phase separation on a large scale is not possible due to the presence of covalent bonds linking the different segment blocks. Thus, at the microphase separation transition (MST), an energy minimization of the order–disorder process occurs. In the special case of materials produced by RIM, therefore, the molecular mixture formed comprises block copolymers and homopolymers, possessing the characteristics of polymer blends, but with the covalent bonds across phase interfaces that are associated with block copolymers.

The driving force for phase separation to occur in RIM copolymers is the difference in solubility between the hard- and soft-segment blocks. Most simply, thermodynamic considerations of phase behaviour are dealt with in terms of the balance between the enthalpic (ΔH_v) and the entropic (ΔS_v) contributions to the free energy of mixing. The basic thermodynamic equations for polymers in a solvent are usually derived from consideration of a liquid–lattice model, originally developed by Flory and Huggins [45]. Thus, the entropy of mixing per unit volume is given as

$$\begin{aligned}\Delta S_v &= -R\left[\frac{\phi_1}{V_1}\ln(\phi_1) + \frac{\phi_2}{N_2 V_1}\ln(\phi_2)\right] \\ &= -R[n_1 \ln(\phi_1) + n_2 \ln(\phi_2)] \qquad (5.2)\end{aligned}$$

where R is the gas constant, n_1 $(= \phi_1/V_1)$ and n_2 $(= \phi_2/N_2V_1)$ are the numbers of moles of solvent and polymer, respectively, and ϕ is volume fraction. V_1 is the solvent and polymer-segment molar volume, with polymer chains comprising N_2 segments. The free energy of interaction per unit volume $\Delta G_v^{\rm int}$ is given as

$$\Delta G_v^{\rm int} = RT\chi_{12}\phi_1\phi_2 \tag{5.3}$$

where χ_{12} characterizes the interaction energy between a solvent molecule and a chain segment (divided by kT) and is commonly defined as the Flory–Huggins interaction parameter. The Gibbs energy change per unit volume is obtained from equations (5.2) and (5.3), giving

$$\begin{aligned} \Delta G_v &= \Delta G_v^{\rm int} - T\,\Delta S_v \\ &= -RT\left\{ n_1 \ln(\phi_1) + n_2 \ln(\phi_2) + \chi_{12}\frac{\phi_1\phi_2}{V_1} \right\} \end{aligned} \tag{5.4}$$

For a phase-separated polymer containing hard (H) and soft (S) segment blocks, equation (5.4) may be rewritten as

$$\Delta G_v = RT\left\{ n_{\rm H} \ln(\phi_{\rm H}) + n_{\rm S} \ln(\phi_{\rm S}) + \chi_{\rm HS}\frac{\phi_{\rm H}\phi_{\rm S}}{V_{\rm r}} \right\} \tag{5.5}$$

where the subscripts 1 and 2 have been replaced by H and S. The volume fractions $\phi_{\rm H}$ and $\phi_{\rm S}$ are related to the total volume V of the phase-separated polymer according to

$$V = V_{\rm r}\{n_{\rm H}N_{\rm H} + n_{\rm S}N_{\rm S}\} \tag{5.6}$$

where $n_{\rm H}$ and $n_{\rm S}$ are the numbers of moles of hard- and soft-segment blocks (assumed to be randomly mixed), having degrees of polymerization $N_{\rm H}$ and $N_{\rm S}$, respectively, and $V_{\rm r}$ is the common (average) molar volume of the structural units of the two blocks. Qualitatively, for phase separation to occur, the Gibbs energy of mixing per unit volume must be positive, and the boundary between stable and unstable states is determined by the condition $\Delta G_v = 0$. In fact, ΔG_v being positive is not sufficient as a criterion for phase separation and the second derivative of the free energy expression being zero defines the stability limit in terms of the free energy versus composition (ϕ) curve. Thus, two conditions must be satisfied for microphase separation to occur, namely

$$\Delta G_v = \Delta G_v^{\rm int} - T\,\Delta S_v > 0 \tag{5.7}$$

and

$$\left(\frac{\delta^2 \Delta G_v}{\delta\phi^2}\right)_{T,P} \leq 0 \tag{5.8}$$

The second differential of the Gibbs energy of mixing with respect to volume fraction defines the spinodal. In most cases, a critical value $\chi_{\rm c}$ of the temperature-dependent Flory–Huggins interaction parameter is used to characterize

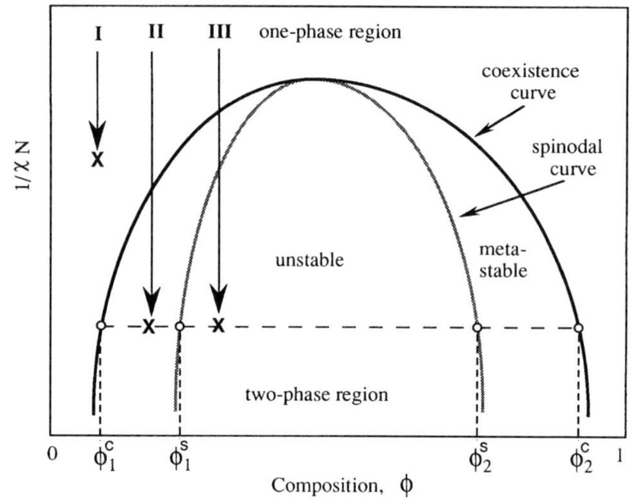

Figure 5.7 A generalized phase diagram, in terms of $1/(\chi N)$ versus composition for a polymerizing mixture, showing the coexistence and spinodal curves. The vertical lines I, II and III represent quenching routes from the miscible region into the stable (single-phase), the metastable and the unstable (two-phase) regions of the phase diagram. The equilibrium concentrations of the two phases formed are ϕ_1^c and ϕ_2^c (coexistence) and ϕ_1^s and ϕ_2^s (spinodal).

phase-separation processes such that

$$\chi_c = \frac{1}{2}\left\{\frac{1}{N_H n_H^{1/2}} + \frac{1}{N_S n_S^{1/2}}\right\}^2 \tag{5.9}$$

Thus, in the original Flory–Huggins model for a polymer–solvent mixture, assuming $N_H n_H \to \infty$ for the polymer and $N_S n_S = 1$ for the solvent, the critical value of χ_c is 0.5. More specifically for polymer systems, as distinct from polymer–solvent systems, it is the product χN [42], where $N = (N_H + N_S)$, which is the critical molecular parameter defining phase separation.

A generalized phase diagram for a mixture of polymeric liquids or a block copolymer [46] is shown in Figure 5.7. Normally, the ordinate of a phase diagram would be defined in terms of temperature but, in this case for a polymerizing system, the product of molar mass and the interaction parameter dominates the phase behaviour and an effective 'temperature' may be defined as $1/(\chi N)$. Thus polymerization, which changes χ and increases significantly the value of N, is parametrically equivalent to quenching a polymer mixture by decreasing temperature. Two curves are shown in the generalized phase diagram in Figure 5.7. The upper curve is the coexistence curve, defining the compositions of phases in equilibrium if a separation process occurs. The lower curve is the spinodal, below which a homogeneous system is thermodynamically unstable and must microphase separate into an energetically more favoured state. The critical point is where the binodal and spinodal converge, i.e. when $(\delta^3 G_v/\delta\phi^3)_{T,P} = (\delta^2 G_v/\delta\phi^2)_{T,P} = 0$. The area between the two curves defines the metastable region and, for phase separation to occur, the reactive system must overcome the energy barrier to the formation of nuclei. The position of a material in phase space may be changed by quenching, either by decreasing the temperature T or by chemical reaction $1/(\chi N)$ and, in the case of the latter, the use of the phase diagram may be exemplified by considering three types of quenching or polymerization routes, marked as I, II and III in Figure 5.7. These routes represent changes in position of a polymerizing mixture from the one-phase region of the phase diagram to: (i) another position in the one-phase region I; (ii) a position in the metastable region II; (iii) a position below the spinodal curve III. No phase separation occurs via route I, whereas polymerization via route II **may** result in phase separation by nucleation and growth and, in the case of polymerization via route III, phase separation **will** occur by spinodal decomposition.

The nucleation and growth process [47] is shown schematically in Figure 5.8(a) and occurs when a mixture in the metastable region is stable to infinitesimal (thermal) concentration fluctuations but is unstable to sufficiently large concentration fluctuations where nuclei are formed. Once the energy barrier has been overcome by the formation of nuclei, randomly in space and time, these nuclei then grow. Note that the concentration of nucleated species is depleted around the nucleus and the molecules or chain segments that feed the new phase follow the ordinary transport phenomena with a positive diffusion coefficient (downhill diffusion). The early stages of the phase separation process are characterized by the existence of a continuous-phase dispersed-phase morphology. Only if nuclei grow to such an extent that they reach the percolation limit and coalesce, will they achieve continuity and, unless phase separation is arrested either by vitrification or by crystallization, two layers will eventually form.

The spinodal decomposition process [47] is shown schematically in Figure 5.8(b). In the unstable region of the phase diagram (beneath the spinodal curve) the mixture is unstable to infinitesimal concentration fluctuations and there is no thermodynamic barrier to phase growth, so that phase separation occurs by a continuous and spontaneous process. Since the mixture is initially uniform in composition, this spontaneous process must occur by a diffusional flux against the concentration gradient. Diffusion of a species from a region of low concentration to a region of high concentration is unusual and is termed uphill diffusion and, when this process occurs, negative diffusion coefficients can be measured. The spinodal decomposition process is characterized by a bicontinuity of phases in the early stages, although this feature may be lost through Ostwald ripening [47] as the system seeks to minimize its free energy by minimizing the interfacial area between the phase domains. As in the case of nucleation and growth, unless phase separation is arrested, either by vitrification or by crystallization, two layers will eventually form.

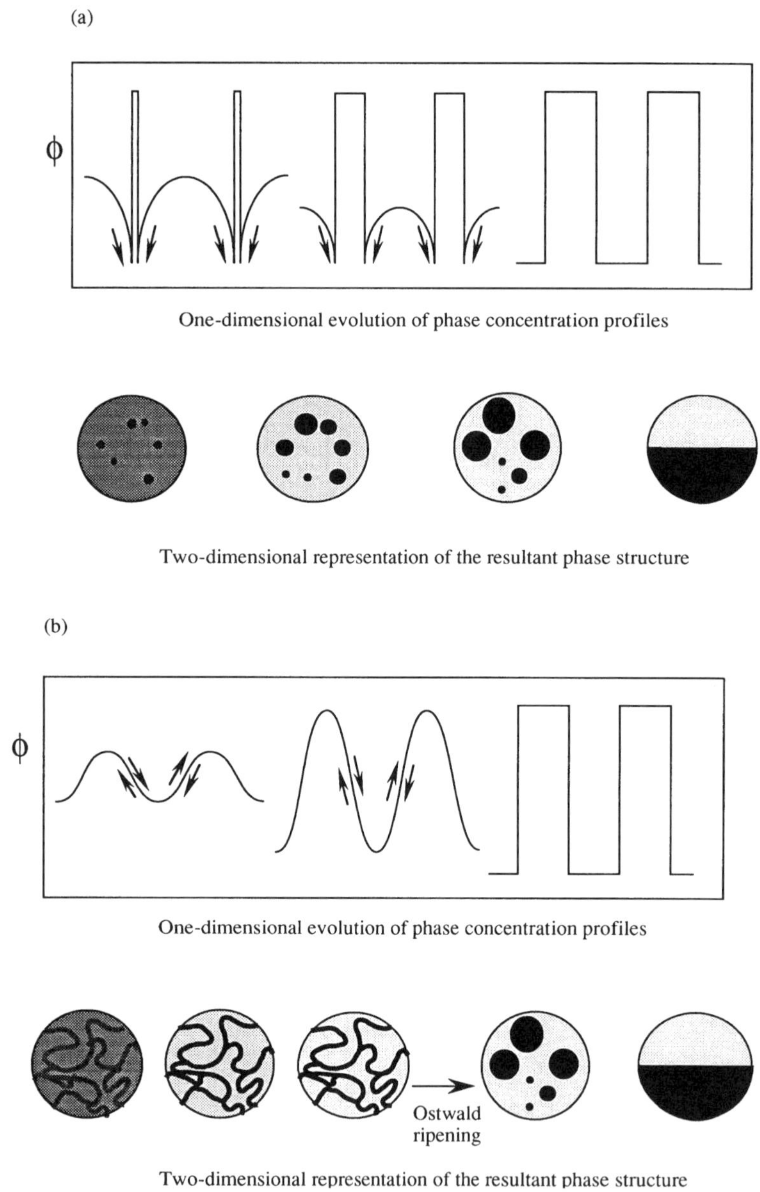

Figure 5.8 Schematic representations of (a) the nucleation and growth process showing the evolution of (sharp) concentration profiles for the two, continuous and discontinuous phases being formed, and (b) the spinodal decomposition process showing the evolution of (sinusoidal) concentration profiles for the two co-continuous phases being formed.

The minimization of interfacial area causes Ostwald ripening in a system undergoing spinodal decomposition and coalescence of particles in systems undergoing nucleation and growth – both mechanisms, given time, eventually form two layers. Morphologies are generally fixed by the classic solidification mechanisms of vitrification, crystallization and network formation, and it is nearly impossible to discern from the final morphology produced which phase-separation mechanism was operative if the early stage structure is lost. Nevertheless, it is possible in some cases to observe relics of a phase-separation mechanism.

A good example of spinodal decomposition in copolymer network formation via RIM is provided by the data shown in Figure 5.9 for a copoly(urea–isocyanurate) (PUrI) [48]. The microphase separation occurring by spinodal decomposition has been arrested prematurely by vitrification of the polyisocyanurate-rich phase to produce a material with a non-equilibrium co-continuous morphology. The representative data in Figure 5.9 were determined from *a posteriori* studies on the as-moulded and postcured PUrI materials. The TEM micrographs in Figure 5.9(b) show that high-temperature annealing improves the phase contrast of the microphase-separated structure, an observation confirmed by the increases in both the peak intensity q^* and relative invariant Q' (the area under the Iq^2 versus q plots) in the static SAXS data shown in Figure 5.9(a). These data illustrate clearly the co-continuous morphology, with a characteristic length-scale ≈ 100 Å associated with the relic of spinodal decomposition, so characteristic of very fast phase-separating RIM systems. The fine-scale morphology created within the postcured PUrI leads to the description of such RIM materials as rubber-modified (thermoset) resins. The materials exhibit two glass-transition temperatures, one at $-40\,°C$ associated with the polyether-rich rubber phase, and the other at $\sim 180\,°C$ associated with the network polyisocyanurate-rich phase. The PUrI materials show only small modulus-temperature dependences between $-20\,°C$ and $250\,°C$, with flexural moduli of ~ 2.0 and 0.5 GPa at $20\,°C$ and $200\,°C$, respectively. In addition, such PUrI materials have been used successfully to produce, in cycle times under 60 s, structural hybrid composites [10] comprising up to 40% by weight of pre-placed fibre mats.

The driving force for phase separation is the large increase in χN with copolymerization. Sequence length build-up, especially in non-linear systems, causes the value of N to increase sharply near the chemical gel point. Additionally, the conversion of functional groups to urethane and urea groups causes a change in the mutual solubility characteristics of the developing hard and soft segments. The onset of phase separation as χN increases can be predicted using the Flory–Huggins equation by calculating a critical value of the interaction parameter, according to equation (5.9). To calculate values of χ requires a knowledge of the partial molar free-energy per unit volume which, in the case of polymer solutions, may be determined

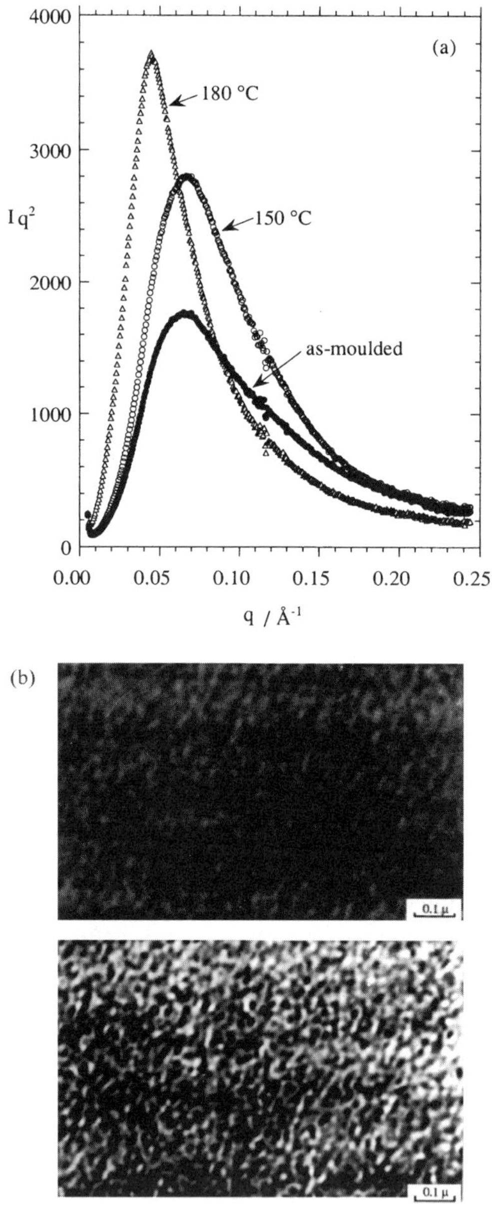

Figure 5.9 Characterization by (a) SAXS and (b) TEM of the co-continuous morphology typical of microphase-separated copolymer networks formed by RIM. The data are for a copoly(urea–isocyanurate) network formed from a POP diamine ($M_n \approx 2000\,\mathrm{g\,mol^{-1}}$), liquefied MDI and a trimerization catalyst. Reactant temperature = 35 °C; mould temperature = 90 °C. Copolymers comprise ~50:50% v/v polyether rubber:polyisocyanurate glass. The effects of annealing are shown by the increased intensity of the SAXS peaks and the enhanced phase-contrast in the TEM from the as-moulded PUrI (top TEM) to the PUrI annealed at 150 °C for 1 h (bottom TEM).

approximately from the heat of mixing per unit volume according to the Hildebrand–Scott equation [49],

$$\Delta H_v = \phi_1\phi_2(\delta_1 - \delta_2)^2 \tag{5.10}$$

where δ_1 and δ_2 are the solubility parameters of solvent and polymer. Combining equations (5.9) and (5.10) gives the relationship between the interaction parameter and solubility parameter difference as

$$\chi_{12} = \frac{V_1}{RT}(\delta_1 - \delta_2)^2 \tag{5.11}$$

As before, equation (5.11) may be written in terms of a polymer mixture, comprising hard and soft segments:

$$\chi_{\mathrm{HS}} = \frac{V_\mathrm{r}}{RT}(\delta_\mathrm{H} - \delta_\mathrm{S})^2 \tag{5.12}$$

where χ_{HS} characterizes the interaction energy between hard and soft segments whose solubility parameters are, respectively, δ_H and δ_S. Thus, phase separation can occur when χ_{HS} exceeds the critical value χ_c.

There have been significant advances in the theoretical and experimental understanding of block copolymers, although much of the knowledge has been restricted to monodisperse, diblock (AB) and triblock (ABA) materials. In simple terms, a block copolymer may be regarded as an assembly of macromolecules, each of which has a chain of one polymer type, joined to a chain of the other polymer type, thereby making the pair more compatible, and the assembly may be treated theoretically to predict microphase separation. For example, Liebler [50] has shown, using a mean-field approach, that a fluctuation-induced microphase (spinodal) transition (MST) occurs in monodisperse (AB) block copolymers at a critical value of χN and that, depending on copolymer composition, different morphologies including a body-centred cubic array of spheres, a hexagonal array of rods or lamellae should exist after the MST. The different morphologies in relation to the phase diagram for the polystyrene–polyisoprene diblock copolymer system have been determined experimentally via an elegant combination of SAXS, SANS, TEM and rheology [51], in comparison with predictions of the Liebler theory. Liebler's theory was extended by Benoit and Hadziiannou [52] to cover a wider range of block copolymers. Calculations based on the two theories are represented in terms of a universal phase diagram, as shown in Figure 5.10(a), which allows the determination of the critical (spinodal) values of the interaction parameter χ_c for both AB-type block copolymers [51, 52] and $(\mathrm{AB})_n$-type block copolymers [52] from a knowledge of their degrees of polymerization N $(= N_\mathrm{A} + N_\mathrm{B})$ and the volume fraction of one component unit, ϕ, namely the copolymer composition.

The thermodynamic principles described previously may be used to predict, at least semi-quantitatively, the phase behaviour of the formulations

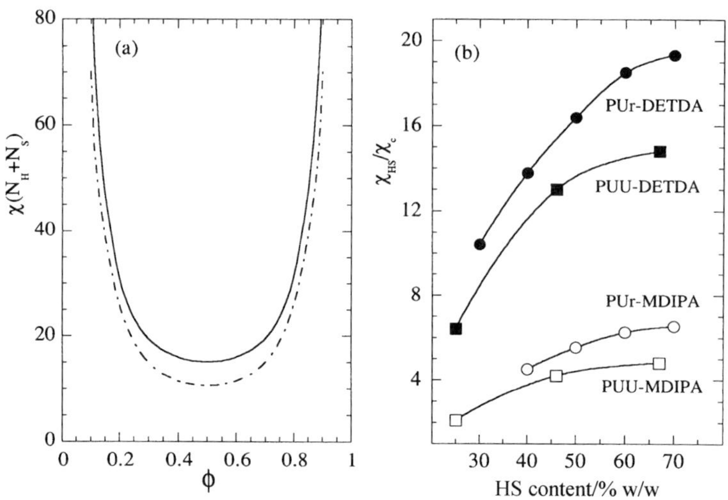

Figure 5.10 Phase separation in segmented block copolymers. (a) The 'universal phase diagram' in terms of the product $\chi(N_H + N_S)$ versus composition for linear multiblock copolymers. The lower (dashed) curve is for monodisperse AB diblocks [50] and the upper (solid) curve is for $(AB)_{n\to\infty}$ multiblocks [50, 52]. (b) Variation of the critical ratio, χ_{HS}/χ_c, with hard-segment content for PUU and PUr copolymer networks showing the effects of hard-and-soft-segment structure on predicted phase separation. PUU-DETDA and PUU-MDIPA comprise MDI/DETDA and MDI/MDIPA hard segments, respectively, and soft segments based on POE-POP triol ($M_n \approx 6000\,\text{g mol}^{-1}$) [42]. PUr-DETDA and PUr-MDIPA comprise MDI/DETDA and MDI/MDIPA hard segments, respectively, and soft segments based on POP triamine ($M_n \approx 5700\,\text{g mol}^{-1}$) [38].

based typically on the reactants shown in Figure 5.3 for the formation of RIM copolymer networks (PU, PUU and PUr). Values of the interaction parameter χ_{HS} have been calculated [42] from solubility parameters (δ_H and δ_S) and compared with the critical value of the interaction parameter, χ_c, estimated from the data of Liebler and Benoit and Hadziiannou represented in the universal phase diagram (Figure 5.10(a)). The calculations are based on the idealized structure of an $(AB)_n$-segmented copolymer shown in Figure 5.11, for which N_S is determined from two arms of the trifunctional polyether soft segment, and the value of N_H is determined from a knowledge of N_S and ϕ, assuming copolymerization has proceeded

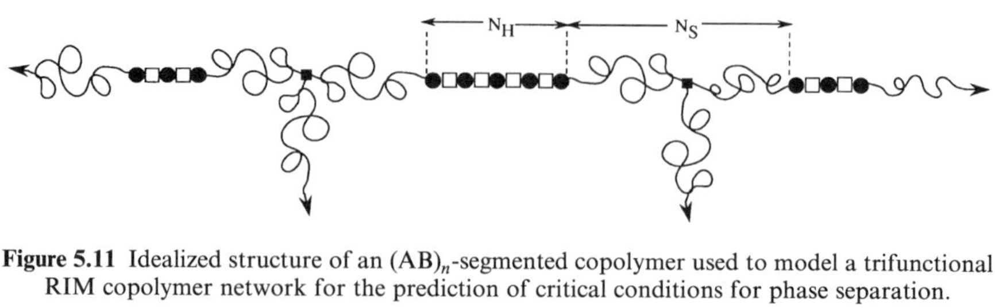

Figure 5.11 Idealized structure of an $(AB)_n$-segmented copolymer used to model a trifunctional RIM copolymer network for the prediction of critical conditions for phase separation.

Table 5.2 Solubility parameters (δ) of hard and soft segments and calculated values of the hard–soft segment interaction parameter (χ_{HS}) at 298 K, normalized to the volume of the soft-segment repeat unit for some RIM copolymer networks[a]

Copolymer network[b]	V_s $cm^3\,mol^{-1}$	δ_H $cal^{1/2}\,cm^{-3/2}$	δ_S $cal^{1/2}\,cm^{-3/2}$	χ_{HS}
PU-EG	53	11.6	8.1	1.14
PUU-DETDA	53	12.4	8.1	1.48
PUU-MDIPA	53	10.6	8.1	0.48
PUr-DETDA	55	12.4	8.1	1.72
PUr-MDIPA	55	10.6	8.1	0.58

[a] $\delta/cal^{1/2}\,cm^{-3/2}$ for monomers: MDI = 10.5, EG = 11.3, DETDA = 12.5, MDIPA = 10.3.
[b] Copolymer networks contain polyether soft segments ($M_n \approx 6000\,g\,mol^{-1}$). A more complete definition of the network systems is given in the caption to Figure 5.3.

to completion. Typical data are given in Table 5.2, in which values of χ_{HS} have been calculated according to equation (5.12), with the reference volume V_r given as the molar mass of the soft-segment repeat unit divided by density. The solubility parameters δ_H and δ_S have been calculated systematically [53] using the chemical structures of the reactants in Figure 5.3. Values of χ_c, estimated by the spinodal values of the product $\chi(N_H + N_S)_s$ at different soft-segment weight fractions (Figure 5.10(a)) and the values of χ_{HS} given in Table 5.2 are then used to determine the critical ratio χ_{HS}/χ_c, which is the parameter used to predict whether or not a particular RIM copolymer-forming system will phase separate. If the value of the critical ratio exceeds unity ($\chi_{HS}/\chi_c > 1$) then phase separation is predicted to occur and the magnitude of the critical ratio gives an indication of the degree of phase separation. Predicted phase-separation behaviour for different RIM copolymer network systems is shown in Figure 5.10(b) as plots of χ_{HS}/χ_c versus hard-segment (HS) content. For all of the systems (PUU and PUr), values of the critical ratio exceed unity and, as expected, the degree of phase separation is predicted to increase with HS content. The driving force for phase separation is related to the magnitude of the critical ratio, and the data presented in Figure 5.10(b) show clearly that copolyurea (PUr) systems will be much more phase-separated than corresponding copoly(urethane–urea) (PUU) systems. For a particular copolymer network system, the effects on phase behaviour of the hard-segment chemical structure, as determined by the hindered amines, DETDA and MDIPA (Figure 5.3) is also clearly evident. Furthermore, the thermodynamic calculations show that the degree of phase separation is predicted to increase in the order PUU-MDIPA < PUr-MDIPA < PUU-DETDA < PUr-DETDA.

The various theoretical models described, that provide predictions of molar mass development, chemical gelation and phase separation in copolymer network-forming systems, may be used as bases for the interpretation and comparison of structure–property relationships in materials produced via RIM. A complete interpretation of such structure–property relationships requires a detailed understanding of the competition between degree of phase separation and reaction kinetics in fast-reacting RIM systems, linked to a comprehensive characterization of materials properties. Thus, as stated previously in Section 5.1, both *in situ* (or real-time) and *a posteriori* techniques are required. A large number of experimental techniques have been utilized in the study of multiphase morphology resulting from microphase separation in copolymers and polymer blends and, more recently, real-time techniques have been developed [9] for the study of structure development during polymer reactive processing such as by RIM. Techniques include DSC [22, 54–56], DMTA [7, 8, 21, 23, 37, 57], FTIR [7, 30, 58], SALS, SAXS and SANS [59, 60], NMR [61] and microscopy [62]. In the following sections, examples are given of some of the techniques used with specific RIM, copolymer network systems.

5.4 Kinetics of phase separation and polymerization during RIM

Examples of studies on the reaction kinetics and the dynamics of microphase separation during RIM copolyurethane formation were discussed in Section 5.2. In these studies, independent techniques were used to establish the factors controlling structure development. More recently, developments in the use of simultaneous techniques, such as time-resolved SAXS and FTIR [63], have enabled significant advances to be made in the understanding of structure development during the bulk copolymerization of linear segmented copolyurethanes formed by RIM. Simultaneous techniques are particularly attractive since the acquisition of multiple experimental data that can be derived from a single sample avoids the effects of anomalies, such as differences in temperature or thermal history, and inaccuracies in time correlations between separate experiments, that are introduced by combining otherwise independent measurements on separate samples. In the reactive processing of polymers, such as by RIM, understanding the development of structure and properties during copolymerization requires a transformation of data from the experimental time domain into the molecular conversion domain, since the reactive process involves a chemical quench from the one-phase (liquid) region to a multiphase region of the phase diagram. By monitoring processing in the conversion domain, the evolution of structure is related to increases in molar mass and to gelation phenomena which, therefore, locate a material in a phase diagram independent of chemical kinetics.

5.4.1 Model system: segmented copolyurethanes

An example of simultaneous studies of reaction kinetics and structure development in a model RIM system [63] is shown by the collected plots in Figure 5.12. The system studied comprised ~30% by weight hard segment (formed from MDI and pentane-1,5-diol) and a soft segment based on a monofunctional POE-POP (monol prepolymer, $M_n \approx 2000\,\mathrm{g\,mol^{-1}}$). The kinetics of microphase separation were derived from SAXS data (Figure 5.12(a)), while FTIR provided data on both reaction kinetics (Figure 5.12(d)) and hydrogen-bonding dynamics (Figure 5.12(c)). The importance of this work is twofold. First, it demonstrates, for the first time, the use of the simultaneous SAXS/FTIR technique and, second, it shows clearly that, in a phase-separating copolyurethane system, the onset of microphase separation precedes that of hydrogen bonding. The latter is shown by the faster growth in the SAXS invariant (scattering power) compared with the concentration of hydrogen-bonded urethane carbonyl moieties as functions of reaction time (Figure 5.12(b)). The order in which these phenomena occur has been a source of controversy for many years. Hydrogen bonding will not in itself drive a phase transition. The mixture phase-separates first, creating regions containing high local concentrations of urethane structures thereby increasing the rate of hydrogen bonding. The important conclusion drawn from this study is that, in the characterization of multiphase copolyurethane-forming systems, the use of a single technique should not be relied upon for the detection of the onset of a phase transition. The fact that there is no indication of hydrogen-bonded urethane or urea carbonyl moieties, or both, does not simply imply that the reacting system is still in a single phase.

5.4.2 *Experimental systems: copoly(urea–isocyanurate) networks*

The model copolyurethane system described in the previous section, in terms of commercial RIM, is a very slowly reacting system. In a more recent study [64], a typically fast copoly(urea–isocyanurate) PUrI system, based on that discussed in Section 5.3.2, has been used in a synchrotron SAXS investigation of structure development during RIM. PUrI copolymers are formed from isocyanurate-based network-forming systems that involve multiple competitive reactions, tailored to provide controlled processability during the reactive processing of structural composites [10, 48, 65]. For example, copoly(urea–isocyanurate)s (PUrI) have been formed by reacting (via RIM) a polyether diamine ($M_n \approx 2000\,\mathrm{g\,mol^{-1}}$) with a large stoichiometric excess of MDI in the presence of an organic trimerization catalyst [48]. Copolymerization occurs by a two-stage process involving firstly the rapid formation of isocyanate-tipped poly(ether–urea) oligomers and, secondly, the trimerization of isocyanate-functionalized moieties to form the final glassy, PUrI resin of high junction-point density. The polymerization kinetics of such rapid RIM systems may be monitored using the adiabatic temperature rise (ATR) technique. Typical ATR kinetics data and the chemical reactions involved in the two-stage copolymerization process are shown in Figure 5.13(a). Such a versatile network-forming system is very

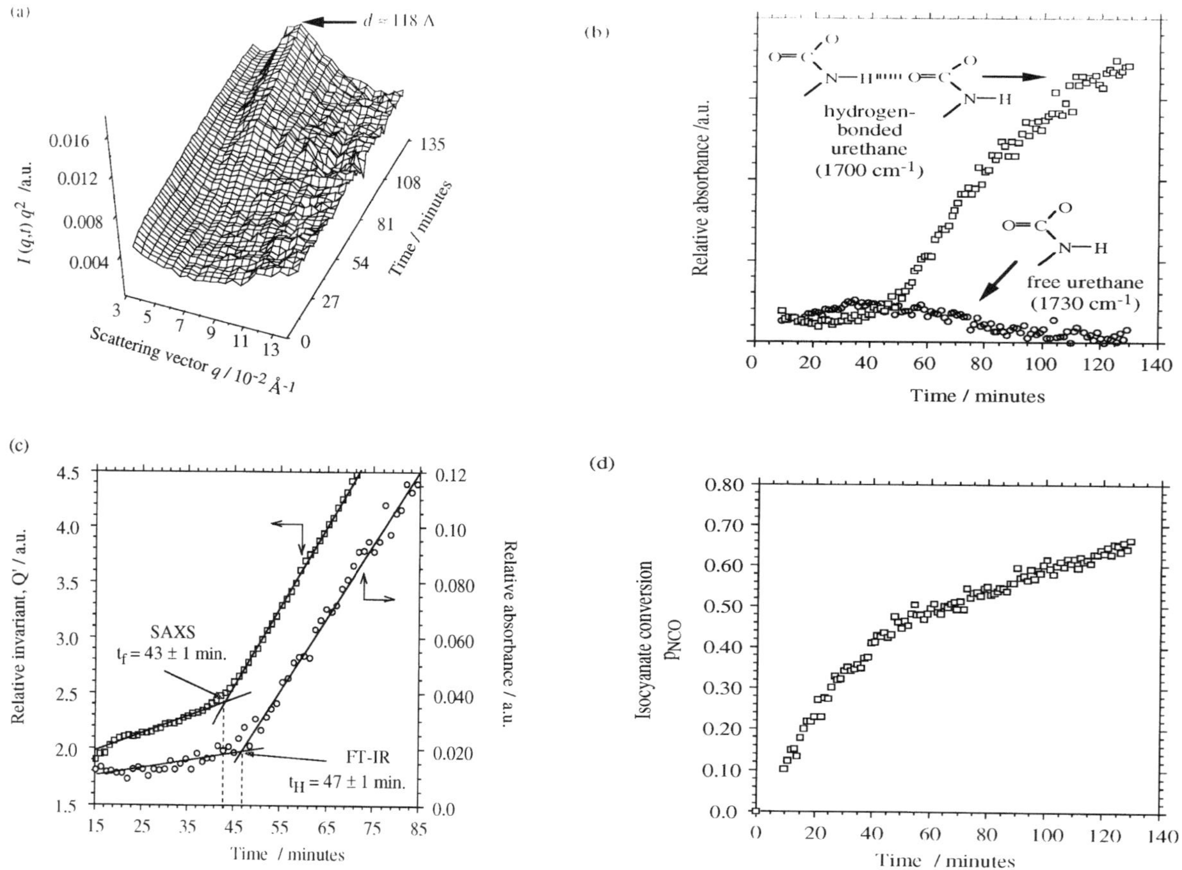
(a)
d ≈ 118 Å
I (q,t) q2 /a.u.
0.016
0.012
0.008
0.004
3
5
7
9
11
13
Scattering vector q / 10-2 Å-1
0
27
54
81
108
135
Time / minutes
(b)
hydrogen-
bonded
urethane
(1700 cm-1)
free urethane
(1730 cm-1)
Relative absorbance /a.u.
0
20
40
60
80
100
120
140
Time / minutes
(c)
SAXS
tf = 43 ± 1 min.
FT-IR
tH = 47 ± 1 min.
Relative invariant, Q' / a.u.
4.5
4.0
3.5
3.0
2.5
2.0
1.5
Relative absorbance / a.u.
0.12
0.10
0.080
0.060
0.040
0.020
0.0
15
25
35
45
55
65
75
85
Time / minutes
(d)
Isocyanate conversion
pNCO
0.80
0.70
0.60
0.50
0.40
0.30
0.20
0.10
0.0
0
20
40
60
80
100
120
140
Time / minutes

attractive, therefore, for the rapid production of structural composites via RIM, since mould-filling viscosity, gel time and resin formation and final properties are readily controlled by systematically varying the molecular structure and molar ratios of monomers and polyether, and using appropriate catalyst concentrations and temperatures.

Both *in situ* and *a posteriori* studies have been carried out on the formation and properties of PUrI networks, and on their use as matrices in RIM structural composites. In the former case [64], the kinetics of microphase separation were monitored using time-resolved synchrotron SAXS. A micro-RIM apparatus was used to meter and mix the two reactants (MDI and POP diamine), and to inject the reactant mixture into a mould cell positioned in the optical-bench assembly of the synchrotron beamline. Typical scattering data for a copolymerization carried out at 90 °C are shown in Figure 5.14(a), in which the evolution of a microphase-separated structure is illustrated by the development of a maximum in the scattering profile at $q^* \approx 0.065\,\text{Å}^{-1}$ that increases in intensity as copolymerization progresses from ≈38 s to 68 s. The interdomain spacing (d) of this structure was calculated from the scattering maxima to give a value of ≈97 Å, which agrees well with post-moulding values of d determined via static SAXS and TEM measurements on this system [48]. PUrI formation was studied at different temperatures in the range 70–135 °C, and values of Q' (the relative invariant) were determined from SAXS data as a function of reaction time. At each temperature, Q' versus time is characterized by an induction period (negligible growth), a period of rapid growth (microphase separation) and a plateau region (growth arrest). Representative data are shown in Figure 5.14(b) for a PUrI formed at 70 °C for which the induction period (0–74 s) and the microphase-separation region (74–128 s) are clearly evident. The onset times for phase separation decrease with increasing temperature following an Arrhenius-type rate dependence, indicating that the MST occurs at the same degree of isocyanate conversion, independent of temperature. At all reaction temperatures, the scattering-peak intensity increases exponentially with time, indicative of spinodal decomposition as predicted by the linearized theory of Cahn and Hilliard [66]. As expected for a block copolymer, the size-scale of the growing structure is independent of temperature, and the molecular connectivity between the polyether- and polyisocyanurate-rich phases restricts the interdomain spacing to values in

Figure 5.12 Simultaneous SAXS-FTIR data showing the kinetic competition between polymerization, microphase separation and hydrogen bonding occurring during the formation of a model linear copolyurethane from a polyether monol ($M_n \approx 2000\,\text{g mol}^{-1}$), 1,5-pentane diol and MDI. The hard-segment content was 30% by weight and the reaction temperature 20 °C. (a) SAXS patterns at different polymerization time; (b) FTIR data showing the formation of free (○) and hydrogen-bonded (□) urethane groups; (c) relative rates of phase separation (SAXS) and hydrogen bond formation (FTIR, normalized to isocyanate conversion); (d) FTIR polymerization kinetics data in terms of isocyanate conversion versus time. (a.u. = arbitrary units.)

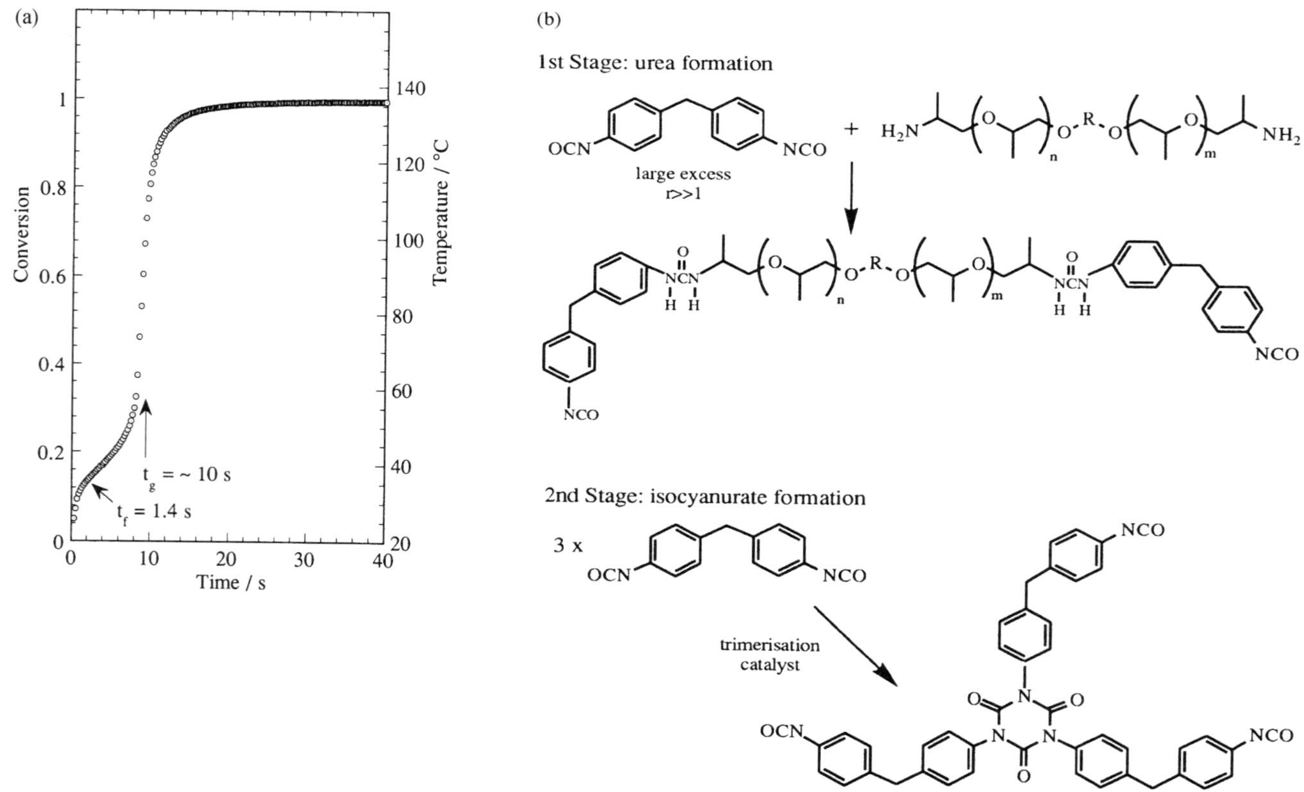

Figure 5.13 Polymerization kinetics, determined using the adiabatic temperature rise (ATR) technique, for the RIM copoly(urea–isocyanurate) (PUrI) described in Figure 5.9. The basic chemical reactions (illustrated using the reactants) involved in the two-stage polymerization process are shown on the right. Ultimately, all NCO groups are reacted into isocyanurate rings to form the network. Mould-fill time (t_f) and gel time (t_g) are indicated by the arrows.

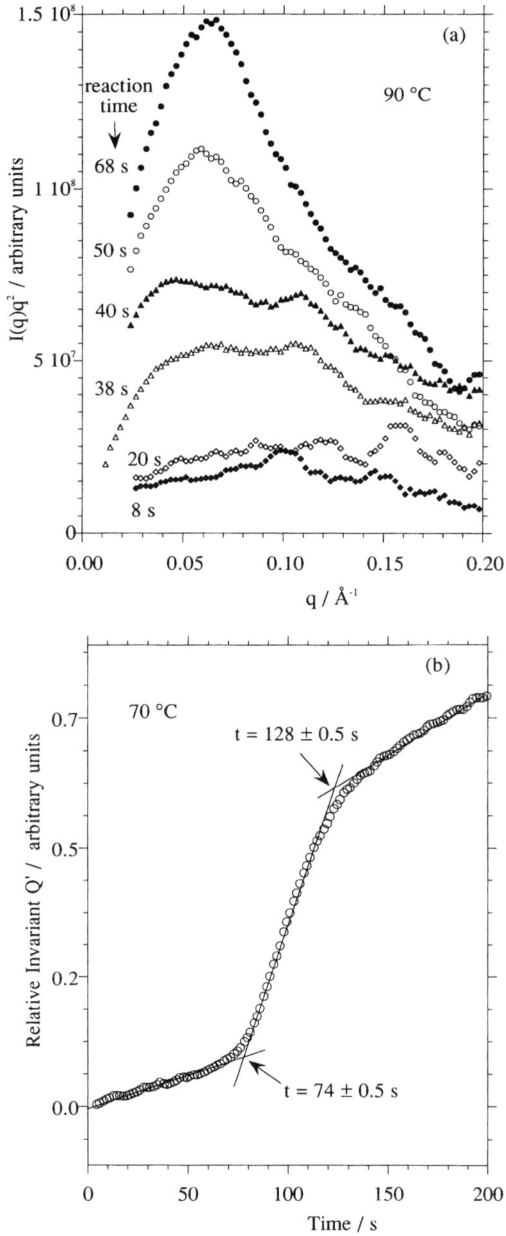

Figure 5.14 Phase-separation kinetics of the RIM-PUrI shown in Figure 5.13, determined from time-resolved SAXS data. (a) Lorentz-corrected intensity $I(q)q^2$ versus scattering vector $\boldsymbol{q}$ at increasing reaction times after mixing at a cell mould temperature of 90 °C. (b) Variation of the SAXS relative invariant Q' with reaction time for the copolymerization carried out using a cell mould temperature of 70 °C.

the range $2 \leq (q^* \cdot R_g) \leq 3$, as predicted by Leibler [50]. The radius of gyration, R_g, of combined (hard-block)–(soft-block) segments in the PUrI is not known, although values of R_g have been determined from SANS data for a polyurethane block copolymer formed using a POP prepolymer of similar molar mass ($\approx 2000\,\mathrm{g\,mol^{-1}}$). The radius of gyration $(R_g)_{pop}$ for POP chains was measured as 13.5 Å in the disordered (relaxed) state, and 16 Å in the ordered (stretched) state. Using these values for the PUrI, for which the volume fraction of POP is $f_{pop} = 0.43$ and R_g is approximated by $(R_g)_{pop}/f_{pop}$, gives values of $q^* R_g$ in the range 2.0–2.4. These values are typical of a block copolymer. Thus, values of d calculated using $(R_g)_{pop} = 16$ Å, lie in the range 74–115 Å and are in reasonable agreement with the experimental value of 97 Å determined from the SAXS data [64]. These studies lead to the somewhat surprising conclusion that a complex copolymerization, producing ultimately a microphase-separated glassy polymer network, shows the dynamics of simple liquid–liquid phase separation. Microphase separation in these RIM-PUrI network-forming systems is arrested prematurely by vitrification of the polyisocyanurate phase, and leads to materials with non-equilibrium co-continuous morphologies. The structure and properties of such RIM network materials is discussed in Section 5.3.2.

5.5 Properties of RIM copolymer networks

In previous sections, the basic principles of RIM, the associated copolymerization kinetics and the thermodynamics of phase separation, were used to define the links between reactive processing, the development of molecular structure and the creation of materials morphology in fast polymer-forming systems. These correlations, explaining the transformation from reactive, liquid state to solid, multiphase material, are fundamentally important for the interpretation of the physical properties of the moulded materials.

The properties of RIM copolymers depend strongly on their morphological structures. In the case of PU materials, however, despite the vast amount of work reported on the subject, the complex microdomain morphology is still not completely resolved and is often confusing, particularly when comparing PU copolymers formed in solution with those formed in bulk. Several studies, using electron microscopy on solution-polymerized PU copolymers [67–69], show their morphologies depend on the amount of crystallizable hard segments. Below 30% w/w of hard segments, a random dispersion of short, cylindrical domains in a continuous soft-segment matrix was observed compared with discrete, cylindrical domains of soft segments in a hard-segment matrix at ~68% w/w hard-segment content. At intermediate compositions (42–61% w/w hard-segment content), co-continuous morphologies were observed comprising relatively short, randomly oriented lamellae with widths of 11–14 nm and lengths that increased with increasing hard-segment content. In contrast, bulk-polymerized PU copolymers formed under heterogeneous reaction conditions that induce macrophase (rather than microphase) separation, produce materials of low molar mass with coarse morphologies and poor mechanical properties. During RIM, such reaction conditions occur either when low catalyst concentrations [29, 70] and poor impingement-mixing conditions [71] are used, or at the surfaces of relatively cold moulds [28, 70]. In addition, large temperature gradients developed within RIM-PU materials result in non-uniform distributions of hard-segment domains and crystallinity. Premature vitrification and phase separation of nascent hard-segment moieties occur when their glass transition temperature exceeds that of the mould surfaces. Under these conditions, favouring macrophase separation, globular and spherulitic hard-segment domains are formed. Several studies using microscopy on bulk-polymerized PU copolymers (see, for example [67, 70, 72, 73]) have shown the existence of spherulites with a radial, fibrillar structure. Often, the spherulites are relatively large, 2–15 μm in diameter [72], within which exist fibrils (50–200 nm) comprising stacked hard-segment lamellae of mean length 190 nm and mean thickness 7.3 nm. The observed lamellae dimensions are in good agreement with those predicted by chain-folded conformation models [74] based on the lengths of structural repeat units of urethane hard segments formed typically in PU copolymers. Processing by RIM using high catalyst concentrations and mould temperatures [30], however, increases the rate of copolymerization and microphase separation becomes the dominant process. This results in the formation of RIM-PUs with higher molar masses and finer domain morphologies within which crystallinity is diffusion controlled, leading to the creation of paracrystalline structures. The short time-scales typical of RIM, however, limit the development of appreciable crystallinity for which further high-temperature postcuring is required.

In contrast to RIM-PU materials, PUU and PUr materials, formed typically from the reactants given in Figure 5.3, are completely amorphous with no evidence of crystallinity detectable by WAXS, DSC or TEM [42]. The relative rates of copolymerization and phase separation in the PUU and PUr materials are such that structure development occurs via spinodal decomposition, giving microphase separation, and the morphologies, produced as relics of the spinodal decomposition process, are co-continuous with characteristic size scales greater than 50 nm for PUU and between 10 and 30 nm for PUr [64].

The following subsections give a comparative discussion of structure–property relationships in PU, PUU and PUr materials and uses, as specific examples, RIM copolymer networks comprising soft segments formed from formulations comprising trifunctional polyether prepolymers (Figure 5.3). Thermal properties (from DSC and DMTA), tensile stress–strain properties and fracture behaviour are presented for chemically related RIM materials, mainly PU, PUU and PUr containing ~50% w/w hard segment. The various copolymers are designated according to the chemical acronym given previously to define the short-chain diol or hindered-amine monomer used to form the hard segment (Figure 5.3). For example, PUU-DETDA defines a copoly(urethane–urea) comprising polyether soft segments, based on a POE-POP triol ($M_n \approx 6000\,\mathrm{g\,mol^{-1}}$), and polyurea hard segments formed from MDI and DETDA.

5.5.1 Thermal behaviour

The thermal properties of RIM copolymers are readily determined using differential scanning calorimetry (DSC) and dynamic-mechanical-thermal analysis (DMTA), which provide data on thermal transitions, phase-separation ratios (*PSR*) and modulus-temperature behaviour [7, 8, 21–24]. From DSC data, values of the glass-transition temperature T_g^S and the heat capacity change ΔC_p^S associated with the soft-segment phase can be related to those ($T_g^{S_0}$ and $\Delta C_p^{S_0}$) of the corresponding isolated (pure) soft-segment material. Thus, normalized values of the PSR $(= (\Delta C_p^S/\Delta C_p^{S_0})/w_{SS})$, where w_{SS} is the mass fraction of soft segments in the copolymer, can be used to define the dependence of degree of phase separation on copolymer composition for the various RIM copolymers [22, 24, 38, 42]. The *PSR* value is a measure of the amount soft segment contributing to the heat capacity at T_g^S and does not, therefore, include soft segments trapped either within the hard segment phase or in interphase regions between hard- and soft-segment phases. Typical data are shown in Figure 5.15 as plots of the *PSR* versus the percentage weight fraction of hard segment for different PUU [22, 42] and PUr [38, 75] materials. The plots in Figure 5.15(a) for DETDA-based materials show the expected increases in *PSR* with increasing hard-segment content, and these data are in qualitative agreement with theoretical predictions in terms of χ_{HS}/χ_c, the critical ratio of interaction parameters given in Section 5.3.2 and Figure 5.10. Also, clearly evident in Figure 5.15(a), is the pronounced increase in degree of phase separation resulting from high-temperature postcuring (200 °C for 1 h) of the PUr-DETDA materials [75]. However, the relative position of the PUU-DETDA [42] and PUr-DETDA curves in Figure 5.15(a) and the decreases in *PSR* with increasing hard-segment content for MDIPA-based materials, shown in Figure 5.15(b), are contrary to the trends with chemical structure and composition predicted by simple thermodynamic considerations.

The experimental observations show the PUr copolymers to be more phase mixed than corresponding PUU copolymers, which may be explained in terms of the difference in rates of copolymerization, relative to those of phase separation, between the two systems. During the processing of PUr, the much faster rate of reaction of the isocyanate groups on MDI towards

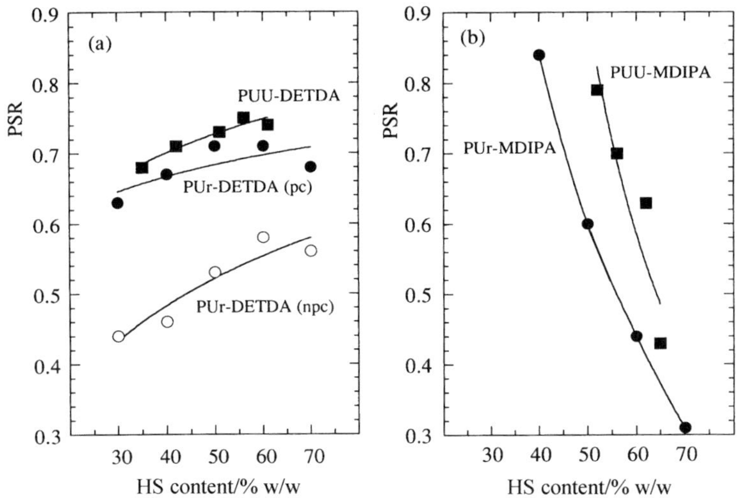

Figure 5.15 Plots of phase-separation ratio (*PSR*) determined from DSC data, versus hard-segment content for the trifunctional RIM copolymer networks defined in Figure 5.10. (a) PUU-DETDA and PUr-DETDA systems (npc = non-postcured; pc = postcured). (b) PUU-MDIPA and PUr-MDIPA systems (postcured).

the aliphatic amine groups on POP triamine, compared with the aromatic amine groups on the hindered diamines, leads to the rapid formation of isocyanate-tipped polyether oligomers. The aromatic amine groups then react competitively with isocyanate groups and hard-segment sequence lengths develop until vitrification occurs. In the case of PUU materials, preformed hard-segment oligomers with isocyanate end groups react with hydroxy groups on polyether prepolymers to give a block copolymer network, and this reaction must be sufficiently catalysed to ensure that its rate is fast in relation to the phase-separation rate, so that covalently-bonded microphases are produced. In contrast, the sequential reaction of the aliphatic and aromatic reactants in PUr systems ensures that there is good connectivity between the hard and soft microphases, even if the vitrification process, that occurs as a consequence of phase separation, precedes chemical gelation. Thus, in simplistic terms, PUU and PUr copolymer-forming systems develop heterogeneous macronetworks via two different routes. More importantly, superimposed on the microphase-separated structure, in both systems, is the subsequent development of strong hydrogen-bonding within the hard-segment domains. Hydrogen bonding is much more extensive in DETDA-based copolymers, compared with MDIPA-based copolymers, due to the much higher density of urea groups per mole of repeat units in the former (5.14 *cf.* 3.57 $mol\,m^{-3}$). The experimental data show that, with increasing hard-segment content, there is increasing dissolution of short (isolated) hard segments within the soft-segment phase of

MDIPA-based copolymers, despite the development of longer, average sequence lengths of hard-segment blocks with higher glass-transition temperatures (T_g^H) [38, 42]. Generally, as copolymerization proceeds and average hard-segment sequence length develops, T_g^H increases to a level that exceeds the temperature within the mould, even allowing for the reaction exotherm, and vitrification occurs rapidly to produce materials with non-equilibrium morphologies. Both the development of hydrogen bonding and the non-equilibrium nature of the rapid RIM process are factors not accounted for in the simple thermodynamic model used to predict phase separation, and these factors therefore explain some of the inconsistencies between theory and experiment.

Additional structure–property information is provided by DMTA, particularly with respect to hard-segment glass transition behaviour and modulus-temperature dependence. Typical DMTA data for RIM materials are shown in Figure 5.16 for PUr-DETDA copolymers [75]. Generally, two major transitions are observed, assigned to the soft- and hard-segment glass transitions at temperatures T_g^S and T_g^H, respectively, and are further confirmation of the microphase-separated block copolymer structure of the RIM materials. To a first approximation, the intensities of the two transitions are in proportion to the relative amounts of each phase. Also, as the hard-segment content increases, the magnitudes of the modulus decrease around T_g^S and the modulus-temperature dependences above T_g^S are clearly reduced. The broad, low-intensity damping region, ranging from ~50–100 °C, between the two major transitions, is attributed to cooperative molecular relaxations arising from the breakdown of intersegmental hydrogen bonding between mixed soft- and hard-segments within interphase regions [22, 76]. Also apparent in Figure 5.16 are the effects of postcuring, which improve modulus and intensify the higher-temperature relaxation behaviour in the region around T_g^H.

Comparative DMTA data for PU [77], PUU [42] and PUr [38] materials containing ~50% w/w hard-segment contents are shown in Figure 5.17, and a summary of the phase-separation and thermal properties, including both DSC and DMTA data, is given in Table 5.3. In all series of copolymers (PU, PUU and PUr), there is little variation in the value of T_g^S with hard-segment content, although T_g^S is raised by 5–10 °C compared with the values (−60 °C to −67 °C) characteristic of the corresponding polyether networks formed as solely soft-segment materials [24]. The increased glass-transition temperatures, indicative of some dissolution of hard segments within the soft-segment phases of the copolymers, and the broadening of the peaks at T_g^S, due to the diffuse nature of the interphase regions, are consistent with the *PSR* data obtained from DSC. In contrast, the hard-segment glass-transition behaviour is strongly influenced by copolymer composition (except in the case of the PU materials in which the hard-segment domains are present as semicrystalline phases). For both the PUU-DETDA [42] and PUr-DETDA materials [38], the value of

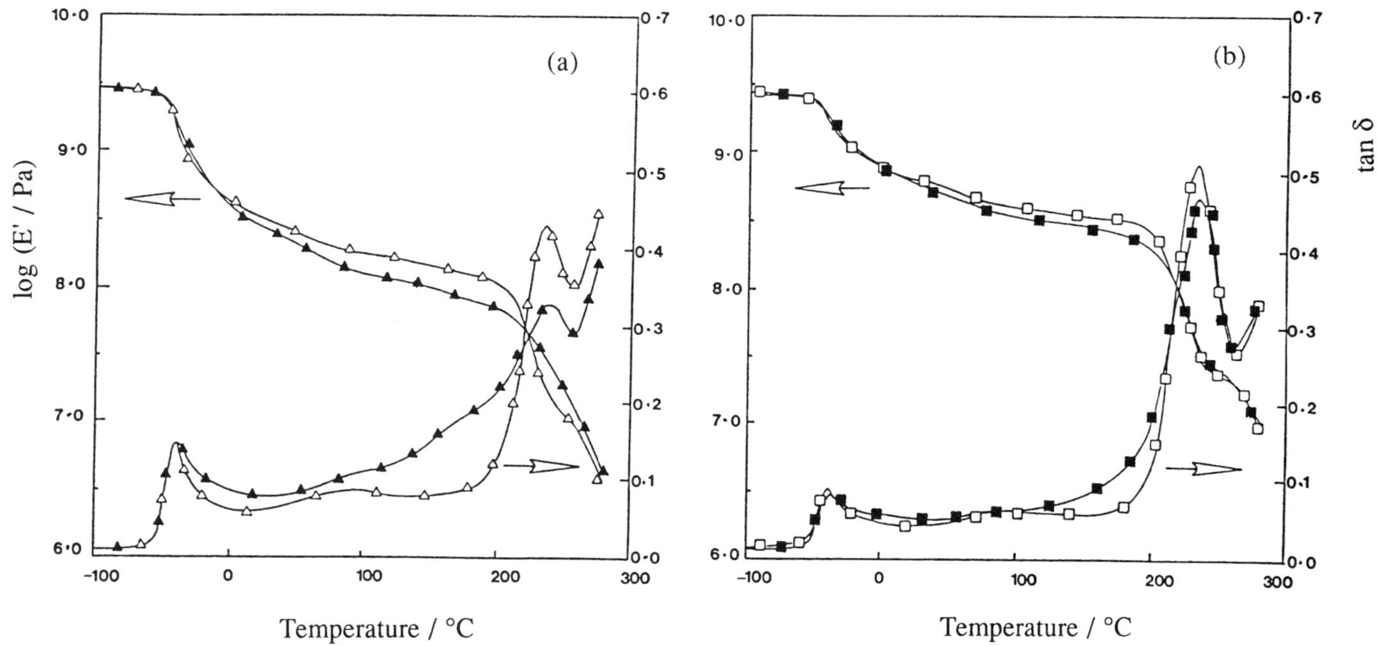

Figure 5.16 Dynamic-mechanical-thermal behaviour at 1 Hz of trifunctional PUr-DETDA networks comprising soft segments based on POP triamine ($M_n \approx 5700\,g\,mol^{-1}$) [75]. Comparative plots of storage flexural modulus (E') and mechanical damping ($\tan\delta$) versus temperature showing materials, non-postcured (closed symbols) and postcured (open symbols), with different hard-segment contents: (a) 40% w/w and (b) 60% w/w.

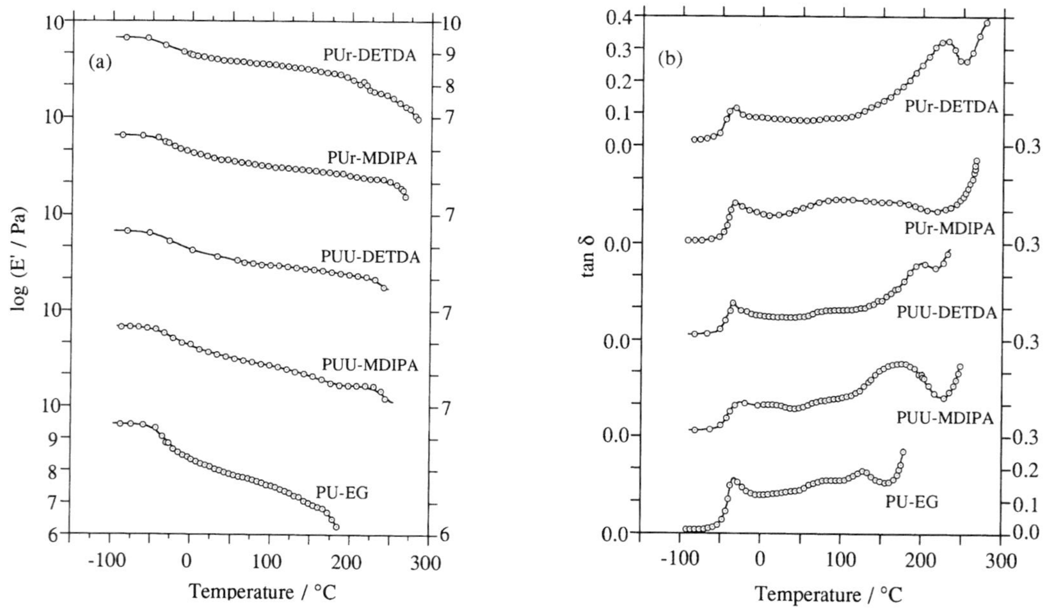

Figure 5.17 Dynamic-mechanical-thermal behaviour at 1 Hz of the trifunctional RIM-copolymer networks defined in Figure 5.10. Also included are data for PU-EG, a copolyurethane network comprising soft segments based on POE-POP triol ($M_n \approx 6000\,g\,mol^{-1}$), and MDI/EG hard segments [77]: (a) flexural storage modulus (E') versus temperature; (b) mechanical damping ($\tan\delta$) versus temperature. For clarity, curves are displaced along the ordinates.

Table 5.3 Phase separation, thermal and tensile stress–strain properties of RIM copolymer networks containing ≈50% w/w hard segments. (For definition of the copolymer systems see Table 5.2 and Figure 5.3.)

Copolymer network	N_n [a]	χ_{HS}/χ_c [b]	T_g^S [c] /°C	PSR [d]	T_g^S [e]		T_g^H [e]		E [f] /MPa	σ_u [f] /MPa	ε_u [f] /%	U_t [f] /MJ m^{-3}
					/°C	tan δ	/°C	tan δ				
PU-EG	14.0	1.2	−63	0.61	−35	0.17	129	0.21	131	22	305	58
PUU-DETDA	9.0	13.2	−58	0.73	−46	0.12	194	0.41	497	26	187	59
PUU-MDIPA	6.2	4.8	−50	0.79	−35	0.11	176	0.24	483	18	163	35
PUr-DETDA	9.0	16.4	−60	0.71	−39	0.11	233	0.42	565	30	98	20
PUr-MDIPA	6.2	5.5	−59	0.60	−35	0.13	171	0.13	332	23	163	26

[a] N_n is the number-average hard-segment sequence length calculated from copolymer stoichiometry.
[b] χ_{HS}/χ_c is the critical ratio of interaction parameters determining the condition for phase separation.
[c] T_g^S is the soft-segment glass-transition temperature determined by DSC.
[d] *PSR* is the phase separation ratio determined from DSC data for T_g^S.
[e] T_g^S and T_g^H are the soft-segment and hard-segment glass-transition temperatures determined by DMTA.
[f] Tensile properties: E is Young's modulus, σ_u is tensile strength, ε_u is failure strain, U_t is toughness.

T_g^H increases linearly with increasing hard-segment content, in accord with the predictions of the Flory–Fox equation [78],

$$T_g^H = T_g^{H\infty} + \frac{B}{N_n} \tag{5.13}$$

in which N_n is the number-average sequence length of the hard segments determined by copolymer stoichiometry, B is a polymer-specific constant, and $T_g^{H\infty}$ is the limiting value of the glass-transition temperature for hard segments of infinite molar mass. Typical data for the trifunctional PUU-DETDA [42] and PUr-DETDA [38] materials give extrapolated values of 200 °C and 249 °C, respectively, for $T_g^{H\infty}$. The difference is due principally to the obvious differences in chemical structures of the hard segments in terms of chain flexibility, urea-group densities and the relative degrees of hydrogen bonding discussed earlier. In addition, as the data in Table 5.3 show, the variations in transitional behaviour are in qualitative agreement with the values of N_n and the predictions of degree of phase separation in terms of the parameter χ_{HS}/χ_c for the different RIM copolymer networks.

5.5.2 *Tensile stress–strain behaviour*

Overall, the tensile stress–strain and fracture properties of RIM materials are determined principally by copolymer composition and morphology. The variations of tensile stress–strain behaviour with hard-segment content for a particular RIM copolymer system, and with copolymer structure for different RIM materials with constant (~50% w/w) hard-segment content, are shown in Figures 5.18(a) and 5.18(b), respectively. Tensile behaviour is also strongly affected by postcuring [75] (which improves tensile properties as shown by comparing the dashed and solid curves in Figure 5.18(a)) and, in the case of the highly catalysed PU and PUU systems, upon the concentration of catalyst [37] used during RIM processing (as shown by the comparative curves and tensile property data given in Figure 5.19).

The tensile curves in Figure 5.18(a) represent the range of materials behaviour, from soft rubbery through ductile plastic to rigid and relatively brittle resins, expected as the hard-segment content is increased, typically, from 30 to 70% w/w in such RIM systems. In general terms, the RIM materials exhibit two characteristic regions of deformation behaviour. At low strains the materials have high modulus and the initial slopes are steep. At higher strains (>10%), all materials show yielding with extensive post-yield drawing under relatively small increases in nominal stress. Such stress–strain behaviour, typical of materials possessing co-continuous morphologies, is interpreted [79] by associating the initial part of the curve with the deformation and yielding of the continuous hard-segment phase, and associating the post-yield region to the deformation of the much more extensible rubbery, network phase. The expected increases of modulus and

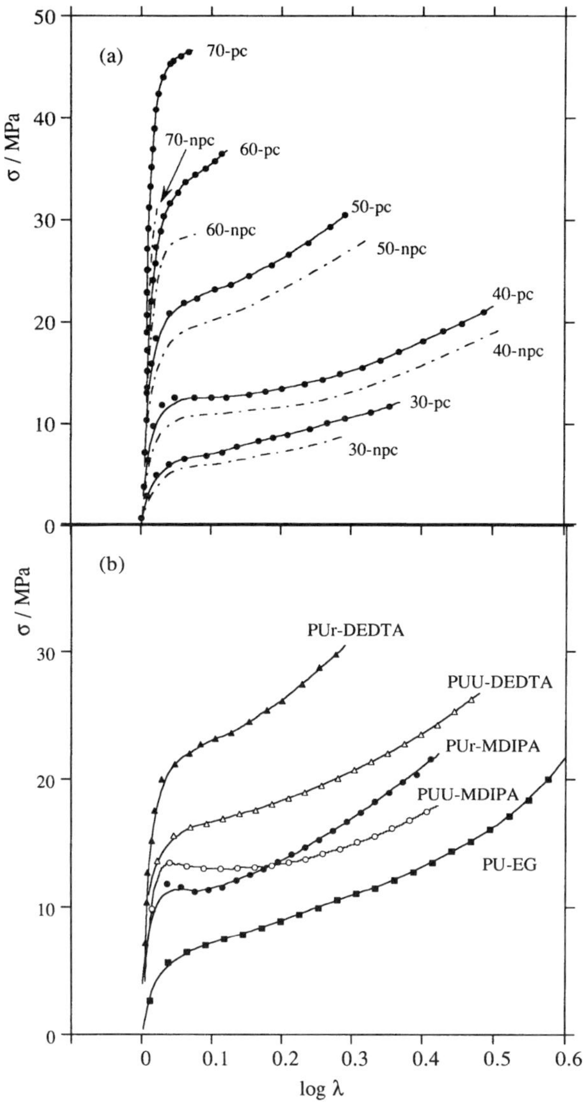

Figure 5.18 Representative tensile stress (σ) versus extension ratio (λ) curves for trifunctional RIM-copolymer networks, showing the effects of (a) hard-segment content for PUr-DETDA materials [75] comprising soft segments based on POP triamine ($M_n \approx 5700\,\mathrm{g\,mol^{-1}}$), and (b) soft-segment and hard-segment structure (at 50% w/w hard-segment content) for the PU, PUU and PUr materials defined in Figures 5.10 and 5.17. In (a) the solid curves are for postcured (pc) samples and the dashed curves are for non-postcured (npc) samples. Numbers against the curves are values of hard-segment content (% w/w).

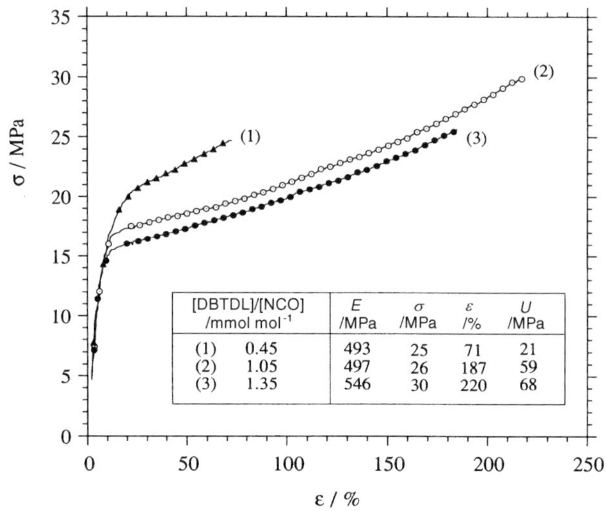

Figure 5.19 Representative tensile stress (σ) versus strain (ε) behaviour of trifunctional PUU-DETDA networks (51% w/w hard-segment content), showing the effects of varying the concentration of dibutyltin dilaurate (DBTDL) catalyst used during processing by RIM [37]. Catalyst concentrations and derived tensile properties are shown in the table inset.

tensile strength, and concomitant decrease in failure strain, with increasing hard-segment content, are shown in the compositional plots of Figure 5.20.

In broad mechanical terms, RIM materials may be considered as composite materials comprising soft and rigid phases, and several composite theories exist (see, for example, reviews by Nielsen [80] and Dickie [81]) to predict the elastic properties of multiphase polymer systems. Theoretical models have been developed for the low-strain elastic behaviour of materials comprising continuous–discontinuous and co-continuous morphologies, in terms of predictive equations for composite moduli. In order to elucidate the morphologies of RIM materials, experimental moduli are compared with three such predictive equations in Figure 5.20(a). In the first, the upper and lower bounds of the modulus for continuous–discontinuous composites are given by the modified Kerner equation [82]:

$$G_c = G_1 \left\{ \frac{\phi_1/A + \phi_2 G_2/B}{\phi_1/A + \phi_2 G_1/B} \right\} \tag{5.14}$$

where $A = 15(1 - \nu_1)$ and $B = (7 - \nu_1)G_1 + (8 - 10\nu_1)G_2$, and G is shear modulus, ν Poisson's ratio, ϕ is the volume fraction, and subscripts 1 and 2 refer to the continuous and discontinuous phase domains, respectively. The second equation due to Budiansky [83] is a special case of the Kerner equation which assumes that both components are continuous, but that the material is isotropic and macroscopically homogeneous. The composite

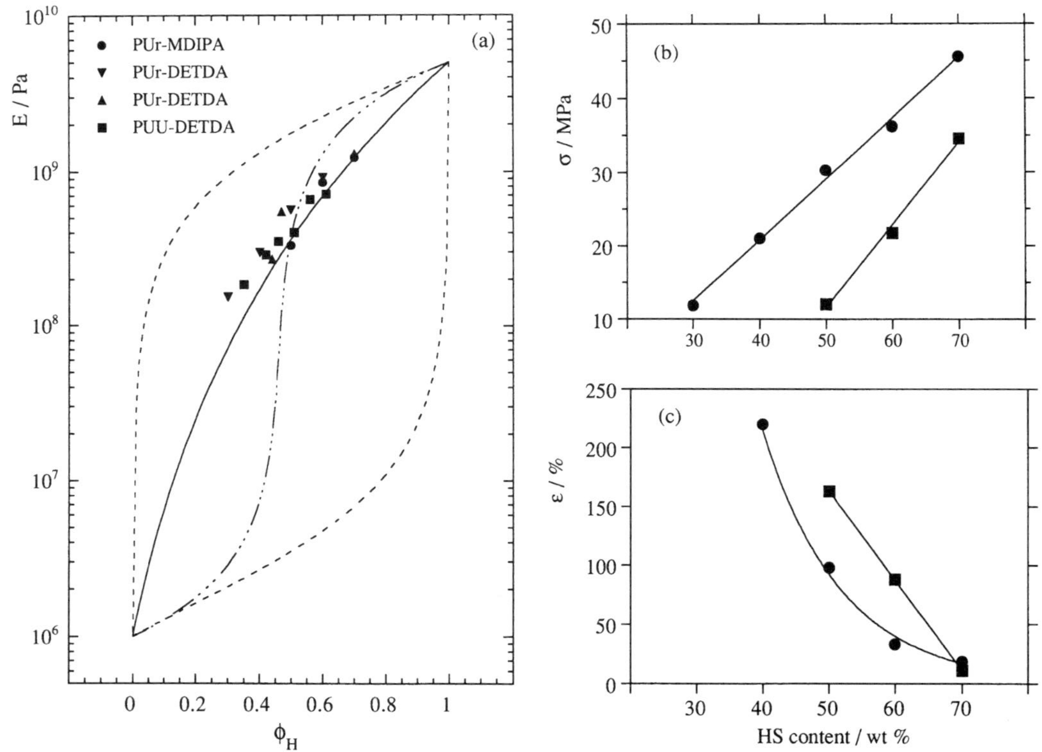

Figure 5.20 Variations of tensile properties with hard-segment content for trifunctional RIM copolymer networks. (a) Semi-log plot of Young's modulus (E) versus hard-segment volume fraction (ϕ_H) outer, dashed curves (Kerner) upper and lower bounds, equation (5.14); central, dashed curve (Budiansky), equation (5.15); solid curve (Davies), equation (5.16). (b) Tensile strength (σ_u) versus hard-segment content. (c) Failure strain (ε_u) versus hard-segment content.

modulus is obtained from a solution of the quadratic equation

$$\phi_1(G_1 - G_c)(G_c + \alpha G_2) + \phi_2(G_2 - G_c)(G_c + \alpha G_1) = 0 \qquad (5.15)$$

where $\alpha = 2(5 - 4\nu_c)/(7 - 5\nu_c)$ and ν_c is the Poisson's ratio of the composite and, in this case, is taken as the weighted average of Poisson's ratios of the two components. The third equation used, due to Davies [84], is based on an expansion of the theoretical treatment of the dielectric constant of two-phase materials and gives an expression with no adjustable parameters to predict the shear modulus of materials with two interpenetrating, continuous phases:

$$G_c^{1/5} = \phi_1 G_1^{1/5} + (1 - \phi_1)G_2^{1/5} \qquad (5.16)$$

According to Davies, the theory requires the existence of a perfect bond between the two phases, so that there is continuity of displacement across phase boundaries, and that the phases are incompressible.

The theoretical predictions of equation (5.14) to (5.16) are shown in Figure 5.20(a) as semi-log plots of Young's modulus versus volume fraction of hard segment, together with experimental values of the modulus derived from tensile stress–strain curves for all of the different RIM copolymer network materials. The data in Figure 5.20(a) show clearly that, in the composition range studied and to a first approximation, the moduli of the RIM copolymers show much closer agreement with the predictions of the Davies model for co-continuous materials than with either of the predictions based on the models of Kerner and Budiansky. Similar observations were reported [85] for well-bonded interpenetrating networks comprising polyurethane and poly(methyl methacrylate) over a wide range of compositions. Despite the apparent success of this type of macroscopic mechanical analysis, it should be noted that the modulus values in Figure 5.20(a) are plotted on an insensitive (logarithmic) scale and there is some scatter in the experimental data about the Davies curve. The Davies equation makes no adjustment for the size-scale of the microphases or of the interphase regions, which must change with both the chemical structure and content of hard segment. Notwithstanding, the modulus-composition data in Figure 5.20(a) provide further confirmation of the characteristic co-continuous morphologies (as illustrated for example by the TEM in Figure 5.9(b)), produced in these rapidly formed RIM copolymers.

The more detailed chemical and morphological structural differences between the PU, PUU and PUr copolymers are clearly illustrated by the comparative stress–strain curves shown in Figure 5.18(b) and the corresponding derived tensile properties given in the last four columns of Table 5.3 for RIM materials containing ~50% w/w hard-segment content. The variations in the mid- to high-strain regions of the stress–strain curves are enormous, given the apparently small chemical differences between the groups (either urethane or urea) linking soft and hard segments, and between the hard segments (based on either EG, DETDA or MDIPA). The more significant structural factors (phase separation and morphology) accounting for the differences in tensile behaviour have already been described in detail, and *all* of the data summarized in Table 5.3 need to be considered in their entirety to explain rigorously the structure–property relationships in these complex RIM-copolymer networks.

5.5.3 *Fracture behaviour*

The determination of absolute fracture properties of RIM materials, with such wide-ranging mechanical properties as those discussed in the previous section, presents very difficult experimental and analytical problems [37, 86]. Values of stiffness ranging from 150 to ~1500 MPa, failure strains from ~20 to ~320% and tensile toughness (estimated from the areas under stress–strain curves) from ~6 to ~48 MJ m^{-3} are illustrative of the spectrum of materials behaviour, spanning that of soft, ductile elastomers to stiff, brittle resins.

The RIM materials of low hard-segment content may be treated as elastomers and their fracture behaviour determined experimentally and analytically according to the tearing criteria of Rivlin and Thomas [87]. Based on these criteria, tearing energies of elastomeric materials may be quantified in terms of the materials parameter G_{Ic}, the critical strain-energy release rate, which gives a direct measure of the resistance to crack propagation in an elastic material. The RIM materials of high hard-segment content are rigid and relatively brittle and their fracture behaviour is more amenable to analysis based on linear-elastic fracture-mechanics (LEFM) [88] to give a fracture-toughness parameter K_{Ic}, defined as the critical stress-intensity factor. Several experimental tests are reported [89] for the evaluation of the fracture parameters G_{Ic} and K_{Ic}, and one of the most convenient is that using single edge-notch tensile (SENT) specimens such as that shown schematically in Figure 5.21. The critical strain-energy release-rate for the tearing of soft RIM elastomers in the form of SENT specimens is given by the relationship

$$2kW_{\mathrm{c}}a = G_{\mathrm{Ic}} \tag{5.17}$$

where a is the initial crack length, W_{c} is the stored elastic-strain-energy density, which is released during fracture, and k is a strain-dependent function. An approximate expression for k has been derived [90] as $k = \pi/\lambda_{\mathrm{c}}^{1/2}$, where λ_{c} is the extension ratio of the SENT specimen at the onset of crack propagation. For the rigid RIM materials, the critical stress-intensity factor for brittle fracture is given by the relationship [88]

$$\sigma_{\mathrm{c}} Y a^{1/2} = K_{\mathrm{Ic}} \tag{5.18}$$

where σ_{c} is the maximum applied stress causing fracture and Y is the shape factor for the SENT specimen geometry, defined by the fourth-order

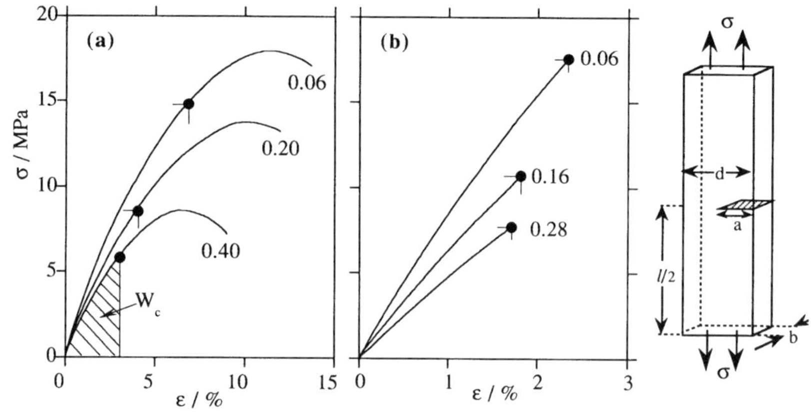

Figure 5.21 Representative single-edge notch (SEN) fracture data for trifunctional PUU-DETDA networks [37]. (a) Stress–strain (tearing) curves for ductile materials with low hard-segment contents (<50% w/w). (b) Stress–strain (elastic) curves for rigid materials with high hard-segment contents (>50% w/w). Typical SEN specimen used (right) has dimensions $l \approx 130\,\text{mm}$, $b \approx 3\,\text{mm}$, $d \approx 25\,\text{mm}$ and notch-to-depth ratios $0.05 \leq a/d \leq 0.60$. The numbers against the fracture curves are the values of a/d used.

polynomial expression [88]

$$Y = 1.99 - 0.41(a/d) + 18.70(a/d)^2 - 38.48(a/d)^3 + 53.85(a/d)^4 \quad (5.19)$$

Typical experimental stress–strain curves [37], obtained using SENT specimens with different initial crack lengths, are shown in Figures 5.21(a) and 5.21(b), respectively, for soft and rigid RIM materials. The data in Figure 5.21(a) illustrate clearly the curvature associated with the tearing of the elastomeric copolymers and how the values of W_c are obtained for analysis according to equation (5.17). In contrast, the plots in Figure 5.21(b) for the rigid copolymers are linear up to fracture. Data from such fracture curves were analysed according to equations (5.17) and (5.18) to give plots of $2kW_c$ versus a^{-1}, for soft RIM materials [37], and $\sigma_c Y^2$ versus a^{-1} for the rigid materials [75], as shown by the representative plots in Figures 5.22(a) and 5.22(b). Both analyses produce linear plots through the origin in accordance with theory and enable consistent values of the fracture parameters to be determined for each type of RIM copolymer. The variations of G_{Ic} and K_{Ic} with hard-segment contents for various PUU and PUr materials are shown in Figures 5.22(c) and 5.22(d), respectively. As expected, G_{Ic} shows monotonic decreases with increasing hard-segment content, with only small differences between the two types of copolymer. In the case of PUr materials, the effects of postcuring are only really evident at higher hard-segment contents (i.e. with the rigid, brittle materials). In contrast, the fracture toughness (K_{Ic}) shows a more complex dependence on hard-segment content. The apparent increases in fracture toughness are

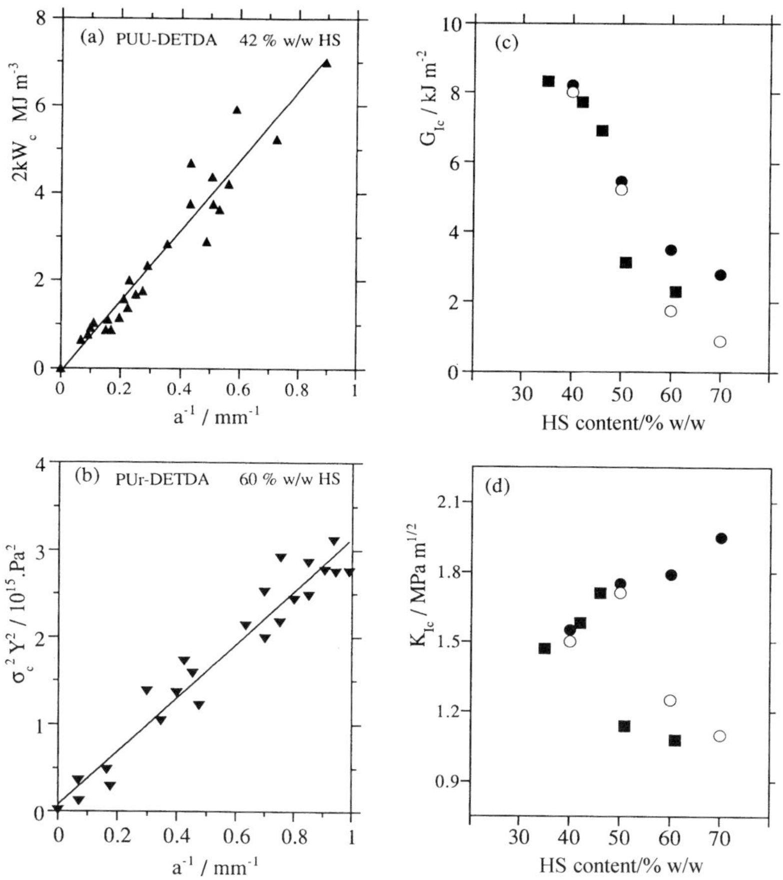

Figure 5.22 SEN fracture behaviour of trifunctional RIM-copolymer networks. (a) Ductile materials with low hard-segment content: typical tear-strength plot of stored energy at fracture ($2kW_c$) versus reciprocal crack length (a^{-1}) for the determination of G_{Ic}. (b) Rigid materials with high hard-segment content: typical LEFM plot of fracture stress ($\sigma_c^2 Y^2$) versus reciprocal crack length (a^{-1}) for the determination of K_{Ic}. (c) Variation of G_{Ic} with hard-segment content. (d) Variation of K_{Ic} with hard-segment content. (■, PUU-DETDA [37]; ○, PUr-DETDA (non-postcured) [75]; ●, PUr-DETDA (postcured) [75].)

explained in terms of the relationship between G_{Ic} and K_{Ic}, given by the fracture mechanics expression [88]

$$G_{Ic} = K_{Ic}^2/E \tag{5.20}$$

which defines the competing effects on K_{Ic} of simultaneous changes in materials stiffness and fracture energy. Comparison of the data in Figures 5.20(a) and 5.22(c) shows clearly that initially, with increasing hard-segment content, the increases in the values of E are greater than the corresponding

Table 5.4 Effect of hard-segment structure on the fracture properties of RIM copolymer networks containing ~50% w/w hard segments. (For definition of the copolymer systems see Table 5.2 and Figure 5.3.)

Copolymer network	E /MPa	σ_y [a] /MPa	G_{Ic} [b] /kJ m^{-2}	K_{Ic} [c] /MPa m$^{1/2}$	b_{min} [d] /mm
PU-EG	131	5 (ex)	3.0	6.3	35
PUU-DETDA	497	15 (ex)	3.1	1.2	17
PUU-MDIPA	483	15 (ex)	6.2	1.7	33
PUr-DETDA	565	16 (ex)	2.6	1.1	12
PUr-MDIPA	332	12 (in)	4.3	1.2	25

[a] σ_y is the tensile yield stress determined either from intrinsic (in) yield points or extrinsic (ex) yield points using Considere's construction [89].
[b] G_{Ic} is the critical strain-energy release rate.
[c] K_{Ic} is the stress intensity factor.
[d] b_{min} is the minimum thickness of SEN specimen calculated from equation (5.21).

decreases in values of G_{Ic}. However, at ~50% w/w hard-segment content, both the PUU-DETDA and non-postcured PUr-DETDA copolymers show a reversal in trend of K_{Ic}, due to the extremely poor fracture resistance of the more rigid materials outweighing their increased stiffness. The data in Figure 5.22(d) also show the advantages obtained by postcuring these non-equilibrium RIM copolymers as demonstrated by the K_{Ic} data for postcured PUr copolymers, which show significant enhancement in fracture toughness that increases progressively with hard-segment content.

Although the interpretation of the fracture behaviour of RIM materials using fracture mechanics enables internal comparisons to be made showing the trends in materials properties, the analyses represented by equations (5.17) and (5.18) are not strictly valid since they are derived on the assumption that the materials are deformed under conditions of plane strain. Plane-strain conditions are normally associated with thick specimens, and, despite the linearity of the plots in Figure 5.22, the values of G_{Ic} and K_{Ic} determined for the RIM materials are not strictly plane-strain values since the thicknesses of SENT specimens used to obtain fracture data, are less than the minimum thickness b_{min} required for conditions of plane strain. The minimum thickness criterion [88] for plane-strain conditions is given as

$$b > 2.5(K_{Ic}/\sigma_y)^2 \tag{5.21}$$

where σ_y is the yield stress of the material. Table 5.4 summarizes fracture properties of various PU, PUU and PUr materials containing ~50% w/w hard segments and gives calculated values of b_{min}. The thicknesses of SENT specimens used to obtain the fracture properties are determined by the thickness (3.5 mm) of moulded RIM plaques and are clearly less than the b_{min} values, so that the values of G_{Ic} and K_{Ic} determined for the various

Table 5.5 Effects of soft-segment functionality (f_n) on the fracture properties of RIM copolymers comprising 50% w/w of DETDA hard segments and polyether soft segments ($M_n \approx$ 6000 g mol^{-1}). (Properties as described in Table 5.4.)

Copolymer	f_n	E /MPa	σ_y /MPa	G_{Ic} /kJ m^{-2}	K_{Ic} /MPa m$^{1/2}$	b_{min} /mm
PUU (diol)	2	436	15 (in)	3.7	1.3	18
PUU (triol)	3	337	13 (ex)	3.5	1.1	17
PUU (tetrol)	4	296	12 (ex)	3.1	1.0	16
PUr (diamine)	2	389	15 (ex)	1.8	0.8	7.8
PUr (triamine)	3	496	16 (in)	2.6	1.3	12

copolymers exceed those expected from ideal plane-strain conditions. Also included for comparison, in Table 5.5, are tensile and fracture properties of RIM PUU and PUr copolymers, containing 50% w/w hard segments but formed using soft-segment prepolymers of different functionalities f_n, as defined in the chemical structures of Figure 5.3. The PUU materials are generally tougher than corresponding PUr materials. However, there appears to be little effect of soft-segment functionality on the fracture toughness of the PUU copolymers and only modest increases in PUr toughness, despite the differences in molar-mass development, phase-separation kinetics and resulting morphologies that are known to occur during RIM processing of the materials.

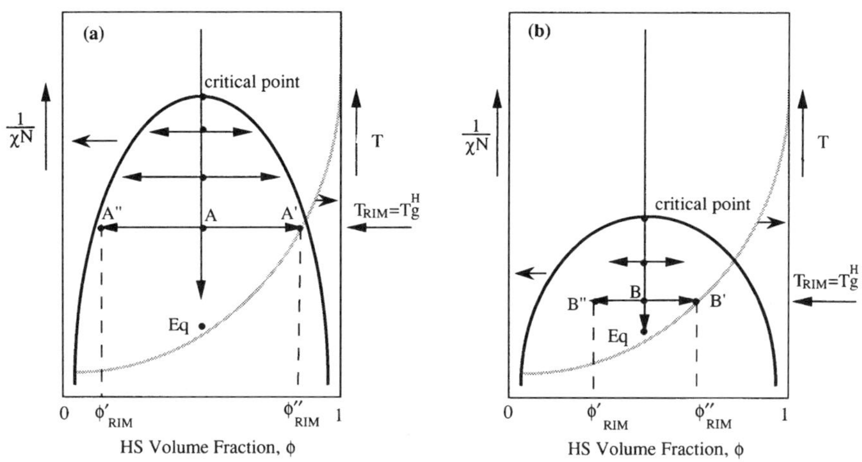

Figure 5.23 Schematic phase diagrams representing coexistence curves (solid) for (a) PUU systems and (b) PUr systems undergoing competitive polymerization and phase separation during RIM. Superimposed on each diagram is a schematic T_g versus composition curve (lighter). The vertical arrow through the critical point represents the extent of reaction of the system, and the horizontal double arrows represent the degrees of phase separation achieved as the polymerization proceeds.

5.5.4 *Summary*

The structure–property relationships discussed in the previous sections depend on the copolymer morphologies formed, as a result of the kinetic competition between copolymerization and microphase separation, as the solidifying system is driven rapidly through phase space during the RIM process. This kinetic competition is represented schematically on phase diagrams in Figures 5.23(a) and 5.23(b) for the RIM-PUU and the RIM-PUr systems, respectively. The diagrams are based on two simplifying assumptions, namely, that the coexistence curve is positioned symmetrically about $\phi = 0.5$ and that the process is isothermal. Thus, the ordinate is defined only in terms of the parameter $1/\chi N$, defined in Section 5.3. Superimposed on each phase diagram is a T_g versus composition curve assuming complete reaction. Under initially homogeneous reaction conditions, copolymerization proceeds rapidly, acting as a thermodynamic quench deep into the unstable region (represented by the vertical arrow through the critical point). Large values of N are quickly achieved, leading to the two-phase equilibrium position marked as 'Eq'. The phase-separation process (represented by the horizontal double arrows) lags behind the reaction and is ultimately arrested by the vitrification of the hard-segment-rich phase, which attains a composition with T_g^H equivalent to the in-mould reaction temperature T_{RIM}. This occurs at a position known as the Berghmans point [91] (shown as A' and B' in Figures 5.23(a) and 5.23(b) and no further phase separation is possible. Thus, the values ϕ'_{RIM} and ϕ''_{RIM} define the compositions of phase domains in the final RIM material. In the phase diagrams, the interphase region is considered to be infinitesimally small although, in real RIM materials, significant interphase regions are formed in which there are gradual changes in composition between the two dominant phases ϕ'_{RIM} and ϕ''_{RIM}.

The differences between the PUU and PUr systems in terms of the different rates of copolymerization, relative to those of phase separation, described previously in Section 5.5.1, account for relative positions of the coexistence curves in Figure 5.23. The higher critical point for the RIM-PUU system results in macrophase separation of nascent hard-segment sequences at a relatively early stage in the polymerization. Macrophase separation and copolymerization continue at similar rates until both processes are arrested by the onset of vitrification at the Berghmans point. Materials formed under these conditions exhibit coarse, highly phase-separated, co-continuous morphologies but possess little strength due to the lack of interphase connectivity. Consequently, organotin catalysts must be used to increase the rate of the hydroxy–isocyanate reaction, and to improve the demould toughness and ultimate mechanical properties of the material. In RIM-PUr, the intrinsically fast reaction of the soft-segment prepolymer ensures that structure development occurs in a manner similar to that of a block copolymer system undergoing microphase separation. The high degree of connectivity between the soft and hard segments decreases the driving force for phase separation, and the rapid development of M_w (and hence viscosity) impedes molecular mobility and reduces the rate of phase separation, such that the system is essentially single-phase until deep into the unstable region. Copolymerization in RIM-PUr systems proceeds rapidly, driving the system towards the two-phase equilibrium position (Eq). Microphase separation, however, lags well behind and if low values of T_{RIM} (i.e. low mould temperatures) are used, vitrification arrests the process at a relatively early stage to produce a highly phase-mixed, brittle material. Thus, significantly higher mould temperatures (100–120 °C), compared to RIM-PUU systems (~70 °C), are used which effectively raise the coexistence curve with respect to the T_g curve, allowing a greater degree of microphase separation to occur prior to vitrification.

5.6 Conclusions

Reactive polymer processes have been reviewed, particularly those defined as reactive extrusion (REX) and reaction injection moulding (RIM). Particular emphasis has been placed on RIM in terms of the correlations between processing, structure and properties of the materials produced. Specific examples of RIM-PU, PUU and PUr copolymer networks, formed from MDI and mixtures of functionalized polyether prepolymer (OH or NH_2 functional groups) and short-chain monomer (EG, DETDA or MDIPA), were used to illustrate the importance of both *in situ* (time-resolved) and *a posteriori* measurements for the determination of such correlations.

The transformation from reactive-liquid state to solid, multiphase material was shown to occur via spinodal decomposition, and the physical properties of the RIM materials were related to the kinetic competition between molar-mass development and network formation (copolymerization kinetics and statistics) and phase separation (thermodynamics). The rapid, bulk copolymerization in RIM was interpreted using phase diagrams to represent the transformation of a reactive mixture, in a thermodynamic stable state, to a blend of copolymer and homopolymers, in an unstable state, by changes in χ and dramatic increases in N.

Structure development in RIM systems does not proceed to equilibrium and the materials formed were shown to possess fine-scale co-continuous, random morphologies. Morphological pinning was shown to occur by network formation (chemical gelation) or vitrification (physical gelation), or both, and the interdomain spacing of the morphological structure was directly related to the molecular dimensions of the soft-segment polyether prepolymer.

The RIM copolymers were shown to exhibit a wide range of materials behaviour, spanning that of soft, ductile elastomers to stiff, brittle plastics.

Examples of thermal, tensile and fracture properties were related to RIM copolymer composition (the relative proportions of soft- and hard-segments), the degree of phase separation and to thermal history.

Acknowledgements

The author wishes to acknowledge the contributions to the studies on RIM made by all former research graduate students and postdoctoral workers. Particular thanks are extended to the collaborations and encouragement of Dr A.J. Ryan, Dr R.H. Still and Dr A.N. Wilkinson.

Glossary of symbols and abbreviations

a	crack length
b_{min}	specimen minimum thickness for plane-strain in fracture
d	interdomain spacing in a phase-separated polymer; specimen depth in fracture measurements
f, f_{OH}, f_{NH_2}	chemical functionality, hydroxyl functionality, amine functionality
f_n	number-average functionality
f_{pop}	volume fraction of POP chains in a copolymer
k	strain-dependent function in fracture
l	specimen length in fracture measurements
n	degree of polymerization
n_1, n_2	numbers of moles of solvent and polymer
n_H, n_S	number of moles of hard-segment and soft-segment blocks in a copolymer sample
p	extent of reaction (conversion)
$p_{g,c}$	conversion at chemical gel
$p_{g,v}$	conversion at physical gel
$p_{p,s}$	conversion at phase-separation
$\mathbf{q}$	scattering vector
q^*	magnitude of scattering vector at maximum intensity of scattered radiation
t	time
t_c	cure time
t_d	demould time
t_f	mould-fill time
$t_{g,c}$	chemical gel time
$t_{p,s}$	phase-separation time
$t_{g,v}$	physical (vitrification) gel time
w, w_{SS}	weight fraction, weight fraction of soft segments in a copolymer

ΔC_p^{S}, $\Delta C_p^{\mathrm{S}_0}$	heat-capacity changes associated with the glass–rubber transitions of the soft-segment phase in a copolymer and in pure soft-segment material
E	Young's modulus
E'	flexural storage modulus
G', G''	elastic shear-modulus, loss shear-modulus
G_1, G_2	shear moduli of components 1 and 2 in a composite material
G_{c}	shear modulus of a composite material
G_{d}	demould shear modulus
G_{Ic}	critical strain-energy release rate
$\Delta G_v^{\mathrm{int}}$	change of Gibbs energy of interaction per unit volume
ΔG_v	Gibbs energy change per unit volume
ΔH_v	heat of mixing per unit volume
$I(q)q^2$	Lorentz-corrected scattering intensity
K	ratio of rate constants
K_{Ic}	critical stress-intensity factor
M_{n}, M_{w}	number-average and weight-average molar masses
M_{w_0}	molar mass of initial reactant mixture
N	degree of polymerization; extruder-screw speed of rotation
N_2	number of segments in a polymer chain
N_{n}, N_{w}	number-average and weight-average hard-segment sequence lengths
N_{H}, N_{S}	degrees of polymerization of hard-segment and soft-segment blocks in a copolymer
ΔP	pressure drop along an extruder barrel
PSR	phase-separation ratio
Q	extruder volumetric throughput (flow rate)
Q'	relative scattering invariant
R	gas constant
R_{g}	molecular radius of gyration
$(R_{\mathrm{g}})_{\mathrm{POP}}$	R_{g} of a POP chain
ΔS_v	entropy of mixing per unit volume
T	temperature
T_{c}	cure temperature
T_{g}	glass–rubber transition temperature
$T_{\mathrm{g}}^{\mathrm{H}}$	hard-segment T_{g} in a copolymer
$T_{\mathrm{g}}^{\mathrm{H}_\infty}$	$T_{\mathrm{g}}^{\mathrm{H}}$ for hard segments of infinite molar mass
$T_{\mathrm{g}}^{\mathrm{S}}$	soft-segment T_{g} in a copolymer
$T_{\mathrm{g}}^{\mathrm{S}_0}$	soft-segment T_{g} in pure soft-segment material
T_{m}	melting temperature
T_0	reactant(s) temperature

T_{mould}, T_{RIM}, T_{w}	mould temperature
T_{reaction}	reaction temperature
U_{t}	tensile toughness
V	total volume
V_1, V_2, V_{r}	molar volumes of solvent and polymer, and average molar volume of the blocks in a block copolymer
W_{c}	stored elastic-strain-energy density
Y	specimen shape factor
α, β	extruder design parameters
δ_1, δ_2	solubility parameters of solvent and polymer
$\tan\delta$	loss tangent
ε	tensile strain
ε_{u}	ultimate tensile strain (failure strain)
λ	extension ratio
λ_{c}	value of λ at the onset of crack propagation
σ	tensile stress
σ_{c}	maximum applied stress causing fracture
σ_{u}	ultimate tensile stress (tensile strength)
σ_{y}	yield stress
ν_1, ν_2	Poisson's ratios of components 1 and 2 in a composite material
ν_{c}	(average) Poisson's ratio of a composite material
ϕ_1, ϕ_2, ϕ_{H}, ϕ_{S}	volume fractions of solvent, polymer, hard-segment blocks and soft-segment blocks
ϕ_1^{c}, ϕ_2^{c}	phase compositions of conjugate points on a binodal
ϕ_1^{s}, ϕ_2^{s}	phase compositions of conjugate points on a spinodal
ϕ'_{RIM}, ϕ''_{RIM}	compositions of phase domains in a material formed by RIM
χ_{12}	Flory–Huggins solvent–polymer interaction parameter
χ_{c}	critical value of χ_{12} for phase separation
χ_{HS}	Flory–Huggins interaction parameter for hard and soft segments
a.u.	arbitrary unit
A, B, C	composition (copolymer), chemical functional group
AB	diblock copolymer
ABA	triblock copolymer
$(\text{AB})_n$	multiblock copolymer
ATR	adiabatic temperature rise
BDO	1,4-butane diol
CM	compression moulding
CTT	cure-transformation-time diagram
DBTDL	dibutyltin dilaurate
DETDA	3,5-diethyltoluene diamine
DMTA	dynamic-mechanical-thermal analysis

DP	degree of polymerization
DSC	differential scanning calorimetry
EDO, EG	1,2-ethane diol, ethylene glycol
FTIR	Fourier transform infra red
HS	hard-segment (content)
IPN	interpenetrating polymer network
LEFM	linear-elastic fracture-mechanics
MAD	moulding-area diagram
MDI	4,4′-diphenylmethane diisocyanate
MDIPA	methylene-bis-2,6-diisopropylamine
MST	microphase separation transition
NCO	isocyanate
NH_2	amine
NG	nucleation and growth
OH	hydroxyl
PDCP	polydicyclopentadiene
PDO	1,5-pentane diol
POE	poly(oxyethylene)
POP	poly(oxypropylene)
PU	copolyurethane
PUU	copoly(urethane–urea)
PUr	copolyurea
PUrI	copoly(urea–isocyanurate)
RA_{fc}, $R'B_{fb}$, $R''C_{fc}$	reactants having f_a reactive A groups, f_b reactive B groups and f_c reactive C groups
REX	reactive extrusion
RIM	reaction injection moulding
RTM	resin-transfer moulding
SALS	small-angle light scattering
SANS	small-angle neutron scattering
SAXS	small-angle X-ray scattering
SD	spinodal decomposition
SEM	scanning electron microscopy
SENT	single edge-notch tensile (specimen)
SRIM	structural RIM (composites)
TEM	transmission electron microscopy

References

1. Xanthos, M. (ed.) (1992) *Reactive Extrusion Principles and Practice*, Hanser, Munich.
2. Lambla, M. (1997) in *Processing of Polymers* (ed. H.E.H. Meijer), VCH, Weinheim, Ch. 6, pp. 345–400.
3. Charrier, J.-M. (1990) *Polymeric Materials and Processing. Plastics, Elastomers and Composites*, Hanser, Munich, pp. 441–524.

4. Elias, H.-G. (1993) *An Introduction to Plastics*, VCH, Weinheim, pp. 212–238.
5. Becker, W.E. (ed.) (1979) *Reaction Injection Moulding*, Van Nostrand Reinhold, New York.
6. Kresta, J.E. (ed.) (1982) *Reaction Injection Moulding and Fast Polymerization Reactions*, Plenum, New York.
7. Macosko, C.W. (1989) *RIM: Fundamentals of Reaction Injection Moulding*, Hanser, Munich.
8. Ryan, A.J. and Stanford, J.L. (1989) in *Comprehensive Polymer Science*, Vol. 5, Step Polymerization (eds G.C. Eastmond, A. Ledwith, S. Russo and P. Sigwalt), Pergamon, Oxford, Ch. 25, pp. 427–454.
9. Stanford, J.L., Elwell, M.J.A. and Ryan, A.J. (1997) in *Processing of Polymers* (ed. H.E.H. Meijer), VCH, Weinheim, Ch. 9, pp. 465–512.
10. Ryan, A.J., Stanford, J.L. and Tao, X.Q. (1995) *Plastics, Rubbers and Composites Processing and Applications*, **23**, 151.
11. van der Loos, L.J.M. and van Geenen, A.A. (1985) in *Reaction Injection Moulding* (ed. J.E. Kresta), American Chemical Society, Washington, DC, Ch. 12, pp. 181–191.
12. Ishida, H. and Scott, C. (1986) *J. Polymer Eng.*, **6**, 201.
13. Klosiewicz, D.W. (1983), US Patent 4,400,340 (Hercules).
14. Geer, R.P. and Stoutland, R.D. (1985) *Society of Plastics Engineering and Technology Papers*, **30**, 1232–1235.
15. See Flory, P.J. (1953) *Principles of Polymer Chemistry*, Cornell University Press, New York, pp. 347–354.
16. Winter, H.H. and Chambon, F. (1986) *J. Rheol.*, **31**, 683.
17. Saunders, J.H. and Frisch, K.C. (1962) *Polyurethane Chemistry and Technology*, Wiley-Interscience, New York.
18. Wright, P. and Cummings, A.P.C. (1969) *Solid Polyurethane Elastomers*, MacLaren and Sons, London.
19. Lyman, D.J. (1966) *Reviews in Macromol. Chem.*, **1**, 191.
20. Reegan, S.L. and Frisch, K.C. (1971) *Advances in Urethane Science and Technology*, **1**, 1.
21. Stanford, J.L., Barksby, N., Dunn, D., Kaye, A. and Stepto, R.F.T. (1985) in *Reaction Injection Moulding* (ed. J.E. Kresta), American Chemical Society, Washington, DC, Ch. 6, pp. 83–96.
22. Ryan, A.J., Stanford, J.L. and Still, R.H. (1988) *Brit. Polymer J.*, **20**, 77.
23. Stanford, J.L., Wilkinson, A.N., Lee, D.-K. and Ryan, A.J. (1990) *Plastics and Rubber Processing and Applications*, **13**, 111.
24. Stanford, J.L., Still, R.H. and Wilkinson, A.N. (1995) *Polymer*, **36**, 3555.
25. Kirkland, C. (1986) *Plastics Technology*, **32**(3), 83.
26. Camargo, R.E., Macosko, C.W., Tirrell, M. and Wellinghoff, S.T. (1982) *Polymer Eng. Sci.*, **22**, 719.
27. Camargo, R.E., Macosko, C.W., Tirrell, M. and Wellinghoff, S.T. (1983) *Polymer Comm.*, **24**, 314.
28. Camargo, R.E., Macosko, C.W., Tirrell, M. and Wellinghoff, S.T. (1985) *Polymer*, **26**, 1145.
29. Camargo, R.E., Macosko, C.W., Tirrell, M. and Wellinghoff, S.T. (1981) in *Reaction Injection Moulding and Fast Polymerization Reactions* (ed. J.E. Kresta), Plenum, New York, pp. 95–118.
30. Yang, W.P. and Macosko, C.W. (1989) *Makromol. Chemie Macromol. Symposia*, **25**, 23.
31. Blake, J.W., Anderson, R.D., Yang, W.P. and Macosko, C.W. (1987) *J. Rheol.*, **31**, 1236.
32. Perry, S.J., Castro, J.M. and Macosko, C.W. (1985) *J. Rheol.*, **29**, 19.
33. Ryan, A.J., Willkomm, W.R., Bergstrom, T.B., Macosko, C.W., Koberstein, J.T., Yu, C.C. and Russell, T.P. (1991) *Macromolecules*, **24**, 2883.
34. Speckhard, T.A., Miller, J.A. and Cooper, S.L. (1986) *Macromolecules*, **19**, 1558.
35. Speckhard, T.A., Miller, J.A. and Cooper, S.L. (1986) *Macromolecules*, **19**, 1568.
36. Dominguez, R.J.G. (1981) *Polymer Eng. Sci.*, **21**, 18.
37. Ryan, A.J., Stanford, J.L. and Still, R.H. (1991) *Polymer*, **32**, 1426.
38. Ryan, A.J., Stanford, J.L. and Birch, A.J. (1993) *Polymer*, **34**, 4874.
39. Lopez-Serrano, F., Castro, J.M., Macosko, C.W. and Tirrell, M. (1980) *Polymer*, **21**, 263.
40. Pannone, M.C. and Macosko, C.W. (1988) *Polymer Eng. Sci.*, **28**, 660.
41. Wang, K.J., Huang, Y.J. and Lee, L.J. (1990) *Polymer Eng. Sci.*, **30**, 654.

42. Ryan, A.J., Stanford, J.L. and Still, R.H. (1990) *Plastics and Rubbers Processing and Applications*, **13**, 99.
43. Peebles, L.H. Jr. (1974) *Macromolecules*, **7**, 872.
44. Peebles, L.H. Jr. (1976) *Macromolecules*, **9**, 58.
45. See Flory, P.J. (1953) *Principles of Polymer Chemistry*, Cornell University Press, New York, pp. 495–540.
46. Binder, K. (1987) *Colloid Polymer Science*, **265**, 273.
47. Olabisi, O., Robeson, L.M. and Shaw, M.T. (1979) *Polymer–Polymer Miscibility*, Academic Press, New York, pp. 26–29.
48. Ryan, A.J., Stanford, J.L. and Tao, X.Q. (1993) *Polymer*, **34**, 4020.
49. Hildebrand, J.H. and Scott, R.L. (1950) *The Solubility of Non-Electrolytes*, 3rd edn, Reinhold, New York.
50. Liebler, L. (1980) *Macromolecules*, **13**, 1602.
51. Khandpur, A.K., Forster, S., Bates, S.F., Hamley, I.W., Ryan, A.J., Bras, W., Almdal, K. and Mortensen, K. (1995) *Macromolecules*, **28**, 8796.
52. Benoit, H. and Hadziiannou, G. (1988) *Macromolecules*, **21**, 1449.
53. Ryan, A.J., Stanford, J.L. and Still, R.H. (1988) *Polymer Comm.*, **29**, 196.
54. Schollenberger, C.S. (1979) in *Multiphase Polymers* (eds S.L. Cooper and G.M. Estes), American Chemical Society, Washington, DC.
55. Camberlin, Y. and Pascault, J.P. (1983) *J. Polymer Sci. Polym. Chem. Ed.*, **21**, 415.
56. Camberlin, Y. and Pascault, J.P. (1984) *J. Polymer Sci. Polym. Phys. Ed.*, **22**, 1835.
57. Ryan, A.J., Stanford, J.L. and Wilkinson, A.N. (1987) *Polymer Bulletin*, **18**, 517.
58. West, J.C. and Cooper, S.L. (1977) *J. Polymer Sci. Polym. Symposia*, **60**, 127.
59. Higgins, J.S. (1988) *Makromol. Chemie Macromol. Symposia*, **15**, 201.
60. Kostorz, G. (1988) *Makromol. Chemie Macromol. Symposia*, **15**, 131.
61. Bovey, F.A. (1988) *Makromol. Chemie Macromol. Symposia*, **20**, 105.
62. Sawyer, L.C. and Grubb, D. (1987) *Polymer Microscopy*, Chapman and Hall, London.
63. Bras, W., Derbyshire, G.E., Bogg, D., Cooke, J.S., Elwell, M.J.A., Komanschek, B.U., Naylor, S. and Ryan, A.J. (1995) *Science*, **267**, 996.
64. Wilkinson, A.N., Naylor, S., Elwell, M.J., Draper, P., Komanschek, B.U., Stanford, J.L. and Ryan, A.J. (1996) *Polymer Comm.*, **37**, 2021.
65. Vespoli, N.P. and Alberino, L.M. (1985) *Polymer Process. Eng.*, **3**, 127.
66. Cahn, J.W. and Hilliard, J. (1958) *J. Chem. Phys.*, **28**, 258.
67. Briber, R.M. and Thomas, E.L. (1983) *J. Macromol. Sci. (Physics)*, **22**, 509.
68. Blackwell, J. and Lee, C.D. (1984) *J. Polymer Sci. (Physics)*, **22**, 759.
69. Li, C., Goodman, S.L., Albrecht, R.M. and Cooper, S.L. (1988) *Macromolecules*, **21**, 2367.
70. Fridman, I.D., Thomas, E.L., Lee, L.J. and Macosko, C.W. (1980) *Polymer*, **21**, 393.
71. Kolodziej, P., Yang, W.P., Macosko, C.W. and Wellinghoff, S.T. (1986) *J. Polymer Sci. Part B (Physics)*, **24**, 2359.
72. Foks, J., Michler, G. and Nauman, I. (1987) *Polymer*, **28**, 2195.
73. Fridman, I.D. and Thomas, E.L. (1980) *Polymer*, **21**, 388.
74. Koberstein, J.T. and Stein, R.S. (1983) *J. Polymer Sci. Polym. Phys. Ed.*, **21**, 2181.
75. Birch, A.J. (1991) PhD Thesis, Victoria University of Manchester.
76. Turner, R.B., Spell, H.L. and Vanderhider, J.A. (1982) *Polymer Sci. Tech.*, **18**, 63.
77. Wilkinson, A.N. (1990) PhD Thesis, Victoria University of Manchester.
78. Flory, P.J. and Fox, P. (1950) *J. Appl. Phys.*, **21**, 581.
79. Bengston, B., Feger, C., MacKnight, W.J. and Schneider, N.S. (1985) *Polymer*, **26**, 895.
80. Nielsen, L.E. (1978) *Predicting the Property of Mixtures*, Marcel Dekker, New York.
81. Dickie, R.A. (1978) in *Polymer Blends* (ed. D.R. Paul), Academic Press, New York, Ch. 8.
82. Kerner, E.H. (1956) *Proc. Phys. Soc. Lond. (B)*, **69**, 808.
83. Budiansky, S. (1965) *J. Mech. Phys. Sol.*, **13**, 223.
84. Davies, W.E. (1971) *J. Phys. (D) Appl. Phys.*, **161**, 318, 1176, 1325.
85. Allen, G., Bowden, M.J., Todd, S.M., Blendell, D.J., Jeffs, G.M. and Davies, W.E. (1974) *Polymer*, **15**, 28.
86. Stanford, J.L., Still, R.H. and Wilkinson, A.N. (1996) *Polymer Int.*, **41**, 283.
87. Rivlin, R.S. and Thomas, A.G. (1953) *J. Polymer Sci.*, **10**, 291.
88. Williams, J.G. (1984) *Fracture Mechanics of Polymers*, Ellis Horwood, Chichester.

89. Kinloch, A.J. and Young, R.J. (1985) *Fracture Behaviour of Polymers*, Applied Science, London.
90. Lake, G.J. (1970) *Proceedings of Yield, Deformation and Fracture in Polymers*, Institute of Physics, London, p. 531.
91. Callister, S., Keller, A. and Hikmet, R.M. (1990) *Makromol. Chemie Macromol. Symposia*, **39**, 19.

6 Interpenetrating polymer network (IPN) materials

H.L. FRISCH, Y. DU and M. SCHULZ

6.1 Introduction

Interpenetrating polymer networks (IPNs) are a type of polymer blend consisting of two or more network polymers [1–6]. They are more or less intimate mixtures of two or more distinct polymer networks held together by permanent entanglements with only accidental covalent bonds between the polymers, i.e. they are catenated polymers [7–9]. The entanglements in IPNs must be of a permanent nature and are made by homo-network formation of the two or more polymers.

There are many different ways to synthesize IPNs. Brief descriptions of some important IPN materials and their formation are summarized as follows [5, 10]:

1. *Simultaneous interpenetrating network (SIN)*: The monomers or pre-polymers, crosslinkers, initiators and sometimes solvent are mixed in liquid form. The non-interfering crosslinking and polymerization reactions are carried out simultaneously.
2. *Sequential IPN (seqIPN)*: Polymer network I is made. The reactants to form network II, for example, monomer II, crosslinker and initiator, are swollen into network I and reacted *in situ*. In general, the sequential IPNs include many possible materials where the synthesis of one network occurs after the initial synthesis of the other.
3. *Gradient IPN (grad.IPN)*: Gradient IPNs can be prepared by the sequential method, by starting to swell a host network with the reactants forming the guest network and stopping at a certain stage before equilibrium is reached. Polymerization is then carried out to produce the guest network, whose concentration in the host network has a gradient resulting from diffusion related to the swelling process. Gradient IPNs can be treated as a sequence of an infinite number of layers of IPNs, whose compositions and properties vary gradually through a sample [11]. For example, a film can be made with network I predominantly on one surface and network II on the other surface.
4. *Latex IPN*: Latex IPNs are prepared by seeded emulsion polymerization (multistage emulsion polymerization) [5, 12]. The preparation involves creation of a seed latex of network I. Reactants for a second network are then introduced into the system and the second

network formed. No fresh emulsifier is added during the second-stage polymerization. Therefore, the growth of network II occurs on the established latex particles forming a core–shell structure. A variation is to mix two different latexes and then form a film which crosslinks both polymers; this variation is sometimes called an interpenetrating elastomer network (IEN) [13–17].

5. *Thermoplastic IPN*: Thermoplastic IPN materials involve physical rather than chemical junction points. These materials flow at elevated temperature, and at in-use temperature they are networks and behave like IPNs. Types of junction points include glassy domains of block copolymers, ionic clusters in ionomers, or the crystalline region of semi-crystalline polymers.
6. *Semi-IPN (pseudo-IPN, PDIPN)*: Semi-IPNs are polymer-blend materials in which one or more polymers are networks and one or more polymers are linear or branched. These materials can be made either sequentially or simultaneously.

IPNs possess several interesting characteristics in comparison to normal polymer blends. The formation of IPNs is the only way of intimately combining network polymers, the resulting mixture exhibiting only limited phase separation. Due to the usual thermodynamic incompatibility of polymers, normal polymer blends have a multiphase morphology. However, if mixing is accomplished simultaneously with network formation, phase separation may be kinetically controlled by the permanent interlocking of entangled chains [18, 19].

6.2 Theoretical description of IPNs

The theoretical description of IPNs includes two parts. The first part is the investigation of IPN formation as a result of the combination of chemical network growth and diffusion controlled dynamics of the monomers and pre-gelled network units. Especially in the case of a repulsive interaction between the network components, the final structure of the network is determined by the competition between the irreversible chemical reactions and the reversible – but progressively frozen – phase separation. The second part is the description of the equilibrium properties of an IPN of a given topology. Although the junction points and entanglements of the network chains are irreversibly fixed, the compositional structure depends strongly on the temperature or, equivalently, on the Flory–Huggins interaction parameter $\chi(T)$. Note that a blend of polymer chains undergoes phase separation if this χ parameter becomes large enough. In comparison, the topological restrictions of an IPN (permanent mutual entanglements as a result of the irreversibly formed junction points) prevent the total separation of phases. This behaviour is similar to that of a melt of block copolymers [20]. Only microphase domains (or microphases), of the order of the mesh size of the network, are created. Furthermore, the quenched stochastic disorder of the topology leads to an additional disturbance of the microphase separation, so that a regular order of the microphases is completely suppressed or at least limited to short-range distances.

Since the theoretical description of the irreversible formation process of an IPN is much more complicated than the analysis of the equilibrium properties, the latter should first be discussed. Furthermore, in both cases only simultaneously synthesized IPNs are investigated. The extension to other types (sequential IPNs and pseudo-IPNs) is possible by using similar theoretical methods and numerical algorithms [21, 22].

A first theory of the equilibrium and phase stability based on phenomenological arguments [21] consists of the construction of a dimensionless effective free-energy density, F, with only mean-field contributions, for a chemically quenched IPN. F is a free-energy density divided by kT. The quenched character of the topological effects (topological structure of the subnetworks and their mutual entanglements) are represented by effective parameters. The IPN is in a stable, homogeneous phase if the interaction between the subnetworks is attractive or weakly repulsive, i.e. the components are compatible. Thus, only the elastic free-energy density of the subnetworks, F_{sub}, and the interaction free-energy density, F_{int}, must be considered for an understanding of the stability regime. In the limit of indefinitely high molar mass between entanglements, the contributions of the mutual entanglements of the subnetworks remains effective outside the stability region because topological restrictions prevent the separation of the subnetworks. Hence, the free-energy density is defined as

$$F = F_{sub} + F_{int} \tag{6.1}$$

In general, the thermodynamic ground state of this phenomenological *Ansatz* is an ideal homogeneous IPN, where each component I $(I = \mathrm{A}, \mathrm{B}, \ldots)$ occupies an average volume fraction f_I, subject to $\sum_I f_I = 1$. The thermodynamic observables of this model are the local volume fractions $\varphi_I(\boldsymbol{r}, t)$ with $\langle \varphi_I(\boldsymbol{r}, t) \rangle = f_I$. For simplicity, only two component IPNs have been studied. Thus, the phenomenological theory is based on one independent thermodynamic observable $\varphi = \varphi_\mathrm{A}$ $(f = f_\mathrm{A})$, because $\varphi_\mathrm{B} = 1 - \varphi$ $(f_\mathrm{B} = 1 - f)$. The free-energy density F_{sub} is a superposition of the elastic free energy densities of the underlying subnetworks, i.e. each summand of F_{sub} is determined by the topological structure of the corresponding subnetwork alone. The free energy of such a network can be expressed in terms of the local deformation tensor [23]. The pre-average over space leads to an isotropic mean-field expression, which depends on the isotropic part of the deformation (i.e. the local volume fraction) only. Hence

$$F_{\text{sub}} = F_{\text{A}}\left(\frac{\varphi}{f}\right) + F_{\text{B}}\left(\frac{1-\varphi}{1-f}\right) \tag{6.2}$$

The effective interaction free-energy between both components is, in the lowest order approximation, a simple linear form in φ_{A} and φ_{B}, which is proportional to the probability of a direct contact between both components, i.e.

$$F_{\text{int}} = \chi\varphi_{\text{A}}\varphi_{\text{B}} = \chi\varphi(1-\varphi) \tag{6.3}$$

The proportionality constant contains the well-known temperature-dependent Flory–Huggins parameter, which is a measure of the effective repulsion ($\chi > 0$) or attraction ($\chi < 0$) between the A and B units.

The free-energy density of an IPN depends on the local volume fraction. Thus, the minimum of F (determined by $\partial F/\partial\varphi = 0$) corresponds to the realized thermodynamic state. Furthermore, this state is stable if $\partial^2 F/\partial\varphi^2 > 0$ and unstable if $\partial^2 F/\partial\varphi^2 < 0$. Thus, the stability limit (spinodal curve) is defined by

$$\partial^2 F/\partial\varphi^2 = 0 \tag{6.4}$$

This equation allows the determination of the Flory–Huggins parameter, χ_{s}, above which the IPN becomes unstable and undergoes spinodal decomposition. χ_{s} depends on the average volume fraction, f, and effective parameters (which are considered in the functional forms of F_{A} and F_{B}). The spinodal curve, $\chi_{\text{s}}(f)$, forms as the plot of f versus χ, the stability border and simultaneously the validity border of the preceding phenomenological theory. If the inequality $\chi < \chi_{\text{s}}(f)$ is fulfilled, thermodynamic fluctuations do not change the homogeneity of the whole IPN. In other words, each thermodynamic fluctuation of the local composition is unstable and decays. However, outside the stable region, i.e. $\chi < \chi_{\text{s}}(f)$, a small thermodynamic fluctuation increases rapidly and the homogeneity of an IPN vanishes – the IPN shows the onset of a spinodal decomposition.

The extension of the phenomenological *Ansatz* is possible through the consideration of thermodynamic composition fluctuations. The approach is the introduction of an additional gradient term $F_{\text{grad}} = a(\nabla\varphi)^2/f(1-f)$ into the expression for the free-energy density. The total free-energy of the whole IPN, $\Gamma = \int F\,\mathrm{d}V$, is expressed now in terms of the corresponding fluctuation modes $\delta\varphi(\boldsymbol{q}, t)$. These are the Fourier-transformed deviations of the local volume fraction $\varphi(\boldsymbol{r}, t)$ from its average f, i.e. $\delta\varphi(\boldsymbol{r}, t) = \varphi(\boldsymbol{r}, t) - f$. One obtains in the lowest non-vanishing order of a Taylor expansion with respect to the composition–fluctuation modes:

$$\Gamma = \sum_{\boldsymbol{q}}\left[\chi_{\text{s}}(f) - \chi + \frac{aq^2}{f(1-f)}\right]|\delta\varphi(\boldsymbol{q}, t)|^2 = \sum_{\boldsymbol{q}} S^{-1}(q)|\delta\varphi(\boldsymbol{q}, t)|^2 \tag{6.5}$$

Note that the expression in square brackets can be interpreted also as the inverse static structure factor. The formula gives a good approximation to the free energy of an IPN in the homogeneous regime. For $\chi \rightarrow \chi_s(f)$ it follows that $S(q = 0) \rightarrow \infty$, i.e. the long-range fluctuations become divergent. This change is characteristic of phase separation into macrophase domains (or macrophases), but is in strong disagreement with the behaviour of IPNs. The cause of this disagreement is the mutual subnetwork entanglements, which are neglected in the theory. These entanglements prevent the complete phase separation of both subnetworks and, therefore, they suppress the long-range composition fluctuations. Entanglements can be taken into account by a second extension of the expression for the free energy, with a phenomenological term proportional to q^{-2} [24, 25]:

$$\Gamma = \sum_{\boldsymbol{q}} \left[\frac{C}{q^2} + (\chi_s(f) - \chi) + \frac{aq^2}{f(1-f)} \right] |\delta\varphi(\boldsymbol{q}, t)|^2 \tag{6.6}$$

The empirical constant C contains the effective length of the entangled network chains. However, one finds now, $S(q = 0) \rightarrow 0$ and a maximum of the static structure factor at $q^2 = \sqrt{C/a}$. This value of q corresponds to a characteristic length, λ of the microphases, of the order of magnitude $\lambda \approx (a/C)^{1/4}$. Furthermore, the static structure factor becomes divergent for $\chi = \chi'_s(f) = \chi_s(f) + 2\sqrt{aC}$, i.e. the spinodal curve is changed as a result of the mutual entanglement of the subnetworks. This behaviour is much closer to that of a real IPN near the spinodal curve than that of the first model (equation (6.5)).

Clearly, the preceding descriptions are still phenomenological theories. A more molecular theory is based on the random phase approximation [26], corresponding to a similar approximation for block copolymers. The central idea is the expansion of the thermodynamic, grand potential Ξ in powers of the average composition–fluctuation field, $\bar{\varphi}(\boldsymbol{r})$. (Note that this field is a constant, i.e. $\bar{\varphi}(\boldsymbol{r}) = f$, in the homogeneous regime and is a spatially dependent quantity in the phase-separated regime):

$$\frac{\Xi}{k_B T} = \sum_{n=2}^{\infty} \frac{1}{n!} \Xi_n(\boldsymbol{r}_1, \boldsymbol{r}_2, \ldots, \boldsymbol{r}_n) \bar{\varphi}(\boldsymbol{r}_1) \bar{\varphi}(\boldsymbol{r}_2) \ldots \bar{\varphi}(\boldsymbol{r}_n) \tag{6.7}$$

The expansion coefficients Ξ_n are now related to the correlation functions of the composition fluctuations $\langle \varphi(\boldsymbol{r}) \varphi(\boldsymbol{r}') \rangle$, $\langle \varphi(\boldsymbol{r}) \varphi(\boldsymbol{r}') \varphi(\boldsymbol{r}'') \rangle, \ldots,$ in the equilibrium state. These correlations can be determined for the case of an IPN without interactions between the two components. The short-range contributions to the correlation functions are determined mainly by the structures of the network chains, whereas the long-range contributions are related to the IPN topology and the mutual entanglement of the subnetworks. This non-interacting IPN can now be used as a reference model for a perturbation theory with respect to the Flory–Huggins parameter χ, allowing the determination of the behaviour of an IPN inside and outside the spinodal curve. Whereas the result is similar to that of the phenomenological theory for $\chi < \chi_s$, leading to a homogeneous stable phase, one obtains now an ordered microphase separated structure for $\chi > \chi_s$, instead of an unstable region. Clearly, the ordered structure is not very long-ranged (the long-range order is destroyed by the quenched, statistically distributed, topological restrictions such as junction points and entanglements). However, the direct environment of a microphase should show an approximately ordered structure of the neighbouring phases, similar to the short-range order of liquids. This behaviour of the inhomogeneous phase domains can also be obtained from numerical Monte-Carlo (MC) simulations [27].

It is well-known that only some special IPNs are homogeneous, i.e. they interpenetrate on the molecular scale of the basic monomers. In general, one finds a separation into regions having an enrichment of one component over the others. It can be shown that such IPNs do not have a homogeneous structure for any value of χ. Moreover, inhomogeneities can occur during formation, i.e. the structure of an IPN is related to the history of the chemical synthesis procedure. This internal separation begins during the growth process of the network clusters and will be frustrated in the completed IPN by topological restrictions. As a result, one obtains microphase-like structures on the nanometer scale. This effect should not be confused with the reversible microphase separation (spinodal decomposition) discussed previously, which occurs in a completed (i.e. chemically quenched) IPN.

In contrast to the described phenomena, the inhomogeneities of nearly all IPNs are the results of irreversible thermodynamic processes. In general, IPNs undergo both nucleation and growth [28] and spinodal decomposition during the syntheses of both subnetworks, with spinodal decomposition often predominating [29]. A possible description of this decomposition can be given by a generalization of the Cahn–Hilliard (mean-field) theory [30], which describes an exponential, time-dependent increase [31] of unstable thermodynamic fluctuations leading to microphase-like structures. The subsequent regime of decomposition can be understood in terms of nucleation and growth [28, 32], with a time-dependent evolution of the phase domains governed by a power law [33]. Although the formation of an IPN is an irreversible process, one can assume that the deviation from equilibrium is not too strong, so that one can work in the linear-response regime. This assumption allows the formulation of a mean-field approximation of the microphase separation in an IPN during the formation of both subnetworks. The separation stops after a finite time and the separated microphases reach an order of magnitude of size of, at most, a few nanometers [34]. Hence, one can assume that phase separation is effected only up to a time t^*, up to when structural motions of the forming subnetworks with a change of the local-order parameter are possible, whereas for $t > t^*$, further separation is prevented by the irreversibly formed junction points of the two subnetworks.

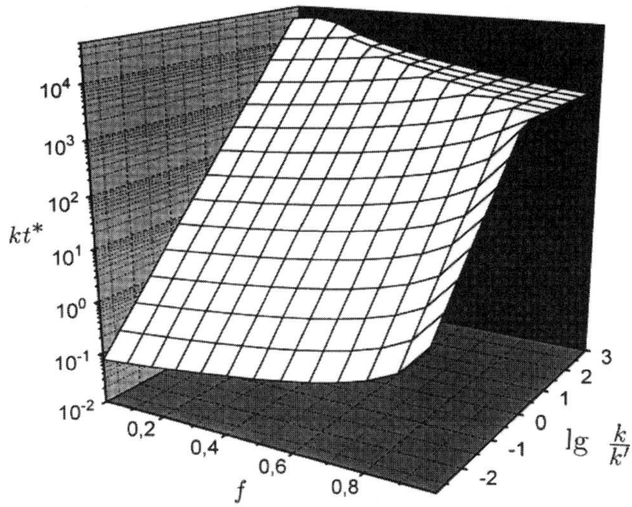

Figure 6.1 Formation of an IPN by the simultaneous polymerizations of tetrafunctional (reactant A, reaction-rate constant k) and trifunctional (reactant B, reaction-rate constant k'). kt^* versus composition, $f(=f_A)$, and the ratio of the reaction rate constants k/k'.

Actually, after network formation is finished one can observe an additional phase separation (i.e. increase in size of the phase domains), but this is a reversible effect (like the microphase separation in a homogeneous IPN) based only on deformations of the mutually entangled network chains of both components [26, 27], which vanish in the high-temperature limit ($\chi \to 0$). Note that the irreversibly separated parts of an IPN are topologically fixed by the interpenetrating network structure and exist also in this limit. The simplest, but mostly sufficient, assumption for a determination of t^* is that t^* is equivalent to the critical time for the formation of an infinite cluster of one or both components [22]. Figure 6.1 shows the time t^* for the formation of an IPN, created by simultaneous polymerization of tetrafunctional (component A, reaction-rate constant k) and trifunctional (component B, reaction-rate constant k') reactants, as predicted by an MC simulation [22]. t^* depends strongly on the composition f and the ratio of the two reaction rate constants k/k'.

The theoretical description of the formation of SIN can be discussed by starting from the dimensionless free-energy density of a polymer mixture [36, 37] of the two components A and B,

$$F(\varphi) = \frac{\varphi}{N_A}\log\varphi + \frac{1-\varphi}{N_B}\log(1-\varphi) + \chi\varphi(1-\varphi) + \frac{a}{f(1-f)}|\nabla\varphi|^2 \quad (6.8)$$

with χ and φ as defined previously and N_A and N_B, respectively, the average number of units of the A and B components. Simultaneously with phase separation starting to occur in the initial mixture, the system changes its equilibrium state by the irreversible polymerization process. If the evolution is sufficiently slow, we can describe the time-dependent structure by a quadratic form with respect to the deviation $\delta\varphi$ (order parameter) from the initial state $\delta\varphi = \varphi - f$:

$$F(\varphi) = F(f) + (\chi_s(f) - \chi)\delta\varphi^2 + \frac{a}{f(1-f)}|\nabla\delta\varphi|^2 \tag{6.9}$$

The time evolution can now be described in a linearization approximation. Note that a Fourier representation is used:

$$\dot{\psi}_q = -\lambda_q\left((\chi_s(f) - \chi) + \frac{aq^2}{f(1-f)}\right)\psi_q \tag{6.10}$$

with a general q-dependent kinetic coefficient λ_q (Onsager coefficient), where $\chi_s(f)$ is the well known spinodal curve, defined now as

$$\chi_s(f) = \frac{1}{2}\left(\frac{1}{N_A f} + \frac{1}{N_B(1-f)}\right) \tag{6.11}$$

Thus, a reasonable kinetic evolution equation is available for the beginning of the phase separation of a system of two components. The main problem in the description of the irreversible phase separation of IPNs is the fact that the parameters of equation (6.10), namely λ_q, N_A and N_B, become time dependent and can no longer be treated as constants as in the case of simple phase transitions. The average sizes of the relevant entities (given by N_A and N_B) continually increase in the course of network formation. Hence, we expect a continuous decrease of the spinodal curve value $\chi_s(f, t)$ for a given mole fraction f. The effective time dependence of N_A or N_B can be determined by the solution of the underlying chemical evolution equations of the IPN formation. Using the time dependence of N_A and N_B one obtains a simple representation of the time-dependent spinodal curve $\chi_s(f, t)$. In the case that at the initial time $t = 0$ the value of χ is less than the spinodal value for the initial mixture ($\chi < \chi_s(f)$), phase separation occurs first if χ_s becomes small enough during the network formation, so that at the time t_{sep} (determined by $\chi = \chi_s(f, t_{sep})$) one can observe the onset of the decomposition into two phases. Note that spinodal decomposition occurs only in the time interval $[t_{sep}, t^*]$ and creates, together with the irreversible chemical reactions, the fixed inhomogeneities of the IPN. Moreover, if $t_{sep} \geq t^*$ no phase separation occurs and the IPN formed is a completely homogeneous structure.

In general, the formation of an IPN is a very versatile process, but which can (at least initially) be simulated by numerical Monte-Carlo methods [22]. As an example, Figure 6.2 shows the mean composition fluctuations of the IPN system created with reaction rate constants $k = k' = 10^{-3}$, as a function of the mole fraction f and the reduced repulsion energy ε ($U(r) = \varepsilon k_B T$ for $r \leq R_0$ and $U(r) = 0$ for $r > R_0$, with the repulsion

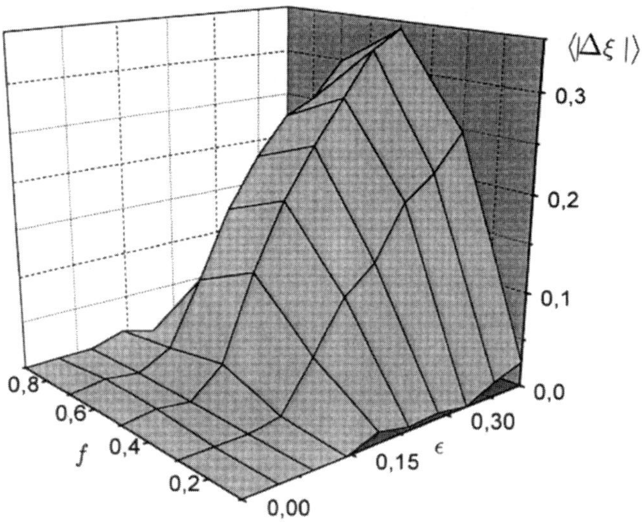

Figure 6.2 Irreversible composition fluctuations as a function of mole fraction f and the effective repulsion energy ε for $k = k' = 10^{-3}$.

radius R_0 of the order of three monomer diameters. (Note that the Flory–Huggins parameter is a function of ε.) Later regimes of the phase decomposition (like nucleation and growth), which lead to characteristic coarsening of the microphases, can be included by a simple extension of the present ideas and a numerical simulation of these effects should be possible for longer computation times.

6.3 Properties of IPN materials

The term 'interpenetrating polymer networks' was proposed by Millar [38] in 1960. Early work by Shibayama and Suzuki [39], Lipatov and Sergeeva [40], Frisch *et al.* [41], and Sperling and Friedman [42] set the basis for modern development of IPNs. Until 1979, the total literature on IPNs consisted of about 125 papers and 75 patents [5]. The present annual production of related papers and patents is more than these numbers. IPNs have many applications, both proposed and in practice. Some commercial materials are shown in Table 6.1.

In general the applications of an IPN material will depend on the glass-transition temperatures, phase continuity, mixing of the two networks, and the sizes of the domains of the component networks. The mechanical properties include stress–strain, lap-shear, peel strength, hardness, thermal degradation etc. Here, we will give a picture of IPN properties compared with polymer blends or homopolymers by reviewing some different IPN systems.

Table 6.1 Commercial IPN materials [5]

Manufacturer	Trade name	Composition[a]	Application
Shell Chemical	Kraton IPN	PSEBS/polyester	automotive parts
Petrarch Systems	Rimplast	silicone rubber/PU	gears, medical
ICI Americas	ITP	PU/(polyester/PS)	sheet moulding
DSM N.V.	Kelbuton	PP/EP, rubber/PP	automotive parts
Shell Research B.V.		rubber/PP	tough plastics
Reichold Chemical	TPR	PEPDM/PP	automotive parts, wire and cable
Rohm & Haas		anionic/cationic	ion exchange resins
Monstanto	Santoprene	PEPDM/PP	tyres, hoses, belts and gaskets
Du Pont	Somel	PEPDM/PP	outdoor weathering
BFGoodrich	Telcar	PEPDM/PP, PE	tubing, liners, wire and cable
Exxon	Vistaon	PEPDM/PP	paintable autoparts
Freeman Chemical	Acpol	acrylic/urethane/PS	sheet moulding
Dentsply	Trubyte		
International	Bioform	acrylic-based	artificial teeth
Hitachi Chemical		vinyl/phenolics	damping composites

[a] PSEBS = poly(styrene-*block*-ethylene/butylene-*block*-styrene); PU = polyurethane; PP = polypropylene; PEP = poly(ethylene-*co*-propylene); PEPDM = poly(ethylene-*co*-propylene-*co*-diene); PE = polyethylene; PS = polystyrene.

6.3.1 *Simultaneous IPNs (SINs)*

Two series of SINs were investigated by Frisch, Zhou and co-workers [43, 79]. In the first series (G), one network was made of poly(2,6-dimethyl-1,4-phenylene oxide) (PPO) and the other network was composed of either polystyrene (PS) [44], poly(methyl methacrylate) (PMMA) [45], polybutadiene (PB) [46], polyurethane/polyacrylate (PUA) [47], or poly(dimethylsiloxane) (PDMS) [48] chains. In the second series (R), one network was a polycarbonate/polyurethane (PCU) and the other one was either PMMA [49], PB [50], poly(4-vinylpyridine) (PVP) [51], or PS [52]. The PCU and PPO were converted into networks by reaction with a biuret triisocyanate and by bromination followed by condensation with a diamine, respectively, which did not interfere with the free-radical reactions used to make all the other networks.

The phase morphologies of the two series of IPNs are summarized and compared with corresponding blends in Table 6.2. All the linear blends in Table 6.2 are immiscible except for the PPO/PS blend. Five of the SINs made from completely immiscible blend components (PPO/PMMA, PPO/PUA, PPO/PB, PCU/PMMA and PCU/PB) are completely miscible over the whole range of composition, while two (PCU/PS and PCU/PVP) are miscible in the upper half range (60–100%) of PCU composition. Only the PPO/PDMS system is microphase-separated, essentially at all compositions with domain sizes of 0.01–0.2 μm [53]. The behaviours correlate well with the differences in solubility parameters [54] and glass-transition temperatures [55] of the linear chains that compose the networks (shown in Table 6.3). Figure 6.3 shows transmission electron micrographs of the SINs, linear-blend and crosslinked components of PCU and PVP.

Table 6.2 Phase morphologies of the G and R series of IPNs and those of the corresponding blends of linear polymers [79]

IPN[a]	Composition range /wt%	Number of distinct T_g values	y_c/wt%[b]	Number of phases in IPN	Number of phases in blends
Series G					
PPO/PS	whole	1	75(PPO)	1	1
PPO/PMMA	whole	1	80(PPO)	1	2
PPO/PUA	whole	1	80(PPO)	1	2
PPO/PB	whole	1	90(PPO)	2	2
PPO/PDMS	whole	2	90(PPO)	2	2
Series R					
PCU/PMMA	whole	1	30(PCU)	1	2
PCU/PB	whole	1	50(PCU)	1	2
PCU/PS	100-60 PCU	1	30(PCU)	1	2
	≤50 PCU or less	2		2	2
PCU/PVP	100-60 PCU	1	70(PCU)	1	2
	≤50 PCU or less	2		2	2

[a] PPO = poly(2,6-dimethyl-1,4-phenylene oxide); PS = polystyrene; PMMA = poly(methyl methacrylate); PUA = polyurethane/polyacrylate; PB = polybutadiene; PDMS = poly(dimethylsiloxane); PCU = polycarbonate/polyurethane; PVP = poly(4-vinyl pyridine).
[b] y_c is the wt% at which a maximum tensile strength to break was observed.

Frisch *et al.* [10] have studied the SINs polyurethane (PU)/acrylic copolymer (AC) containing *n*-butyl methacrylate, ethyl methacrylate and styrene [56]. These SINs exhibit a homogeneous morphology and single glass transition temperatures. The tensile strength and breaking elongations of the IPNs, pseudo-IPNs (PDIPNs) and blends are given in Tables 6.4 and 6.5, respectively. The IPNs show significant enhancements in tensile strength, especially at PU/AC = 80:20. The enhancement is most likely due to the high level of mixing and greater phase-adhesion of the permanent entanglements at the phase boundaries. The elongations show a slight increase from 100% to

Table 6.3 Solubility parameters of the linear polymers of the IPNs in Table 6.2 [79]

Polymer	Solubility parameter $\delta/(\mathrm{cal\,cm^{-3}})^{1/2}$ δ	T_g/°C
PPO		211
PVP		142
PMMA	9.1–12.8	105
PS	8.5–9.3	100
PUA	9.1[a]	
PCU	8.8[a]	−40
PB	8.1–8.6	−95
PDMS	7.3–7.6	−127

[a] δ computed from group contributions [78].

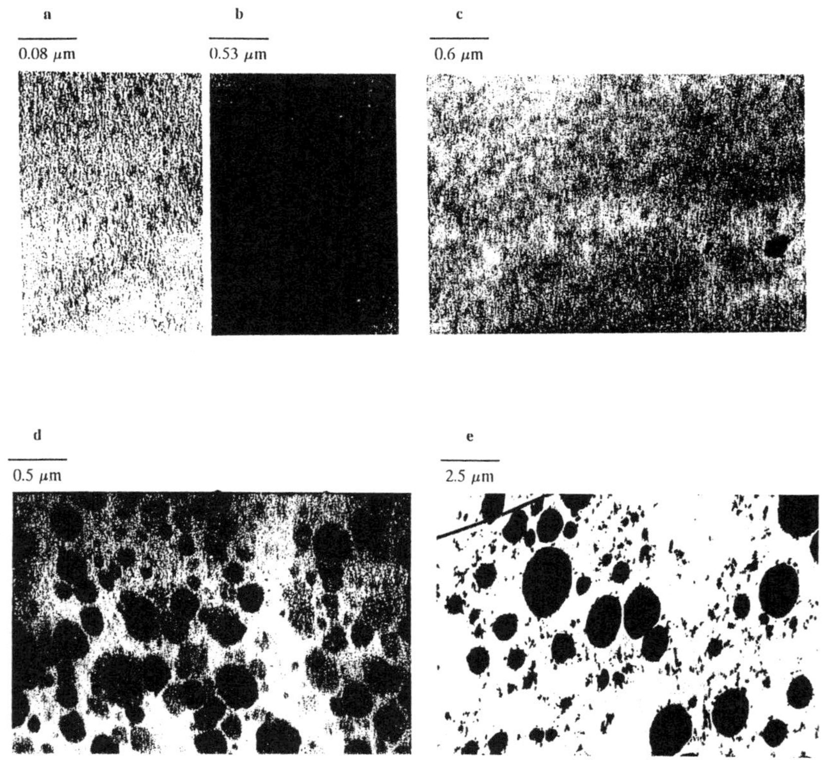

Figure 6.3 Transmission electron micrographs of the SINs, linear blends, and pure network components of PCU and PVP as specified in Table 6.2: (a) pure network PVP (×221 000); (b) pure network PCU (×38 000); (c) SIN of PCU/PVP (70 wt % PCU, ×31 200); (d) SIN of PCU/PVP (50 wt % PCU, ×40 000); (e) linear blend of PCU/PVP (50 wt %, ×7000) [79].

Table 6.4 Tensile strength (/10^4 Pa) SINs, PDIPNs and linear blends of polyurethane (PU) and acrylic copolymer (AC) [10]

PU/%	100	80	60	40	20	0
AC/%	0	20	40	60	80	100
SIN	6100	7100	6525	5000	3715	2571
PDIPN-1[a]	6100	5875	6300	3127	2227	1613
PDIPN-2[b]	5600	4428	5215	3602	3018	2571
Linear blend	5600	1828	3617	3413	2118	1610

[a] PDIPN-1 = network PU/linear AC.
[b] PDIPN-2 = linear PU/network AC.

80% PU in the IPNs, consistent with the exceptional toughness of the IPN. A sharp drop in elongation take places at 60% PU and also rapid increases in Shore A and D hardnesses between 80% and 40% PU (shown in Table 6.6), which indicate that a phase inversion occurs around this composition.

6.3.2 Sequential IPNs (seqIPNs)

The structural features of seqIPNs may be summarized as follows [5, 57, 58]:

1. The two polymer components are fairly well-mixed with each other irrespective of their mutual immiscibility, but at the same time are inevitably phase-separated to some extent.
2. The two polymer components form mutually continuous phase domains, but the phase of the first network usually dominates the properties, unless the composition is highly unbalanced. Thus, a seqIPN of a rubbery first network and a glassy second network may provide a toughened rubber, and one of the reverse combination, a high-impact plastic.
3. The phase-domain sizes are controlled by adjusting junction-point densities. The greater the densities, the smaller the domain sizes of the two phases and higher the degree of interpenetration of the two components.

Kotaka *et al.* [80] have reported some seqIPN systems. Two series of seqIPNs of poly(ethylacrylate) (PEA)/poly(methyl methacrylate) (PMMA) were prepared by a swelling-ratio-controlled method (changing the swelling

Table 6.5 Percentage elongation at break of SINs, PDIPNs and linear blends of Table 6.4 [10]

PU/%	100	80	60	40	20	0
AC/%	0	20	40	60	80	100
SIN	640	780	540	270	180	15
PDIPN-1	640	540	250	138	120	12
PDIPN-2	610	553	520	276	211	15
Linear blend	610	640	357	130	195	12

Table 6.6 Shore A and Shore D hardnesses of the SINs, PDIPNs and linear blends of Table 6.4 [10]

PU/%	100	80	60	40	20	0
PA/%	0	20	40	60	80	100
SIN	68	70	80	92	94	100
	(32)[a]	(35)	(38)	(53)	(63)	(72)
PDIPN-1	68	78	86	94	95	100
	(32)	(33)	(38)	(49)	(55)	(68)
PDIPN-2	73	81	91	93	96	100
	(31)	(37)	(43)	(53)	(65)	(72)
Linear blend	73	75	87	92	95	100
	(31)	(30)	(35)	(48)	(58)	(68)

[a] Hardnesses measured by Shore D( ).

time for different samples, sw-IPNs) and conversion-controlled method (changing the curing time for different samples, cv-IPNs), respectively. Figure 6.4 shows storage modulus, E', loss modulus, E'', and loss tangent, $\tan\delta$, at 11 Hz [59]. The E'' and $\tan\delta$ curves for sw-IPNs exhibit only one broad peak and their E' curves have a one step decrease with increasing temperature. On the other hand, cv-IPNs have a wider transition region than the corresponding sw-IPNs. At the temperature close to the loss-maximum temperatures, T_{max}, of PEA and PMMA, the E'' and $\tan\delta$ curves of cv-10 exhibit double peaks. The T_{max} and E' values for sw- and cv-IPNs show nearly identical dependences on the PMMA composition. Figure 6.5 shows some examples of tensile stress–strain curves obtained at 349 K with a constant strain rate of 3% s^{-1}. Figure 6.6 summarizes the composition dependence of the tensile strength, σ_b, and strain-at-break, ε_b, at 349 K and strain rate of 3% s^{-1}. The differences between the results of the cv- and sw-IPNs suggest that only after the curing proceeds long enough does the second network develop fully interpenetration into the first one and forms a more-or-less heterogeneous but co-continuous nodular structure.

The effect of junction-point densities on the structures and mechanical properties of seqIPNs were tested using polybutadiene(PB)/(styrene/acrylonitrile)copolymer (SANC) seqIPNs [60]. The results show that large SANC domains exist in opaque films, the initial modulus is independent of the junction-point density of the SANC network unless the temperature approaches the glass transition temperature of the SANC and, for IPNs of high SANC proportions, the junction-point densities of the glassy SANC networks influence the mechanical properties significantly, particularly the stress–strain behaviour.

6.3.3 *Gradient IPNs*

The rate of diffusion of the reactants of the guest network in the course of preparing a gradient IPN and the resulting distribution of the guest polymer

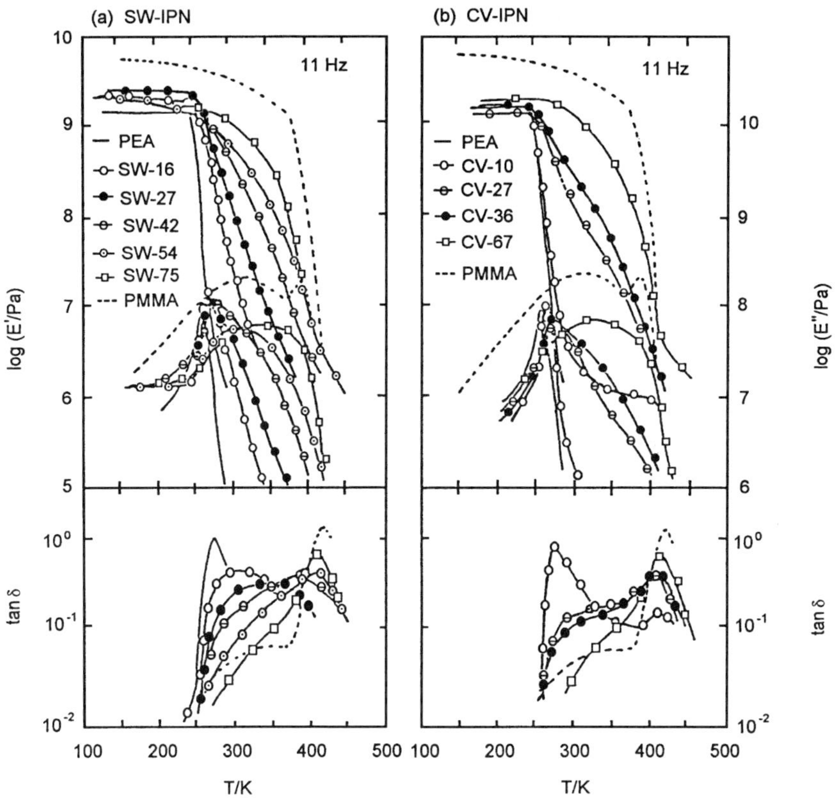

Figure 6.4 Storage modulus, E', loss modulus, E'' and loss tangent, $\tan\delta$, plotted against temperature for (a) sw- and (b) cv-series PEA/PMMA seqIPNs (see text), determined at 11 Hz with a heating rate of $1\,K\,min^{-1}$. Numbers in sample codes indicate wt % PMMA. Curves for the homopolymer networks are also shown for comparison [80].

throughout the host depend on the reactant concentrations in solvent (if used), and the concentration of the crosslinker, initiator etc. Dror *et al.* [61] explored this issue with the system of IPN of poly(ether-urethane) and a gradient of polyacrylamide (PEU/grad.PAM). Gradient IPN samples were prepared by swelling PEU films in an ethanolic solution of acrylamide, tetraethyleneglycol dimethacrylate (branched reactant) and AIBM (initiator), followed by acrylamide polymerization. Increasing the monomer concentration in the swelling solution resulted in regular growth of the PAM concentration in the gradient IPN. Increasing the initiator concentration in the solution also increased the PAM content in the IPN, but only up to concentrations of 0.05–0.09 M, after which its variation ceased to exert an appreciable effect.

Lipatov *et al.* [62] investigated the viscoelastic and mechanical properties of gradient IPNs containing polyurethane with a gradient of butyl methacrylate/

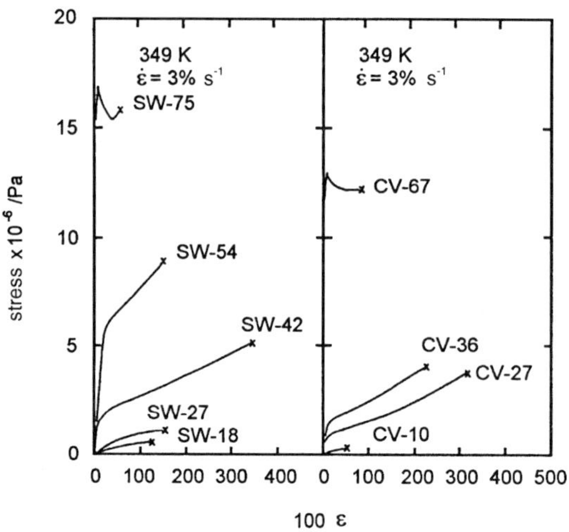

Figure 6.5 Tensile stress–strain curves [80] for (a) sw- and (b) cv-series PEA/PMMA SIPNs (see text) obtained at 349 K with a strain rate of 3% s^{-1}. Numbers in sample codes indicate wt % PMMA. × denotes break point.

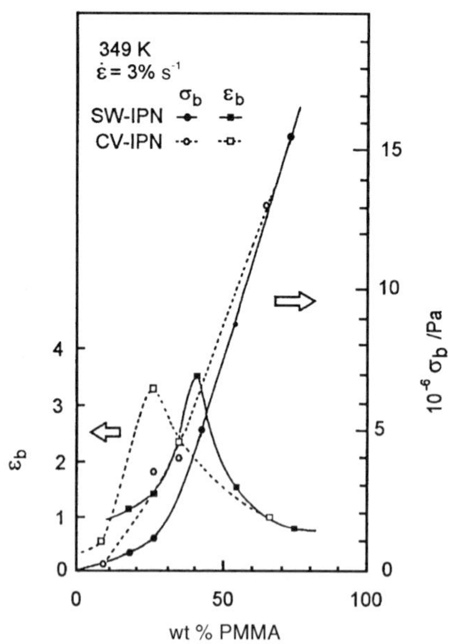

Figure 6.6 Tensile strength, σ_b, and strain-at break, ε_b, versus PMMA content for sw- and cv-series PEA/PMMA SIPNs (see text) determined at 349 K with a strain rate of 3% s^{-1} [80].

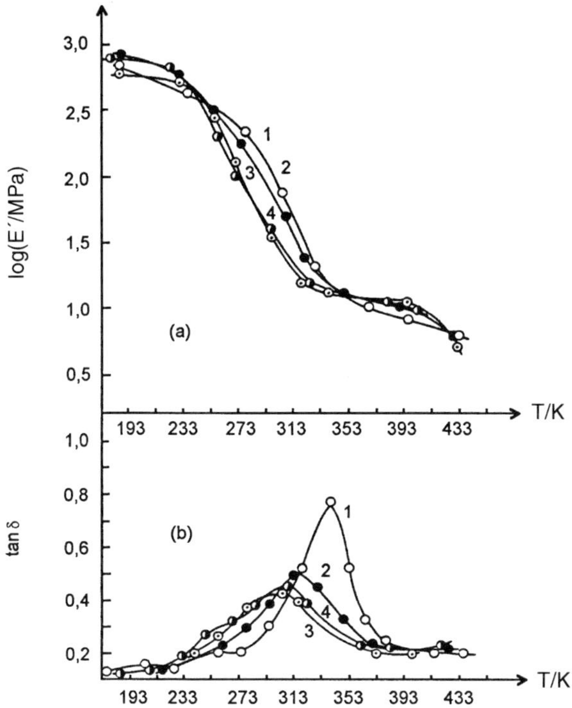

Figure 6.7 Temperature dependence of mechanical loss tangent (b) and elastic modulus (a) in sections of a gradient IPN, PU/grad.P(BMA/TrEGDMA), sliced perpendicular to the (BMA + TrEGDMA) copolymer concentration gradient. Sections are numbered from surface to core of the sample [62].

triethylene glycol dimethacrylate copolymer (PU/grad.P(BMA/TrEGDMA)). The host PU was swollen in a mixture of BMA and TrEGDMA with initiator which started the photopolymerization of the monomers. Sample sections were sliced from a 5 mm thick sample, perpendicular to the copolymer concentration gradient. Four, approximately 0.6 mm-thick sections were cut and the results of measurements of their viscoelastic properties are shown in Figure 6.7. The curve for the first surface layer exhibits peaks characteristic of the glass-transition temperatures of the PU at 253 K, and of the copolymer at 339 K (Figure 6.7(b), curve 1). For the second layer, the copolymer's $\tan\delta$ is lower, broader and shifted towards lower temperatures. For the third and fourth layers, the copolymer's $\tan\delta$ maximum is further shifted towards lower temperatures and a shoulder in the PU glass transition region rises (Figure 6.7(b), curve 4). This behaviour indicates that the copolymer is concentrated mainly in the first layer, its concentration in subsequent layers declining progressively. The temperature dependence of the elastic modulus, E', of the same four sections gives more evidence for this argument. In

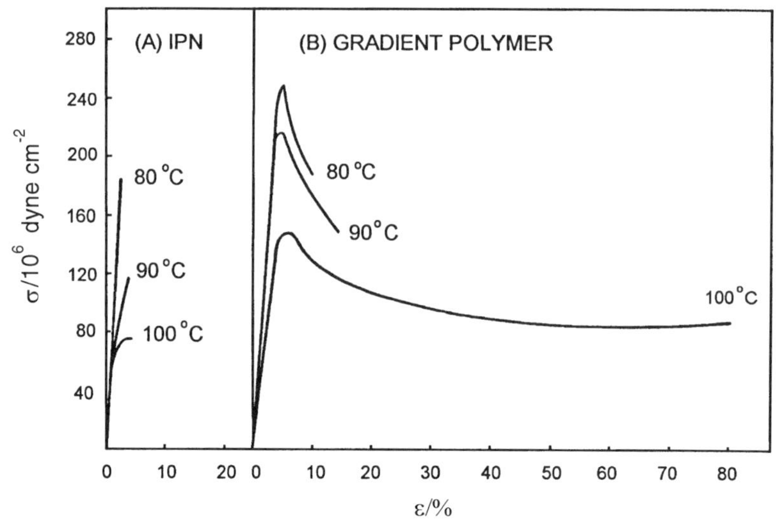

Figure 6.8 Stress(σ)–strain(ε) curves for non-gradient (PMMA/PMA) IPN and gradient (PMMA/grad.PMA) IPN at various temperatures. Total methyl acrylate (MA) content in both samples was 20% [63].

addition, further mechanical measurements demonstrated for the PU/grad.P(BMA/TrEGCDMA) gradient IPNs that their tensile strengths are much higher than those of ordinary IPNs or the individual, component networks. The work of fracture and Young's modulus were both greater than for non-gradient IPN or polyurethane.

Akovali *et al.* [63] studied the mechanical properties of gradient IPNs of poly(methyl methacrylate) and poly(methyl acrylate) (PMMA/grad.PMA) compared with those of non-gradient IPNs. At 373 K pure PMMA fractures at a tensile strength of 2%, and a gradient IPN containing 20% PMA at a strain of almost 80%. The stress–strain curves are shown in Figure 6.8. As can be seen, the tensile strength of gradient IPNs is much higher than that of non-gradient IPNs.

6.3.4 *Latex IPNs*

There are several papers [64, 65, 81] reporting the preparation and properties of latex IPNs based on natural rubber (NR) latex with a synthetic polymer as the second component. Different from the normal latex IPNs, this kind of latex IPN is made from swelling NR particles with the second monomer, followed by polymerization. Hourston *et al.* [81] investigated NR (centrifuged at a solids content of $\approx$60%)/polystyrene (PS) latex IPNs, initiated by an amine-activated hydroperoxide. Figure 6.9 is a typical transmission electron micrograph of the 80NR-20PS material. A complex morphology

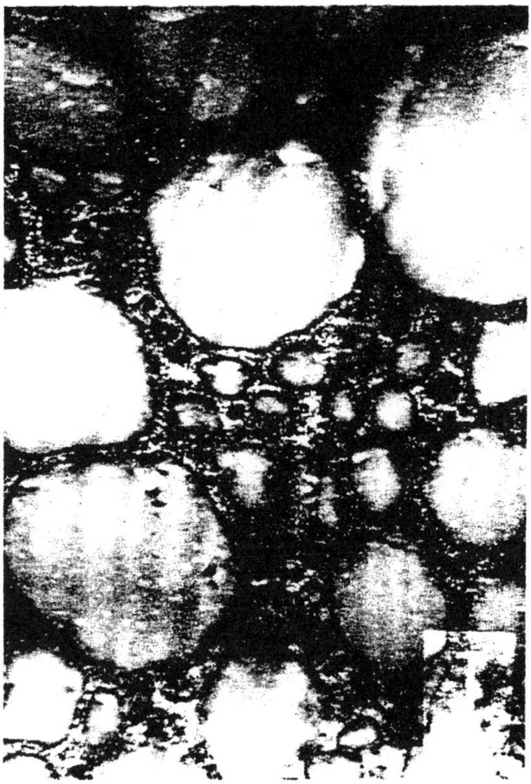

Figure 6.9 Natural rubber(NR)/polystyrene(PS) latex IPN prepared using 1.0 mol% divinylbenzene to styrene. Transmission electron micrograph of an 80 wt% NR–20 wt% PS material. The scale mark represents 410 nm [81].

exists. The large unstained phase domains are of the order of 700–800 nm in diameter and are believed to arise from monomer droplets which have been polymerized as essentially pure PS particles.

Tables 6.7 and 6.8 show dynamic–mechanical data for the NR T_g region and the PS T_g region. It is clear that increasing the percentage of PS in the blends does not shift the T_g significantly nor change its width at half-height. The PS T_g does not show any consistent shift with either composition or junction-point density. These behaviours indicate that there is substantial phase separation of the NR and PS. Table 6.9 and Figures 6.10 and 6.11 give stress–strain data for these latex IPNs. In Figure 6.10, there is an increase in the initial modulus for linear-PS contents between 20 and 30 wt%. The curve for 30 wt % PS is indicative of the presence of continuous PS phase domains. Crosslinking has reduced the ability of the NR to show strain-induced crystallization at high strains. However, with 20% and 30% PS the level of strain hardness is quite remarkable, due to crystallization of the NR and reinforcement by the much-higher-modulus PS phase. The results of tear

Table 6.7 NR latex/PS IPNs. Dynamic–mechanical data at 10 Hz for the NR T_g region [81]. The figures preceding NR and PS in the sample codes are the percentage weights of the two components. The materials prepared in the absence of divinylbenzene are NR latex/PS semi-IPNs

Sample	T_g /°C	Peak width at half-height/°C
CNRL734[a]	−44	12
100NR-0PS	−46	11
90NR-10PS	−44	12
90NR-10PS (1.0)[b]	−44	12
80NR-20PS	−44	12
80NR-20PS (0.25)	−42	12
80NR-20PS (0.5)	−44	12
80NR-20PS (0.75)	−44	12
80NR-20PS (1.0)	−42	17
80NR-20PS (1.5)	−44	12
70NR-30PS	−44	15
70NR-30PS (0.5)	−44	10

[a] As-received natural rubber.
[b] Mol% divinylbenzene to styrene.

tests in a radial direction (Table 6.10) demonstrate that an increase in PS content reduces the tear strength of the materials.

6.3.5 *Thermoplastic IPNs*

Thermoplastic IPNs can be recognized as hybrid materials, demonstrating some of the characteristics of polymer blends and some of the properties of more conventional IPNs. The two major methods that have been employed to prepare these materials are (i) mechanical blending [66, 67] of

Table 6.8 NR latex/PS IPNs. Dynamic mechanical data at 10 Hz for the PS T_g region [81]

Sample[a]	T_g/°C
90NR-10PS	–
90NR-10PS (1.0)	–
80NR-20PS	102
80NR-20PS (0.25)	103
80NR-20PS (0.5)	101
80NR-20PS (0.75)	101
80NR-20PS (1.0)	107
80NR-20PS (1.5)	106
70NR-30PS	102
70NR-30PS (0.5)	106
0NR-100PS	108

[a] Sample-codes as Table 6.7.

Table 6.9 Stress–strain data for the NR latex/PS IPNs [81]

Sample[a]	Initial modulus /MPa	Tensile strength /MPa	Elongation at break /%
CNRL 734	1.08	2.68	653
100NR-0PS	1.97	2.17	740
90NR-10PS	0.72	1.88	814
90NR-10PS (1.0)	0.91	2.50	814
80NR-20PS	1.76	4.00	667
80NR-20PS (0.25)	2.24	4.50	753
80NR-20PS (0.5)	1.98	4.47	780
80NR-20PS (0.75)	3.60	4.45	735
80NR-20PS (1.0)	6.30	2.07	510
80NR-20PS (1.5)	16.00	4.34	560
70NR-30PS	–	12.90	590
70NR-30PS (0.5)	–	6.64	475

[a] Samples codes as Table 6.7.

the two polymers in the melt state and (ii) chemical blending [68] by swelling or dissolving polymer 1 in monomer 2 and then polymerizing polymer 2 *in situ*.

The hydrogenation of elastomeric polybutadiene segments of styrene-butadiene-styrene copolymers (SBSCS, Kratons) produces an olefinic elastomeric block-segment, which is best described as an ethylene/butylene copolymer and the block copolymer is termed a PSEBS block copolymer [69]. Elastomeric properties are achieved by the network-forming capability of the block copolymer. In PSEBS polymers, the solubility parameter

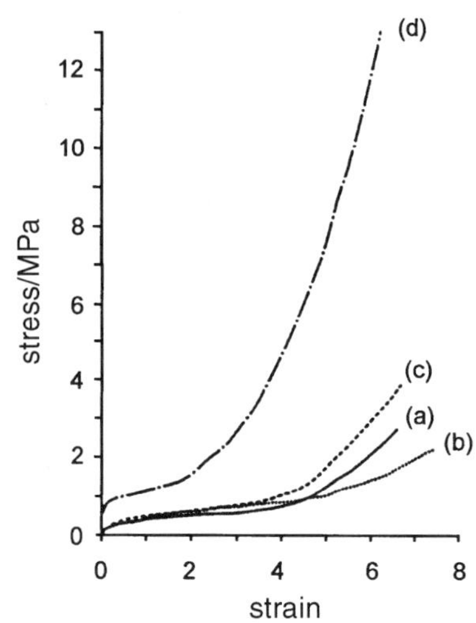

Figure 6.10 NR latex/PS IPNs. Stress–strain curves of (a) the as-received natural rubber, (b) the 100NR-0PS, (c) 80NR-20PS, and (d) the 70NR-30PS materials [81].

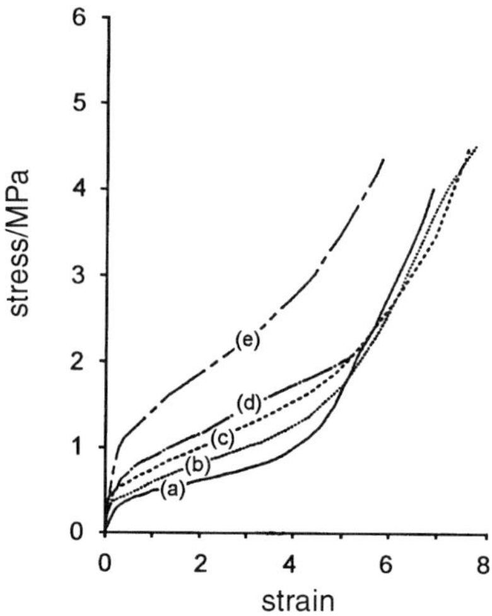

Figure 6.11 NR latex/PS IPNs. Stress–strain curves of the (a) 80NR-20PS, (b) 80NR-20PS (0.25), (c) 80NR-20PS (0.75), (d) 80NR-20PS (1.0), and (e) 80NR-20PS (1.5) materials [81].

difference between segments is large and microphase separation has been found in block copolymers. The commercial products based on PSEBS copolymers have the processing advantage of being very responsive to changes in shear rate. The unique rheology of PSEBS polymers also makes them ideal as a network-forming phase of thermoplastic IPNs. Kratons, thermoplastic IPNs, are available for products from the flexible elastomers preferred for high visibility, exterior, decorated automotive parts. These Kratons contain nylon-6, 11, or 6.6. Some properties are shown in Table 6.11 [70].

Chemically and mechanically blended thermoplastic IPNs have been prepared by Siegfried *et al.* [71] by combining a triblock copolymer with an ionomer made of a statistical copolymer of styrene, methacrylic acid and isoprene (PSMAAI; 90/10/1 by volume) at weight ratios of 25–75%

Table 6.10 Tear-strength data for the NR latex/PS IPNs [81]

Sample[a]	Tear strength ($kN\,m^{-1}$)
CNRL 734	5.40
100NR-0PS	1.62
90NR-10PS	5.76
90NR-10PS (1.0)	3.48
80NR-20PS	3.9
80NR-20PS (0.25)	3.4
80NR-20PS (0.75)	3.6
80NR-20PS (1.0)	18

[a] Sample codes as Table 6.7.

Table 6.11 Some properties of a Kraton IPN [70]

Property	Value
Hardness (Shore D)	38
Flexural modulus of elasticity at 23°C	125 MPa
Flexural modulus of elasticity at 70°C	34 MPa
100% modulus	76 MPa
Tensile strength	17 MPa
Elongation at break	500%

PSEBS. Morphological studies indicated the thermoplastic IPNs go through phase inversions during neutralization of the MAA repeat unit. However, many compositions in the mechanically-blended thermoplastic IPNs, and the chemically blended thermoplastic IPNs with high ionomer contents retain some degree of dual-phase continuity. Stress–strain and dynamic–mechanical analyses demonstrated [72] that the chemically-blended thermoplastic IPNs exhibited greater tensile strengths and higher elongations at break, but lower moduli than the mechanically-blended thermoplastic IPN materials of the same compositions. The results are compared briefly in Table 6.12 [71, 72].

The toughness of certain thermoplastics, plus the high-temperature behaviour and solvent resistance of thermosets, and the best processing characteristics of both are offered by a novel class of plastics called dicyanate-thermoplastic polymers [73–77]. Allied Corporation has investigated these materials [75] in an attempt to combine the strength and toughness of a thermoplastic with the high use-temperature of a thermoset. When dicyanates are used as the thermoset component, dicyanate thermoplastic pseudo-IPNs, having excellent strength, high elongations at break, and good toughness with high softening temperatures are produced (Table 6.13).

The electrical conductivity (particularly the DC conductivity) has been studied for both full IPNs and pseudo-IPNs in which the linear component is more conductive than the surrounding network [82–85]. At sufficiently low temperatures, below 115 K to liquid-He temperature, the conductivity

Table 6.12 Rheological and mechanical behaviour of a 50/50 PSEBS/PSMAAI thermoplastic IPN [71, 72]

Property	Blending methods	
	Chemical	Mechanical
Torque in Brabender at 180 °C/mg	1600	2050
Tensile strength/MPa	29.5	16.6
Percentage strain at break	410	200
Young's modulus/MPa	214	311

Table 6.13 Tensile properties of 50/50 dicyanatobisphenol A/thermoplastic pseudo-IPNs [75]

Thermoplastic	Elongation at break /%	Tensile strength /MPa	Tensile modulus /GPa
Polyester-carbonate copolymer	17.6	84.5	2.14
Polycarbonate	17.3	84.8	2.06
Polysulfone	12.7	73.0	2.05
Poly(ethylene terephthalate)	12.5	76.5	2.45
Poly(ether sulfone)	96	71.7	2.34

change with temperature reverses sign and the conductivity can increase by several orders of magnitude as liquid-He temperature is approached.

Acknowledgement

This work was supported by the National Science Foundation under Grant DMR-90-23541.

Glossary of symbols and abbreviations

f	overall (average) volume fraction of component A ($\equiv f_A$) in a two-component IPN
f_I	overall (average) volume fraction of component I
t^*	time during the formation of an IPN, up to which phase-separation can occur
t_{sep}	time during the formation of an IPN, after which phase-separation can occur
E'	storage modulus
E''	loss modulus
F	dimensionless free-energy density of an IPN
F_{int}	dimensionless interaction free-energy density of an IPN
F_{sub}	dimensionless (elastic) free-energy density of the subnetworks of an IPN
N_A, N_B	average number of units of components A and B
S	static structure-factor of an IPN
T_g	glass-transition temperature
T_{max}	(mechanical) loss-maximum temperature
δ	solubility parameter
$\tan \delta$	loss tangent
ε_b	strain-at-break
σ_b	tensile strength

φ	local volume fraction of component A ($\equiv f_A$) in a two-component IPN
$\bar{\varphi}(\boldsymbol{r})$	average composition-fluctuation field
χ	Flory–Huggins interaction parameter
χ_s	critical value of the Flory–Huggins interaction parameter for the spinodal decomposition of a two-component IPN
Γ	dimensionless free-energy of a two-component IPN
cv-IPN	sequential IPN, prepared by a conversion-control method
grad.	concentration gradient of a component, as in a gradient IPN
pseudo-IPN	polymer-blend material in which one or more polymers are networks and one or more polymers are linear or branched
semi-IPN	synonym for pseudo-IPN
seqIPN	sequential IPN
sw-IPN	sequential IPN, prepared by a swelling-ratio-control method
AC	acrylic copolymer
IPN	interpenetrating network
MAA	methacrylic acid
NR	natural rubber
PAM	polyacrylamide
PB	polybutadiene
P(BMA/TrEGDMA)	(butyl methacrylate)/(triethylene glycol dimethacrylate) copolymer
PCU	polycarbonate/polyurethane
PDIPN	synonym for pseudo-IPN
PDMS	poly(dimethyl siloxane)
PE	polyethylene
PEA	poly(ethyl acrylate)
PEP	poly(ethylene-*co*-propylene)
PEDPM	poly(ethylene-*co*-propylene-*co*-diene)
PEU	poly(ether-urethane)
PMA	poly(methyl acrylate)
PMMA	poly(methyl methacrylate)
PP	poly(propylene)
PPO	poly(2,6-dimethyl-1,4-phenylene oxide)
PS	polystyrene
PSEBS	poly(styrene-*b*-(ethylene/butylene)-*b*-styrene)
PSMAAI	poly(styrene-*co*-methacrylic acid-*co*-isoprene)
PUA	polyurethane/polyacrylate
PVP	poly(4-vinyl pyridine)

SANC	styrene/acrylonitrile copolymer
SBSC	styrene/butadiene/styrene copolymer
SIN	simultaneous IPN

References

1. Klempner, D. and Frisch, K.C. (eds) (1989–1990) *Advances in Interpenetrating Polymer Networks*, Vols I–IV, Technomic, Lancaster, PA.
2. Klempner, D., Sperling, L.H. and Utracki, L.A. (eds) (1994) *Interpenetrating Polymer Networks*, American Chemical Society, Washington, DC.
3. Frisch, K.C. and Klempner, D. (eds) (1989) *Recent Developments in Polyurethanes and Interpenetrating Polymer Networks*, Technomic, Lancaster, PA.
4. Paul, D.R. and Sperling, L.H. (eds) (1986) *Multicomponent Polymer Materials*, American Chemical Society, Washington, DC.
5. Sperling, L.H. (1981) *Interpenetrating Polymer Networks and Related Materials*, Plenum, New York.
6. Lipatov, Yu.S. and Sergeeva, L.M. (1979) *Interpenetrating Polymer Networks*, Maukova Dumka, Kiev.
7. Frisch, H.L. and Klempner, D. (1970) *Adv. Macromol. Chem.*, **2**, 149.
8. Frisch, H.L. and Wasserman, E. (1961) *J. Am. Chem. Soc.*, **83**, 3789.
9. Shill, G. (1971) *Catenanes, Rotaxanes and Knots*, Academic Press, New York.
10. Frisch, K.C., Klempner, D. and Frisch, H.L. (1982) *Polym. Eng. Sci.*, **22**, 1144.
11. Lipatov, Yu.S. and Karabanova, L.V. (1995) *J. Mater. Sci.*, **30**, 1095.
12. Liucheng, Z. (1990) *Interpenetrating Polymer Networks*, Hydrocarbon-Procession Press, Beijing, China, pp. 17–88.
13. Frisch, H.L., Klempner, D. and Frisch, K.C. (1969) *Polym. Lett.*, **7**, 775.
14. Klempner, D., Frisch, H.L. and Frisch, K.C. (1970) *J. Polym. Sci. A-2*, **8**, 921.
15. Matsuo, M., Kwei, T.K., Klempner, D. and Frisch, H.L. (1970) *Polym. Eng. Sci.*, **10**, 327.
16. Klempner, D. and Frisch, H.L. (1970) *Polym. Lett.*, **8**, 525.
17. Klempner, D., Frisch, H.L. and Frisch, K.C. (1971) *J. Elastoplast.*, **3**, 1, 2.
18. Zhou, P. and Frisch, H.L. (1994) *Macromolecules*, **27**, 1788.
19. Zhou, P., Xu, Q. and Frisch, H.L. (1994) *Macromolecules*, **27**, 938.
20. Leibler, L. (1980) *Macromolecules*, **13**, 1602.
21. Binder, K. and Frisch, H.L. (1984) *J. Chem. Phys.*, **81**, 2126.
22. Schulz, M. and Frisch, H.L. (1994) *J. Chem. Phys.*, **101**, 10008.
23. Dušek, K. and Prins, W. (1969) *Adv. Polym. Sci.*, **6**, 1.
24. Bettachy, A., Derouchiche, A., Benhamou, M. and Daoud, M. (1991) *J. Physique I*, **1**, 153.
25. Frisch, H.L. and Grosberg, A.Yu. (1993) *Makromol. Chem. Theo. Simul.*, **2**, 1.
26. Schulz, M. (1992) *J. Chem. Phys.*, **97**, 5631.
27. Schulz, M. and Binder, K. (1993) *J. Chem. Phys.*, **98**, 655.
28. Petschek, R. and Metiu, H. (1983) *J. Chem. Phys.*, **79**, 3443.
29. An, J.H. and Sperling, L.H. (1988) in *Cross-Linked Polymers: Chemistry, Properties and Applications* (eds R.A. Dicke, S.S. Labana and R.S. Bauer), American Chemical Society, Washington, DC.
30. Cahn, J.W. and Hilliard, J.E. (1958) *J. Chem. Phys.*, **28**, 258.
31. Binder, K. (1983) *J. Chem. Phys.*, **79**, 6387.
32. Matsuda, S. (1991) *Polym. J.*, **23**, 435.
33. Hashimoto, T. (1987) in *Current Topics in Polymer Science* (eds R. Ottenbrite, L.A. Utracki and S. Inoue), Carl Hanser Verlag, München.
34. McGarey, B. (1989) in *Advances in IPNs*, Vol. I (eds D. Klempner and K.C. Frisch), Technomic, Lancaster, PA.
35. Stauffer, D. (1977) *Am. J. Phys.*, **45**, 1001.
36. Flory, P.J. (1986) *Principles of Polymer Chemistry*, Cornell University Press, Ithaca.
37. Sariban, A. and Binder, K. (1987) *J. Chem. Phys.*, **86**, 5859.
38. Millar, J.R. (1960) *J. Chem. Soc.*, 1311.

39. Shibayama, K. and Suzuki, Y. (1967) *Rubber. Chem. Tech.*, **40**, 476.
40. Lipatov, Yu.S. and Sergeeva, C.M. (1967) *Russ. Chem. Rev.*, **45**, 1, 63.
41. Frisch, H.L., Klempner, D. and Frisch, K.C. (1969) *J. Polym. Sci., Polym. Phys. Ed.*, **7**, 11, 775.
42. Sperling, L.H. and Friedman, D.W. (1969) *J. Polym. Sci.*, **A-2**, 7, 425.
43. Frisch, H.L., Xu, Q. and Zhou, P. (1994) *Polymer Gels and Networks*, **2**, 257.
44. Frisch, H.L., Klempner, D., Yoon, H.K. and Frisch, K.C. (1980) *Macromolecules*, **13**, 1016.
45. Singh, S., Ghiradella, H. and Frisch, H.L. (1990) *Macromolecules*, **23**, 375.
46. Frisch, H.L. and Hua, Y.H. (1991) *Macromolecules*, **22**, 91.
47. Mengnjoh, P.C. and Frisch, H.L. (1989) *J. Polym. Sci., Polym. Chem. Ed.*, **27**, 3363; Frisch, H.L. and Mengnjoh, P.C. (1989) *J. Polym. Sci., Polym. Chem. Ed.*, **27**, 285.
48. Frisch, H.L., Gebreyes, K. and Frisch, K.C. (1988) *J. Polym. Sci., Polym. Chem. Ed.*, **26**, 2589; Gebreyes, K. and Frisch, H.L. (1989) *J. Polym. Sci., Polym. Chem. Ed.*, **26**, 3391; Huang, W. and Frisch, H.L. (1989) *Macromol. Chem. Phys. Suppl.*, **15**, 137.
49. Frisch, H.L., Zhou, P., Frisch, K.C., Xiao, X.H., Huang, W. and Ghiradella, H. (1991) *J. Polym. Sci., Polym. Chem. Ed.*, **29**, 1031.
50. Frisch, H.L. and Zhou, P. (1992) *J. Polym. Sci., Polym. Chem. Ed.*, **30**, 2794.
51. Zhou, P. and Frisch, H.L. (1992) *J. Polym. Sci., Polym. Chem. Ed.*, **30**, 835.
52. Zhou, P. and Frisch, H.L. (1992) *J. Polym. Sci., Polym. Chem. Ed.*, **30**, 887.
53. Frisch, H.L. and Huang, M.W. (1993) in *Siloxane Polymers* (eds J.A. Semylen and S. Clarson), Prentice-Hall, Englewood Cliffs, NJ.
54. Frisch, H.L. (1990) *Mater. Res. Soc. Symp. Proc.*, **171**, 231.
55. de Barros, G.G., Huang, M.W., Frisch, H.L. (1992) *J. Appl. Polym. Sci.*, **44**, 255.
56. Klempner, D., Yoon, H.K., Frisch, K.C. and Frisch, H.L. (1979) in *Polymer Alloys II* (eds D. Klempner and K.C. Frisch), Plenum Press, New York, p. 185.
57. Manson, J.A. and Sperling, L.H. (1976) *Polymer Blends and Composites*, Plenum Press, New York.
58. Klempner, D. and Frisch, K.C. (1977) *Polymer Alloys*, Plenum Press, New York.
59. Adachi, H. and Kotaka, T. (1982) *Polym. J.*, **14**, 379.
60. Kojima, K. and Kotaka, T. Unpublished results (1988).
61. Dror, M., Elsabee, M.L. and Berry, G.C. (1981) *J. Appl. Polym. Sci.*, **26**, 1741.
62. Lipatov, Yu.S., Sergeeba, L.M., Karabanova, L.V. and Rosovitskij, V.P. (1988) *Mekhanika Kompozitsionnykh Materialov*, **6**, 1028.
63. Akovali, G., Biliyar, K. and Shen, M. (1976) *J. Appl. Polym. Sci.*, **20**, 2419.
64. Almen, G.R., MacKenzie, P., Malhotra, V., Maskell, R.K., McGrail, P.T. and Sefton, M.S. (1987) *20th International SAMPE Technical Conference*, pp. 46–59.
65. Almen, G.R., Maskell, R.K., Malbotra, V., Sefton, M.S., McGrail, P.T. and Wilkinson, S.P. (1988) *33rd International SAMPE Symposium*, pp. 970–980.
66. Kresge, E.N. (1978) in *Polymer Blends*, Vol. 2 (eds D.R. Paul and S. Newman), Academic Press, New York.
67. Bull, A.L. and Holden, G. (1977) *J. Elast. Plast.*, **9**, 281.
68. Siegfrid, D.L., Thomas, D.A. and Sperling, L.H. (1980) *Polymer Preprint*, **21**, 186.
69. Gergen, W.P. (1984) *Kautsch. Gummi.*, **37**, 284.
70. Klempner, D. and Berkowski, L. (1987) *Encyclopedia of Polymer Science and Engineering*, **8**, 279.
71. Siegfrid, D.L., Thomas, D.A. and Sperling, L.H. (1981) *J. Appli. Polym. Sci.*, **26**, 177.
72. Siegfrid, D.L., Thomas, D.A. and Sperling, L.H. (1981) *J. Appli. Polym. Sci.*, **21**, 39.
73. *Polym. News* (1983) **9**, 188.
74. Wertz, D.H. and Prevorsek, D.C. (1984) *Plast. Eng.*, April, 31.
75. Wertz, D.H. and Prevorsek, D.C. (1984) *ANTEC'84*, 483.
76. Prevorsek, D.C. and Chung, D.C. (1979) US Patent 4,157,360.
77. Hsiue, E.S., Miller, R.L. and Segal, L. in *Fifth Int. Conf. Compos. Mater., ICCM-V, Conf. Proc.* (eds C.W. Harrigan, Jr., J. Strife and A.K. Dhingra), Warrendale, PA, Metal Soc. AIME, pp. 1655–1665.
78. Olabisi, O., Robeson, L.M. and Shaw, M.T. (1976) *Polymer-Polymer Miscibility*, Academic Press, New York.
79. Frisch, H.L. and Zhou, P. (1994) in *Interpenetrating Polymer Networks* (eds D. Klempner, L.H. Sperling and L.A. Utracki), American Chemical Society, Washington, DC, p. 269.

80. Kotaka, T., Nishi, S. and Adachi, H. (1990) in *Advances in Interpenetrating Polymer Networks*, Vol. II (eds D. Klempner and K.C. Frisch), Technomic, Lancaster, PA, p. 47.
81. Hourston, D.J. and Romaine, J. (1994) in *Advances in Interpenetrating Polymer Networks* Vol. IV (eds D. Klempner and K.C. Frisch), Technomic, Lancaster, PA, p. 77.
82. Frisch, H.L., Chen, Z.J. and Chen, X. (1994) *Macromol. Symp.*, **81**, 181.
83. Frisch, H.L. and Chen, Z.J. (1994) *J. Polym. Sci. A*, **32**, 1317.
84. Frisch, H.L. and Chen, X. (1993) *J. Polym. Sci. A*, **31**, 3307.
85. Frisch, H.L. and deBarros, G.G. (1992) *J. Polym. Sci. A*, **30**, 937.

7 Molecular aspects of rubberlike elasticity

J.E. MARK and B. ERMAN

7.1 Introduction

The networks of interest in the present context exhibit high deformability with essentially complete recovery, a phenomenon called 'rubberlike elasticity', since it is exemplified by natural rubber. Of central importance in this type of elasticity is the fact that natural rubber itself and other rubberlike materials all consist of relatively long polymeric chains having a high degree of flexibility and mobility, which are joined into network structures [1–4]. The requirement of flexibility and mobility is associated with the very high deformability. As a result of an externally imposed stress, the long chains alter their conformations, a process which can take place relatively easily in the typical case of high chain mobility. The requirement of linking the chains into a network structure is associated with solid-like features, where the chains are prevented from irreversibly flowing relative to one another under external stresses. As a result of these circumstances, a typical rubber or elastomer may be stretched up to about 10 times its original length. Upon removal of the external force, it rapidly recovers its original dimensions in an attempt to regain the entropy it lost in the deformation process, and can generally do this with essentially no residual or non-recoverable strain.

The preceding general features of elastomeric materials have long been known and, in fact, the area of rubberlike elasticity has had one of the longest and most distinguished histories in all of polymer science [1–3, 5]. For example, quantitative measurements of the mechanical and thermodynamic properties of natural rubber and other elastomers go back to 1805, and some of the earliest studies have been carried out by such luminaries as Joule and Maxwell. Also, the earliest molecular theories for polymer properties of any kind were, in fact, addressed to the phenomenon of rubberlike elasticity.

The present review briefly discusses methods for forming networks, and defines the parameters used to characterize the resulting network structures. The theories of rubberlike elasticity are covered, with the focus essentially entirely on theories that are molecular and treat elastomeric properties at equilibrium (i.e. to the exclusion of time-dependent or viscoelastic effects). Theoretical and experimental results are compared, both for the evaluation of theory and for guidance for the development of improved molecular models and more probing experimental investigations.

Much of the material covered is described in much greater detail in the specialized literature, but some more general sources of related information are also readily available. For example, theory and experiment up to approximately 1974 were treated in the now classic book by Treloar [2], and the status of theory has been reviewed by Eichinger [6], by Heinrich *et al.* [7], and by the present authors [4, 8, 9]. More specialized reviews, describing theoretical results obtained with particular models, have also appeared [10–17]. Several recent books have focused on elastomers and rubberlike elasticity [18–20], and several others have reported the proceedings of symposia based on this subject [21–27]. Two general review articles of relevance have been published in encyclopaedias [28, 29], and in several articles in a recent Handbook [30]. Finally, an overall survey has been provided in an introductory book [3] and in a more comprehensive sequel [9] by the present authors.

7.2 Preparation and structure of networks

7.2.1 *Polymerizations with multifunctional monomers (cf. Chapter 2)*

For rubberlike elasticity to be achieved, not only must the material consist of flexible polymer chains, but the chains must be joined (crosslinked or end-linked) into a network structure [1–3, 8]. One way of doing this is by a copolymerization in which at least one of the co-monomers has a functionality ϕ of 3 or larger. This is one of the oldest ways of preparing networks, but has been used mostly to prepare materials with such high densities of junction points as to be relatively hard thermosets rather than highly deformable elastomers [3, 22, 31]. Recent work in this area involves the use of sol fractions and statistical arguments to obtain information on the structures of the sol phases and gel (elastomeric) phases [32–40].

7.2.2 *Physical aggregation (cf. Chapter 9)*

Preparation of elastomeric networks is also possible by causing physical aggregation of some of the chain segments [3, 8, 41]. Examples are the adsorption of chain segments onto filler particles, formation of polymer microcrystallites, condensation of ionic side chains onto metal ions, chelation of ligand side chains to metal ions, and microphase separation of glassy end blocks in an elastomeric triblock copolymer [42]. These materials are now not much used in quantitative studies of rubberlike elasticity. The nature and extent of the junction zones can change with temperature, presence of diluent, and degree of deformation in an uncontrolled manner, and the 'junction points' are frequently so large as to complicate greatly the theoretical analysis of any experimental results.

7.2.3 *Random chemical crosslinking (cf. Chapter 3)*

Polymers can be crosslinked by using chemical reactions that attack a pair of chains at essentially random locations [1, 3]. Examples are the addition of sulfur atoms to the double bonds of diene elastomers, and the attack of free radicals from peroxide thermolysis or high-energy radiation on side chains (frequently unsaturated) or on the chain backbone itself.

The covalent linkages thus formed are generally highly stable, thus avoiding some of the problems with physically crosslinked systems. These networks, however, still present the problem that the crosslinking reaction used is highly uncontrolled in that it is not known how many crosslinks are introduced or where they lie along the chain trajectories. It is thus very difficult to use these networks for quantitative purposes such as the development of structure–property relationships. Obtaining independent measures of their degree of crosslinking (as represented directly by the number density of crosslinks or the number density of network chains, or inversely by the average molecular weight M_c between the crosslinks that mark off a network chain) is virtually impossible. Some recent simulations of the sulfur vulcanization process [43], however, may be of help in this regard.

7.2.4 *Highly specific chemical endlinking (cf. Chapter 2)*

If networks are formed by endlinking functionally terminated chains instead of haphazardly joining chain segments at random, then the nature of this very specific chemical reaction provides the desired structural information [3, 41, 44–75]. Thus, the functionality of the junction points is the same as that of the end linking agent, and the molecular weight M_c between junction points and its distribution are the same as those of the starting chains prior to their being endlinked.

Because of their known structures, such endlinked elastomers are now the preferred materials for the quantitative characterization of rubberlike elasticity. Such very specific endlinking reactions have also been shown to be useful in the preparation of liquid-crystalline elastomers [76, 77]. Additionally, bimodal networks prepared by these endlinking techniques have very good ultimate properties, and there is currently much interest in preparing and characterizing such networks [3, 46, 52, 53, 55–57, 60, 61, 64, 68, 72, 75, 78] and developing theoretical interpretations for their properties [79–83].

7.2.5 *Crosslinking under unusual conditions*

For some investigations, networks have been prepared by partially disentangling the polymer chains prior to their crosslinking by either stretching or dissolution in a solvent [3, 8, 84]. Such networks are of particular interest because they come to elastic equilibrium relatively rapidly, show little stress relaxation, and yield stress–strain isotherms of an unusually simple form. Interpretation of their properties thus constitutes an important challenge to the various molecular theories of rubberlike elasticity.

In a rather novel series of experiments [3, 85], results on the trapping of cyclic molecules during network formation were found to support the notion that 'olympic' or 'chain-mail' networks can be formed in a similar manner. This could be done by threading functionally terminated chains through such cyclics and then difunctionally endlinking them [85–88].

7.2.6 *Network structure*

The structure of a perfect network may be defined by two variables, the cycle rank ξ and the average junction functionality ϕ [3, 8]. Cycle rank is defined as the number of network chains that must be cut in order to reduce it to a tree. The three other parameters used often in defining a network, the number of junction points μ, the number of chains ν, and the molecular weight M_c, of chains between two junction points, may be obtained from ξ and ϕ by the following three relationships:

$$\nu = \xi/(1 - 2/\phi) \tag{7.1}$$

$$\mu = 2\nu/\phi \tag{7.2}$$

$$M_c = \frac{(1 - 2/\phi)\rho N_A}{\xi/V_0} \tag{7.3}$$

where ρ is the density, V_0 is the reference volume of the network, and N_A is the Avogadro constant.

In a typical network, the number of skeletal bonds in a network chain ranges from about 100 to 700 [8, 89]. Networks with chains shorter than 100 bonds have low degrees of extensibilities. Those having chains much larger than 700 bonds may have very high extensibilities but are too weak to serve as load-carrying materials. As mentioned previously, it is possible to prepare bimodal networks, however, by endlinking very short and very long chains in the same network, and these elastomers have greatly increased toughness [8, 72, 90].

7.3 Elasticity experiments

7.3.1 *Mechanical properties*

These are the properties that determine the unique usefulness of elastomeric materials. Not surprisingly, they are the properties that have been most extensively studied [1–3, 19–29]. The relationship of primary interest is the stress–strain isotherm. Measurements to determine it are typically carried out to the rupture point of a sample, thus yielding, as well, the two ultimate properties of the elastomer, namely its ultimate strength and maximum extensibility. In addition to providing practical information, such results are also much used to obtain estimates of the concentration of junction points, and to test the predictions of theory. Most experimental studies have involved simple elongation, largely because of the simplicity of the techniques involved. Results obtained with other deformations [2, 91–101] such as biaxial extension, compression, shear and torsion are, however, particularly valuable for gauging the generality of a particular theory, and more such investigations should be carried out. Some important non-mechanical properties are mentioned in the following three sections.

7.3.2 Swelling

In a swelling experiment the network is typically placed into an excess of solvent, which it imbibes until the dilational stretching of the chains prevents further absorption [1–3, 8, 102]. This equilibrium extent of swelling can be interpreted to yield the concentration of junction points in the network, provided the polymer–solvent interaction parameter χ is known. Conversely, if the concentration of junction points is known from an independent experiment, then the interaction parameter can be determined. The equilibrium degree of swelling and its dependence on various parameters and conditions provide, of course, additional tests of theory.

One important new topic in this area is 'gel collapse', in which a swollen gel can dramatically shrink in response to a relatively minor change in the environment, for example, temperature, pH, ionic strength and exposure to light [103–128]. These materials could be very useful in a variety of applications as mechanical switches, actuators, chemical valves, devices for converting chemical energy into mechanical energy, and artificial muscles.

7.3.3 Optical and spectroscopic properties

An example of a relevant optical property is the birefringence of a deformed polymer network [2, 129–131]. This strain-induced birefringence can be used to characterize chain segmental orientation, both Gaussian and non-Gaussian elasticity, crystallization and other types of chain ordering, and short-range correlations [2, 3, 8]. Other optical and spectroscopic techniques are also important, particularly with regard to segmental orientation. Some examples are fluorescence polarization [132, 133], deuterium NMR [134, 135], and polarized and two-dimensional infrared spectroscopy [136–139].

7.3.4 Scattering

The technique of this type of greatest utility in the study of elastomers is small-angle neutron scattering; for example, from deuterated chains in a non-deuterated host [140–142]. One application has been the determination of the degree of randomness of the chain conformations in the undeformed state, an issue of great importance with regard to the basic postulates of elasticity theory. Of even greater importance is determination of the manner in which the dimensions of the chains follow the macroscopic dimensions of the sample. This relationship between the molecular and macroscopic worlds is one of the central problems in rubberlike elasticity.

Some small-angle X-ray scattering techniques have also been applied to elastomers, an example being the characterization of fillers incorporated into elastomers to improve their mechanical properties [143–145].

7.4 Elasticity theories involving structureless chains

The molecular theory of elasticity has developed along two lines. The first is based on the statistical mechanics of networks, and is mostly concentrated on network topology, determination of stress–strain behaviour, number of effective junction points, contributions from entanglements, etc. Most experimental effort has been on this first aspect. In this approach, the chemical structure of the chains constituting the medium has not been considered important – only the topological nature of the network is relevant. These models are considered in the present section [2, 3, 8].

The second line of theoretical development has not received much attention (except for birefringence and thermoelasticity). In this second approach, the chemical structure and the resulting conformational characteristics of the chains are considered important. Topics are birefringence, thermoelasticity, rotational isomerization upon stretching, strain dichroism, local segmental orientation, local segmental mobility and neutron scattering. Advances in present-day experimental techniques focus more attention on these phenomena. Additionally, as already mentioned, changes in the swelling behaviour of networks resembling first-order transitions are being studied experimentally in several laboratories. Molecular theories of rubberlike elasticity are most often based on the Gaussian distribution of vectorial end-to-end distances $\boldsymbol{r}$ for the network chains. In this approach, the probability $W(\boldsymbol{r})\,\mathrm{d}x\,\mathrm{d}y\,\mathrm{d}z$ that one end of $\boldsymbol{r}$ is at the origin and the other end is in an infinitesimal volume $\mathrm{d}V = \mathrm{d}x\,\mathrm{d}y\,\mathrm{d}z$ is satisfactorily represented by the Gaussian function [146, 147]:

$$W(\boldsymbol{r})\,\mathrm{d}x\,\mathrm{d}y\,\mathrm{d}z = (3/2\pi\langle r^2\rangle_0)^{3/2}\exp(-3r^2/2\langle r^2\rangle_0)\,\mathrm{d}x\,\mathrm{d}y\,\mathrm{d}z \tag{7.4}$$

Here, $\langle r^2\rangle_0$ represents the average of the squared end-to-end vectors, and the subscript zero indicates that the chain is in the unperturbed or the so-called

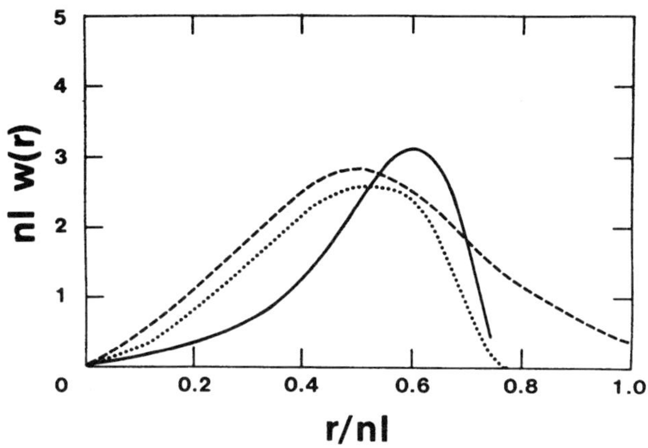

Figure 7.1 Distributions for the end-to-end distance of a poly(dimethylsiloxane) chain having $n = 20$ skeletal bonds of length l [79]. The Fixman–Alben distribution (dotted curve) and that from a Monte-Carlo simulation (solid curve) are compared with the Gaussian approximation (dashed curve).

theta-state [1]. It is now well established that chains in the bulk, undiluted state are in this unperturbed state. Equation (7.4) represents the probability distribution of the vectorial quantity $\boldsymbol{r}$. A less detailed form of representation is the distribution $w(r)$ showing the probability that the magnitude $|r|$ of $\boldsymbol{r}$ has a certain value irrespective of the direction. Thus the probability that the chain end-to-end length is in the range r to $r + \mathrm{d}r$ irrespective of its direction is

$$w(r)\,\mathrm{d}r = (3/2\pi\langle r^2\rangle_0)^{3/2}\exp(-3r^2/2\langle r^2\rangle_0)4\pi r^2\,\mathrm{d}r \qquad (7.5)$$

A schematic representation of the Gaussian function of equation (7.5) is given by the dashed curve in Figure 7.1. The abscissa is normalized by dividing the end-to-end distance by the contour length of the chain and the ordinate is made non-dimensional by multiplying $w(r)$ by the contour length. For chains having fewer than 50 bonds, such as the short chains in a bimodal network, for example, the distribution departs markedly from the Gaussian limit. Among various representations of $w(r)$ for short chains are the Hermite series [146, 147], the Fixman–Alben distribution [148], or Monte-Carlo simulations [79]. The Fixman–Alben distribution is given by

$$w(r)\,\mathrm{d}r \propto \exp(-ar^2 + b^2r^4)4\pi r^2\,\mathrm{d}r \qquad (7.6)$$

where a and b are coefficients. This distribution and the results for Monte-Carlo simulations for a poly(dimethylsiloxane) (PDMS) chain of 20 skeletal bonds are compared with the Gaussian approximation in Figure 7.1. The molecular theories of networks to be presented in the following paragraphs are based on this Gaussian picture of the individual network chain.

The elastic free energy, A_{el}, of a Gaussian chain is related to the probability distribution $W(r)$ by the thermodynamic expression [3]

$$A_{el} = C(T) - kT \ln W(\boldsymbol{r}) \tag{7.7}$$

where $C(T)$ is a function only of temperature T, and k is the Boltzmann constant. Substituting equation (7.4) into equation (7.7) leads to

$$A_{el} = A^*(T) + (3kT/2\langle r^2\rangle_0)r^2 \tag{7.8}$$

where $A^*(T)$ is a function of temperature alone. Equation (7.8) represents the elastic free energy of a Gaussian chain with ends fixed at a separation r. The average force required to keep the two ends at this separation is obtained from the thermodynamic expression [149]

$$f = \left(\frac{\partial A_{el}}{\partial r}\right)_T \tag{7.9}$$

$$= \left(\frac{3kT}{\langle r^2\rangle_0}\right) r \tag{7.10}$$

where equation (7.10) is obtained by substituting equation (7.8) into equation (7.9). The subscript T denotes differentiation at fixed temperature.

Equation (7.10) shows that the single chain behaves like a spring, with a spring constant equal to $3kT/\langle r^2\rangle_0$.

This approach is directly extendable to the chains making up a network structure. Specifically, the total elastic free energy ΔA_{el} of the network relative to the undeformed state is obtained by summing equation (7.8) over the ν chains of the network [3]

$$\Delta A_{el} = \frac{3kT}{2\langle r^2\rangle_0}\sum_{\nu}(r^2 - \langle r^2\rangle_0) \tag{7.11}$$

$$= \frac{3\nu kT}{2}\left(\frac{\langle r^2\rangle}{\langle r^2\rangle_0} - 1\right) \tag{7.12}$$

where $\langle r^2\rangle = \sum r^2/\nu$ is the average square end-to-end distances of chains in the deformed network. Substituting

$$\langle r^2\rangle = \langle x^2\rangle + \langle y^2\rangle + \langle z^2\rangle \tag{7.13}$$

into equation (7.12) and using the fact that chain dimensions are isotropic in the deformed state, i.e.

$$\langle r^2\rangle_0 = \langle x^2\rangle_0 + \langle y^2\rangle_0 + \langle z^2\rangle_0 = 3\langle x^2\rangle_0 = 3\langle y^2\rangle_0 = 3\langle z^2\rangle_0 \tag{7.14}$$

yields

$$\Delta A_{el} = \frac{\nu kT}{2}\left[\frac{\langle x^2\rangle}{\langle x^2\rangle_0} + \frac{\langle y^2\rangle}{\langle y^2\rangle_0} + \frac{\langle z^2\rangle}{\langle z^2\rangle_0} - 3\right] \tag{7.15}$$

The x, y and z coordinates appearing in the above equations represent laboratory-fixed coordinates.

The ratios of mean-squared dimensions appearing in equation (7.15) are molecular quantities. In order to express the elastic free energy of a network in terms of the macroscopic state of deformation, an assumption has to be made relating molecular chain dimensions to macroscopic deformation.

The state of macroscopic deformation may be characterized by considering the deformation of a rectangular prism, with deformation ratios λ_x, λ_y and λ_z along the x, y and z directions, respectively, as

$$\lambda_x = \frac{L_x}{L_{x0}} \qquad \lambda_y = \frac{L_y}{L_{y0}} \qquad \lambda_z = \frac{L_z}{L_{z0}} \tag{7.16}$$

where L_{x0}, L_{y0} and L_{z0} are the sides of the prism before deformation, and L_x, L_y and L_z are the corresponding sides in the deformed state.

Two network models have been used in the elementary theories of elasticity in order to relate the molecular deformation to the macroscopic one: the affine model and the phantom model. Both are described as parts of the following section on classical models in general.

7.4.1 *Classical network theories*

Molecular theories of rubber-like elasticity may be conveniently discussed in two categories: (i) the classical theories, comprising the phantom and the affine network models, introduced between 1941 and 1943, and (ii) the modern theories, introduced after 1975.

The term 'classical' is used in describing the phantom and affine models because of their conceptual simplicity and their status as suitable reference systems for all later treatments of the molecular theory. The characteristic features of the classical theories are now outlined with extensive references to relevant literature, followed by a discussion of the modern theories and their extensions and applications to various phenomena observed by advanced experimental techniques of the present day.

7.4.1.1 *Affine-network model*

One of the earlier assumptions regarding molecular deformation in networks is that the junction points in the networks move affinely with macroscopic deformation. It follows that chain end-to-end vectors deform affinely also, and

$$\langle x^2 \rangle = \lambda_x^2 \langle x^2 \rangle_0 \qquad \langle y^2 \rangle = \lambda_y^2 \langle y^2 \rangle_0 \qquad \langle z^2 \rangle = \lambda_z^2 \langle z^2 \rangle_0 \tag{7.17}$$

Substituting equation (7.17) into equation (7.15) leads to

$$\Delta A_{\text{el,affine}} = \tfrac{1}{2} \nu k T (\lambda_x^2 + \lambda_y^2 + \lambda_z^2 - 3) \tag{7.18}$$

A more rigorous statistical analysis [1, 8] gives an additional volume term, $-\mu kT(V/V_0)$, in equation (7.18). This term does not appear in the simplified derivation presented here.

Theoretical developments after 1975 (Section 7.4.2) have discredited the affine-network model [89, 150–155], mainly because the assumption of embedding each junction securely into the continuum of the network volume was judged to be unrealistic, and was later contradicted by neutron scattering results [156–158]. The phantom network model thus seems to be more appropriate in this sense, but variations of the affine theory still appear in the literature [159], however.

7.4.1.2 Phantom-network model

The theory of James and Guth, which has subsequently been termed the 'phantom-network theory', was first outlined in two papers [160, 161], followed by a mathematically more rigorous treatment [162–164]. More recent work has been carried out by Duiser and Staverman [165], Eichinger [166], Graessley [167, 168], Flory [89], Pearson [169], and Kloczkowski *et al.* [170, 171]. In the phantom network approach, the instantaneous vector $\boldsymbol{r}$ joining two junctions at the extremities of a network chain may be written as the sum of a time-averaged mean $\bar{\boldsymbol{r}}$ and the instantaneous fluctuation $\Delta\boldsymbol{r}$, from this mean, i.e.

$$\boldsymbol{r} = \bar{\boldsymbol{r}} + \Delta\boldsymbol{r} \tag{7.19}$$

According to the phantom network model, the fluctuations $\Delta\boldsymbol{r}$ are independent of deformation and the mean vectors $\bar{\boldsymbol{r}}$ deform affinely with macroscopic strain. Squaring both sides of equation (7.19) and averaging over all chains gives

$$\langle r^2 \rangle = \langle \bar{r}^2 \rangle + \langle (\Delta r)^2 \rangle \tag{7.20}$$

The average of the quantity $\bar{\boldsymbol{r}} \cdot \Delta\boldsymbol{r}$ has been equated to zero in obtaining equation (7.20) inasmuch as this quantity is equally likely to be positive or negative due to the fluctuating term $\Delta\boldsymbol{r}$, and the average over all possible occurrences vanishes. Equation (7.20) is valid both in the deformed and undeformed states. It may be written in terms of λ_x, λ_y and λ_z as

$$\langle r^2 \rangle = \left(\frac{\lambda_x^2 + \lambda_y^2 + \lambda_z^2}{3} \right) \langle \bar{r}^2 \rangle_0 + \langle (\Delta r)^2 \rangle \tag{7.21}$$

$$= \left[\left(\frac{\lambda_x^2 + \lambda_y^2 + \lambda_z^2}{3} \right) \left(1 - \frac{2}{\phi} \right) + \frac{2}{\phi} \right] \langle r^2 \rangle_0 \tag{7.22}$$

Equation (7.14) is used in obtaining equation (7.21), and equation (7.22) is obtained from

$$\langle \bar{r}^2 \rangle_0 = \left(1 - \frac{2}{\phi} \right) \langle r^2 \rangle_0 \qquad \langle (\Delta r)^2 \rangle_0 = \frac{2}{\phi} \langle r^2 \rangle_0 \tag{7.23}$$

These relationships result from the phantom network model, and their derivations are given elsewhere [3, 89].

Using equation (7.1) in equation (7.22) and substituting the resulting expression into equation (7.12) results in

$$\Delta A_{\mathrm{el,phantom}} = \tfrac{1}{2}\xi kT(\lambda_x^2 + \lambda_y^2 + \lambda_z^2 - 3) \tag{7.24}$$

Further information on the theory of phantom networks can be found in a review article by Eichinger [6].

Comparison of the expressions for the elastic free energies for the affine and the phantom network models shows that they differ only in the front factor. Expressions for the elastic free energy of more realistic models than the affine and the phantom network models are given in the following sections.

7.4.1.3 Non-Gaussian network theories

Most molecular theories are based on the Gaussian distribution of end-to-end distances of the network chains. This distribution, however, is not suitable for very short chains, or for any chains stretched to near the limits of their extensibility [2, 3, 28, 79, 172]. In these cases the modulus shows a marked upturn at high elongations, a result that may be of practical as well as fundamental importance.

Non-Gaussian theories developed for such circumstances are either analytical, involving generalization or modification of the Gaussian limit, or simulations, typically based on rotational isomeric-state models. Some of the results thus obtained are described in a later section.

7.4.2 Modern network theories

The term 'modern' refers to theories of rubberlike elasticity introduced after 1975 mainly to account for the disagreement between experiment and the predictions of the phantom or affine network models. All the theories in this category may essentially be regarded as corrections to the phantom network model. Corrections come in the form of contributions of entanglements to the elastic free energy.

7.4.2.1 Constrained-junction model

The constrained-junction model assumes that the fluctuations of junctions are diminished below those of the phantom network due to the presence of entanglements. It is based on initial work carried out by Ronca and Allegra [150], and subsequently developed by Flory and Erman [8, 151, 155]. As indicated by the second part of equation (7.23), the fluctuations in a phantom network are substantial. For a tetrafunctional network, the mean-square fluctuations of junctions amount to as much as half of the mean-square end-to-end distance of the network chains. The strength of the constraints is measured by a parameter κ, defined as

$$\kappa = \frac{\langle(\Delta R)^2\rangle}{\langle(\Delta s)^2\rangle} \tag{7.25}$$

where $\langle(\Delta R)^2\rangle$ and $\langle(\Delta s)^2\rangle$ denote, respectively, the mean-square dimension of junction fluctuations in the phantom network and in the entanglement domain. If the range of fluctuations decreases to zero due to entanglements, κ becomes infinitely large. If the effect of entanglements is nil, then $\kappa = 0$. The elastic free-energy of the constrained-junction model is given by the expression

$$\Delta A_{\text{el}} = \tfrac{1}{2}\xi kT\left\{\sum_{i=1}^{3}(\lambda_i^2 - 1) + \frac{\mu}{\xi}\sum_{i=1}^{3}[B_{\text{t}} + D_{\text{t}} - \ln(1 + B_{\text{t}}) - \ln(1 + D_{\text{t}})]\right\} \tag{7.26}$$

where

$$B_{\text{t}} = \kappa^2(\lambda_{\text{t}}^2 - 1)(\lambda_{\text{t}}^2 + \kappa)^2 \qquad D_{\text{t}} = \lambda_{\text{t}}^2\kappa^{-1}B_{\text{t}} \tag{7.27}$$

This theory has been shown to give results which agree quantitatively with stress–strain swelling data [173], birefringence [174, 175], and segmental orientation [133, 176] in uniaxially stretched networks and with experimental data obtained in shear and multi-axial states of stress [177].

The Ronca–Allegra [150] and Flory–Erman [8, 151, 155] theories are both based on the idea that effects of constraints are local and decrease with increasing strain and swelling. The basic difference between the two theories is that in the Ronca–Allegra theory the fluctuations of junctions become exactly affine as the undeformed state is approached, whereas in the Flory theory they are close to but below those of the affine state.

7.4.2.2 Constrained-chain model

This refinement of the constrained-junction model is based on re-examination of the constraint problem and evaluation of some neutron scattering estimates of actual junction fluctuations [156–158]. It was concluded that the suppression of the fluctuations was overestimated in the theory, presumably because the entire effect of the interchain interactions was arbitrarily placed on the junctions. The theory was therefore revised to make it more realistic by placing the effects of the constraints on the centers-of-mass of the network chains [178–180]. This modification also provided improved agreement between theory and experiment.

7.4.2.3 Diffused-constraints model

This theory attempts an even greater realism, by distributing the constraints continuously along the network chains. In its application to stress–strain isotherms in elongation [181], it has the advantage of having only a single constraint parameter and the values it exhibits upon comparing theory and experiment seem more reasonable than in the earlier models. Applications to strain birefringence [182], however, yield values of the birefringence that are much larger than those in the constrained-junction and constrained-chain theories. Additional comparisons between theory and experiment for a variety of elastomeric properties should be very helpful. Also, neutron scattering measurements carried out on series of networks having different junction functionalities ϕ would be extremely useful in suggesting how to position the constraints along a chain in refining such models, since ϕ should have a pronounced effect on the magnitudes of the fluctuations of the junctions.

7.4.3 *Attempts at greater rigour*

The preceding theories on the effects of constraints on fluctuations start from a detailed molecular model of the real network [4, 8]. The fundamental postulate of the theories states that only strain-dependent contributions to the elastic free energy are of importance, and that these contributions vanish at infinitely large extensions or swelling. Contributions from trapped entanglements, for example, are categorically eliminated when this postulate is accepted. According to this approach, whether the trapped entanglements contribute to stress can only be determined by experiments. A complete statistical mechanical theory that does not depend on physical assumptions of the type stated has been outlined by Freed [183], and a mathematical theory of elasticity of networks with internal constraints has been worked out by Edwards and Freed [184–188]. According to these theories, a set of internal constraints, including knots and entanglements, are assumed to be fixed during the formation of the network. These constraints are conserved under deformation, thus contributing to the elastic properties.

In this area, the slip-link model [184–188] incorporates the effects of entanglements along the chain contour into the elastic free energy. According to the mechanism of the slip link, sketched in Figure 7.2, a link joins two different chains and may slide a distance a along the contour of the chains. The elastic free energy resulting from this model is

$$A_{\mathrm{el}} = \tfrac{1}{2}N_{\mathrm{c}}kT\left\{\sum_{i=1}^{3}\lambda_i^2 + \frac{N_{\mathrm{s}}}{N_{\mathrm{c}}}\sum_{i=1}^{3}\left[\frac{(1+\eta)\lambda_i^2}{1+\eta\lambda_i^2} + \log(1+\eta\lambda_i^2)\right]\right\} \tag{7.28}$$

where N_{c} and N_{s} are the number of chemical crosslinks and slip links, respectively, and $\eta = 0.2343$. The first term on the right-hand-side of equation (7.28) is the contribution to the elastic free energy from the phantom network. The effect of entanglements enter as further contributions and is proportional to the number of slip links.

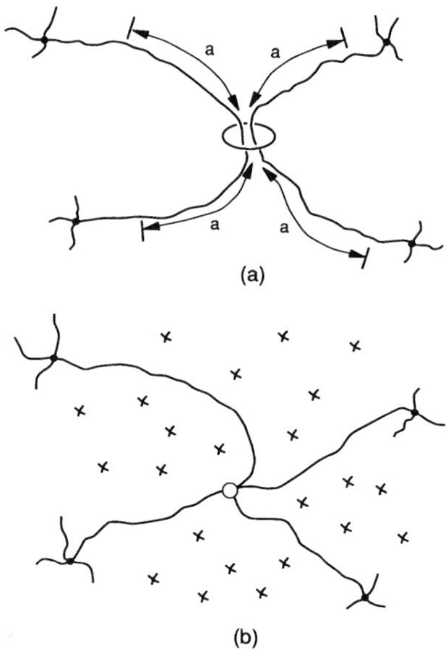

Figure 7.2 (a) Schematic drawing of a slip link, with its possible motions along the network chains specified by the distances a, and (b) its locking into position as a crosslink (junction point) [4].

The slip-link theory has subsequently been treated in a simplified mathematical approach by Edwards and Vilgis [189], and is discussed further in a recent review [12].

7.5 Rotational-isomeric-state chains

7.5.1 Analytical calculations

As will be described, the major applications of rotational-isomeric-state models have been to thermoelasticity [3, 28, 190, 191], strain birefringence [191–195], segmental orientation [196], strain dichroism [197], and rotational isomerization [198].

7.5.2 Simulations

Monte-Carlo simulations have been used to calculate thermoelastic results through the temperature coefficient of the unperturbed dimensions [191]. In the case of networks of the protein elastin, such results were used to evaluate alternative theories for the molecular deformation mechanism for this bioelastomer [199].

Stress–strain isotherms have also been calculated with this approach. Examples are unimodal networks of polyethylene and poly(dimethylsiloxane) [200], polymeric sulfur and selenium [201], short *n*-alkane chains [202], natural rubber [203], several polyoxides [204, 205], elastin [206] and bimodal networks of poly(dimethylsiloxane) [207]. It is possible to include excluded-volume [1] effects [208]. In the case of the partially helical polymer poly(oxymethylene), the simulations were used to resolve the overall distributions into contributions from unbroken rods, once-broken rods, twice-broken rods, etc. [205]. It was also shown how applying stresses to the ends of chains of this type can be used to bias the distributions in the direction of increased helical content and increased average end-to-end distances [205]. In this sense, imposition of a stress has the same effect on the helix-coil equilibrium as a decrease in temperature.

7.6 Phenomenological elasticity theory

The phenomenological approach to elasticity theory is based on continuum mechanics and symmetry arguments rather than on molecular concepts [2, 209–211]. It attempts to fit stress–strain data with a minimum number of parameters, which are then used to predict other mechanical properties of the same material.

The elastic free energy given by the elementary and the more advanced theories in this area are symmetric functions of the three deformation ratios λ_x, λ_y and λ_z. One may also express the dependence of the elastic free energy on strain in terms of three other variables, which are in turn functions of λ_x, λ_y and λ_z. In phenomenological theories of continuum mechanics, where only the observed behaviour of the material is of concern rather than the molecular mechanism, these three functions are chosen to be

$$\begin{aligned} I_1 &= \lambda_1^2 + \lambda_2^2 + \lambda_3^2 \\ I_2 &= \lambda_1^2\lambda_2^2 + \lambda_1^2\lambda_3^2 + \lambda_2^2\lambda_3^2 \\ I_3 &= \lambda_1^2\lambda_2^2\lambda_3^2 \end{aligned} \tag{7.29}$$

Here, I_1, I_2 and I_3 are referred to as the three strain-invariants. The term invariant comes from the property that the value of the expression is independent of the choice of the coordinate frame. The elastic free energies of the phantom and the affine network models now take the simple form

$$\Delta A_{\text{el}} = \mathscr{F}kT(I_1 - 3) \tag{7.30}$$

where $\mathscr{F}$ equates to $\xi kT/2$ and $\nu kT/2$ for the phantom and the affine network models, respectively. In general, the elastic free energy of the network is expressed as

$$\Delta A_{\text{el}} = \Delta A_{\text{el}}(I_1, I_2, I_3) \tag{7.31}$$

The most general form of the elastic free energy may be written as a power series

$$\Delta A_{\mathrm{el}} = \sum_{i,j,k=0}^{\infty} C_{ijk}(I_1 - 3)^i (I_2 - 3)^j (I_3 - 1)^k \tag{7.32}$$

where C_{ijk} are the phenomenological coefficients. The simple cases of the phantom and affine networks are obtained as the first term of the series

$$\Delta A_{\mathrm{el}} = C_{100}(I_1 - 3) \tag{7.33}$$

The elastic free energy of the so-called Mooney–Rivlin solid is obtained from equation (7.32) as

$$\Delta A_{\mathrm{el}} = C_{100}(I_1 - 3) + C_{010}(I_2 - 3) \tag{7.34}$$

This form of the elastic free energy has been used widely in the treatment of data from simple tension experiments on elastomers. It should be noted, however, that although the Mooney–Rivlin expression represents the behaviour of rubbers under tension satisfactorily, it fails utterly in compression.

7.7 Stress–strain relationships

The stress–strain relationships, or the equations of state, may be easily understood by considering the homogeneous deformation of a cube of sides L_0 in the reference state and L_x, L_y and L_z in the deformed state, as shown in Figure 7.3. The reference state is defined as the state in which the

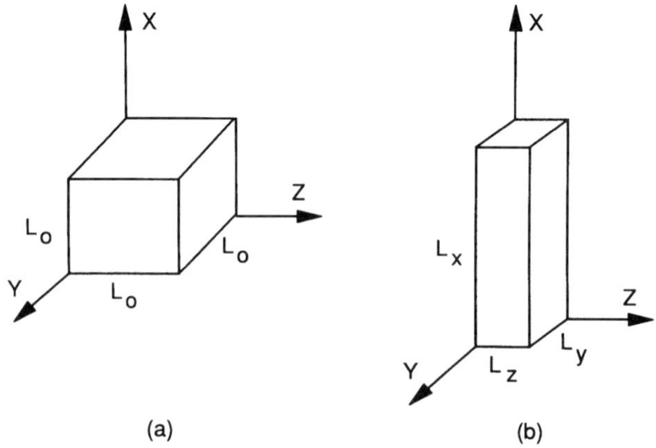

Figure 7.3 Homogeneous deformation of a cube of sides L_0 [4].

network has been formed. The components of the deformation tensor $\boldsymbol{\lambda}$ representing the state of deformation shown in Figure 7.3 are given as

$$\boldsymbol{\lambda} = \begin{bmatrix} \lambda_x & 0 & 0 \\ 0 & \lambda_y & 0 \\ 0 & 0 & \lambda_z \end{bmatrix} = \begin{bmatrix} L_x/L_0 & 0 & 0 \\ 0 & L_y/L_0 & 0 \\ 0 & 0 & L_z/L_0 \end{bmatrix} \tag{7.35}$$

The state of the network during formation is of particular importance in obtaining the correct expression for the stress. If the network is prepared in the presence of diluent, the reference volume $V_0 = L_0^3$ equals the sum of the dry volume V_d of bulk polymer, and the volume of solvent V_s.

The deformation tensor $\boldsymbol{\lambda}$ of equation (7.35) may be written in terms of a distortional and a dilatational part as

$$\boldsymbol{\lambda} = \begin{bmatrix} \alpha_x(v_{2c}/v_2)^{1/3} & 0 & 0 \\ 0 & \alpha_y(v_{2c}/v_2)^{1/3} & 0 \\ 0 & 0 & \alpha_z(v_{2c}/v_2)^{1/3} \end{bmatrix} \tag{7.36}$$

Here, α_x represents the ratio of the length along the x direction in the deformed state to that in the swollen but undistorted state, with similar definitions for the y and z directions. v_{2c} is the volume fraction of polymer during network formation and v_2 is that during the experiment (which is assumed to remain constant during network deformation). This approximation fails, obviously, if the sample is highly compressed, or if the deformation is carried out while the sample is immersed in the solvent. The volume of a dry network remains approximately constant during deformation, according to which

$$\alpha_x\alpha_y\alpha_z = 1 \tag{7.37}$$

Equation (7.37) is a statement of the incompressibility of rubber during distortion. In terms of the components of $\boldsymbol{\lambda}$, using equation (7.37) in equation (7.36), one obtains

$$\lambda_x\lambda_y\lambda_z = V/V_0 \tag{7.38}$$

The equation of state relating the components of the true stress tensor t_i to deformation is given for an incompressible material by the thermodynamic relation

$$t_i = 2V^{-1}\lambda_i^2 \frac{\partial \Delta A_{el}}{\partial \lambda_i^2} \qquad i = 1, 2, 3 \tag{7.39}$$

where the ith component t_i of the true stress tensor is defined as the force along the ith coordinate direction per unit deformed area. It should be noted that the three components t_i $(i = 1, 2, 3)$ constitute the diagonal elements of the stress tensor $\boldsymbol{t}$, corresponding to the normal stresses acting on the three faces of the prism shown in Figure 7.3. The off-diagonal components correspond to shear stresses. In the particular case when the shear stresses equate to zero, the normal stresses are referred to as the principal stresses. In the following sections only simple tension or compression, which is the simplest state of principal stresses, is considered. The reader is referred to Treloar [2] for applications to other states of stress.

Equation (7.39) allows the derivation of the three principal stresses from the elastic free energy of the network. The elastic free energy follows either from the molecular theories or from the phenomenological formulation. In the following, the elastic free energy expression according to the constrained-junction theory will be considered. This model is adopted because it has been widely studied in recent years and its predictions have been compared with experimental data from different polymeric systems.

Substituting the expression for ΔA_{el} from equation (7.26) into the right hand side of equation (7.39), the principal components t_i are obtained as

$$t_i = 2V^{-1}\lambda_i^2 \frac{\partial \Delta A_{\mathrm{el}}}{\partial \lambda_i^2} = \frac{\xi kT}{V}\lambda_i^2\left[1 + \frac{\mu}{\xi}K(\lambda_i^2)\right] \tag{7.40}$$

where

$$\begin{aligned} K(\lambda^2) &= B[\dot{B}(B+1)^{-1} + \kappa^{-1}(\lambda^2\dot{B} + B)(B + \kappa\lambda^{-2})^{-1}] \\ \dot{B} &= \frac{\partial B}{\partial \lambda^2} = B[(\lambda^2 - 1)^{-1} - 2(\lambda^2 + \kappa)^{-1}] \end{aligned} \tag{7.41}$$

The subscript i in equation (7.41) is omitted for simplicity. Equation (7.40) is given for the general state of principal stresses and will now be specialized to simple tension. For simple tension along the x direction, the deformation tensor given by equation (7.36) reduces, with the help of equation (7.37), to

$$\boldsymbol{\lambda} = \left(\frac{V}{V_0}\right)^{1/3}\begin{bmatrix} \alpha & 0 & 0 \\ 0 & \alpha^{-1/2} & 0 \\ 0 & 0 & \alpha^{-1/2} \end{bmatrix} \tag{7.42}$$

The appropriate quantity measured in experiments is the force f along the x direction which is related to the components of the true stress as

$$f = A(t_x - t_y) \tag{7.43}$$

$$= \frac{\xi kT}{L_0}\left(\frac{V}{V_0}\right)^{1/3}\left\{(\alpha - \alpha^{-2}) + \frac{\mu}{\xi}[\alpha K(\lambda_x^2) - \alpha^{-2}K(\lambda_y^2)]\right\} \tag{7.44}$$

In equation (7.43), A is the deformed cross-sectional area. The lateral true stress t_y equates to the hydrostatic pressure in the case of simple tension and has therefore been subtracted from t_x in equation (7.43), inasmuch as

f is defined as the net force acting along the x direction. The reduced stress, $[f^*]$, is defined according to the relationship

$$[f^*] \equiv \frac{f v_2^{1/3}}{A_d(\alpha - \alpha^{-2})} \tag{7.45}$$

where A_d is the dry, undeformed cross-sectional area. Substituting the force given by equation (7.44) into equation (7.45) leads to

$$[f^*] = \left(\frac{\xi kT}{V_d}\right) v_{2c}^{1/3} \left\{1 + \frac{\mu}{\xi}\left[\frac{\alpha K(\lambda_x^2) - \alpha^{-2}K(\lambda_y^2)}{\alpha - \alpha^{-2}}\right]\right\} \tag{7.46}$$

The reduced force may be interpreted as the shear modulus of the network [3]. According to equation (7.46), the reduced force consists of a term due to contributions from the phantom network and another from the constraints. The contribution of constraints, proportional to the term in μ/ξ in the brackets in equation (7.46), decreases as the network is stretched or swollen, an important feature in comparisons of theory and experiment.

7.8 Theoretical versus experimental behaviour

7.8.1 *Stress–strain*

Numerous investigations have been made on stress–strain relationships and have involved a variety of types of deformations [1–3, 8, 189, 212–215]. Typical experimental results for elongation and compression (biaxial extension) for unswollen and swollen networks are shown schematically [3] in Figure 7.4. They are represented in a manner consistent with the Mooney–Rivlin equation [2, 216–221]:

$$[f^*] = 2C_1 + 2C_2\alpha^{-1} \tag{7.47}$$

where $2C_1$ and $2C_2$ are constants, independent of elongation α. The constrained-junction theory successfully describes most of the features of these isotherms. Specifically, the decrease in modulus $[f^*]$ with an increase in elongation is viewed as the deformation becoming more nonaffine as the stretching of the network chains decreases chain-junction entangling; this in turn increases junction fluctuations. The observation that the decrease in $[f^*]$ is less in the case of swollen networks results from the swelling-diluent decreasing some of this entangling even at low elongations. The observation that the isotherms become approximately horizontal in compression is also successfully predicted by the theory [3, 6]. The upturn in $[f^*]$ at very high elongations is a non-Gaussian effect not accounted for in the theory and must be treated separately.

Interpretation of the upturn in $[f^*]$ in terms of the limited extensibility of the network chains first requires demonstration that reinforcement from

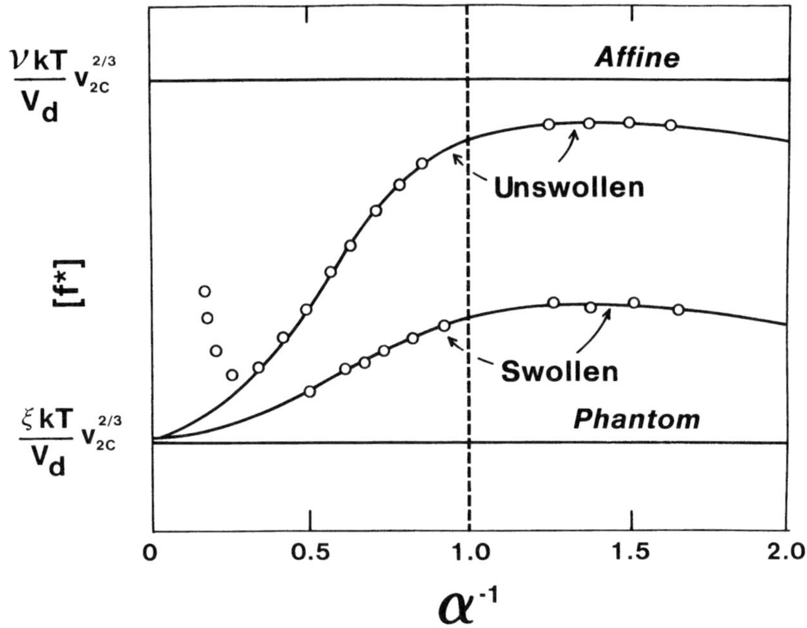

Figure 7.4 The reduced stress or modulus shown schematically as a function of the reciprocal of the relative length of the sample, for both elongation ($\alpha > 1$) and compression ($\alpha < 1$) [3]. The upper and lower lines represent results from affine and phantom network models, respectively. Circles show representative data from experiments, and the curves are from the constrained-junction theory.

strain-induced crystallization is not a significant part of the effect [222]. Failure to test for this by swelling or increase in temperature has caused a great deal of confusion in the literature.

7.8.2 *Swelling*

Experiment and theory are in at least approximate agreement regarding swelling, in that estimates of junction-point concentration from equilibrium swelling are generally in fair agreement with values obtained from mechanical properties [1, 2, 223].

A major unresolved issue concerns the observation that the swelling or dilation modulus frequently goes through an unanticipated maximum with increase in degree of swelling [224–227]. Modifications of existing theories to reproduce this feature quantitatively have not been very successful. In a worst-case scenario, this discrepancy could undermine one of the crucial assumptions in swelling theory, namely that the mixing and elastic contributions to the swelling free energy are separable and simply additive [225].

7.8.3 Birefringence

The reduced birefringence shows some properties that parallel the reduced stress or modulus. In other respects, however, it can be quite different showing, for example, significant increases with increase in degree of swelling. Nonetheless, experimental results and those from the constrained-junction theory seem to be in at least fair agreement [174, 175].

7.8.4 Orientation

Experimental results on the elongation and swelling dependence of the reduced orientation seem to be well-reproduced by theory [133, 176]. Preparing networks by crosslinking a polymer in solution [228] can give very useful orientation results, but it is essential in such cases to account for changes in reference state in the interpretation of the data [229].

7.8.5 Scattering

The most important elasticity result of this type is obtained by small-angle neutron scattering. In particular, such studies yield information on how the radius of gyration of the network chains transforms with elongation [3]. Preliminary comparisons between theory and experiment are quite encouraging [230], but many issues remain unresolved in this relatively new approach to characterizing elastomeric behaviour.

7.9 Concluding comments

The reinforcement of elastomers by the addition of fillers such as carbon black or silica is an extremely important aspect of rubberlike elasticity [231–236]. The subject has not been covered in this review, however, since relatively little is known about the molecular origins of the reinforcement, in spite of attempts going back to the 1930s [237–239]. There is now growing interest in obtaining a better molecular understanding of this phenomenon, both from analytical theory [240] and computer simulations [83, 241–245]. There may well be important advances in this area in the near future.

Acknowledgements

We wish to acknowledge the financial support generously provided by the National Science Foundation, Army Research Office, and Air Force Office of Scientific Research, and fellowship support by the Dow Corning Corporation, IBM-San Jose, and the Raychem Corporation.

Glossary of symbols and abbreviations

f	elastic force exerted on a network
$[f^*]$	reduced uniaxial stress acting on a network
l	skeletal-bond length
n	number of skeletal (main-chain) bonds in a network (polymer) chain
$\boldsymbol{r}$	end-to-end vector of a network (polymer) chain
r	end-to-end distance of a network (polymer) chain ($= \lvert\boldsymbol{r}\rvert$)
$\langle r^2 \rangle$	mean-square value of r
$\langle r^2 \rangle_0$	unperturbed, or θ-state value of $\langle r^2 \rangle$
$\langle \bar{r}^2 \rangle_0$	mean-square of the time-averaged value of r
$\langle (\Delta r)^2 \rangle_0$	mean-square instantaneous fluctuation of r
ν_2	volume fraction of a polymer in a network during deformation
ν_{2c}	volume fraction of a polymer in a network during formation
$w(r)$	radial probability distribution function of end-to-end distances (r) of a network (polymer) chain
A_{el}	elastic free energy of a network chain; elastic free energy of a network in the slip-link model
$A_{el,affine}$	elastic free energy of a network in the affine-network model
$A_{el,phantom}$	elastic free energy of a network in the phantom-network model
I_1, I_2, I_3	strain invariants in the phenomenological theory of rubberlike elasticity
M_c	average molecular weight between junction points (crosslinks)
N_A	Avogadro constant
N_C	number of chemical crosslinks (junction points) in a network in the slip-link network model
N_S	number of slip links in a network in the slip-link network model
V_d	volume of a dry network
V_0	reference volume of a network
V_s	volume of solvent in a network
$W(\boldsymbol{r})$	probability-density distribution function of end-to-end vectors of a network (polymer) chain
$\alpha_x, \alpha_y, \alpha_z$	macroscopic deformation ratios of a network along laboratory-fixed Cartesian axes and measured relative to the actual network volume
κ	parameter describing the strength of constraints on junction-point fluctuations
$\lambda_x, \lambda_y, \lambda_z$	macroscopic deformation ratios of a network along laboratory-fixed Cartesian axes and measured relative to the reference state (volume $= V_0$)
μ	number of junction points in a network
ν	number of elastically active chains in a network
ξ	cycle rank of a network

ρ density of a dry network
ϕ average functionality
χ Flory–Huggins polymer–solvent interaction parameter

References

1. Flory, P.J. (1953) *Principles of Polymer Chemistry*, Cornell University Press, Ithaca, NY.
2. Treloar, L.R.G. (1975) *The Physics of Rubber Elasticity*, 3rd edn, Oxford University Press, Clarendon.
3. Mark, J.E. and Erman, B. (1988) *Rubberlike Elasticity. A Molecular Primer*, Wiley-Interscience, New York.
4. Erman, B. and Mark, J.E. (1994) in *Science and Technology of Rubber*, 2nd edn (eds J.E. Mark, B. Erman and F.R. Eirich), Academic Press, New York, p. 189.
5. Morawetz, H. (1985) *Polymers: The Origin and Growth of a Science*, Wiley-Interscience, New York.
6. Eichinger, B.E. (1983) *Ann. Rev. Phys. Chem.*, **34**, 359.
7. Heinrich, G., Straube, E. and Helmis, G. (1988) *Adv. Polym. Sci.*, **85**, 33.
8. Erman, B. and Mark, J.E. (1989) *Ann. Rev. Phys. Chem.*, **40**, 351.
9. Erman, B. and Mark, J.E. (1997) *Structures and Properties of Rubberlike Networks*, Oxford University Press, New York.
10. Furukawa, J. (1985) *Makromol. Chem. Suppl.*, **14**, 3.
11. Kilian, H.G., Enderle, H.F. and Unseld, K. (1986) *Colloid Polym. Sci.*, **264**, 866.
12. Edwards, S.F. and Vilgis, T.A. (1988) *Rep. Prog. Phys.*, **51**, 243.
13. Schwarz, J. (1989) *Macromol. Symp.*, **30**, 251.
14. Abramchuk, S.S., Nyrkova, I.A. and Khokhlov, A.R. (1989) *Polym. Sci. U.S.S.R.*, **31**, 1936.
15. Altenberger, A.R. and Dahler, J.S. (1990) *J. Chem. Phys.*, **92**, 3100.
16. Kloczkowski, A., Mark, J.E., Erman, B. and Eichinger, B.E. (1991) *J. Chem. Phys.*, **94**, 7015.
17. Dayantis, J. (1995) *Eur. Polym. J.*, **31**, 203.
18. Gent, A.N. (ed.) (1992) *Engineering with Rubber. How to Design Rubber Components*, Hanser Publishers, New York.
19. Mark, J.E. and Erman, B. (eds.) (1992) *Elastomeric Polymer Networks*, Prentice Hall, Englewood Cliffs, NJ.
20. Mark, J.E., Erman, B. and Eirich, F.R. (eds.) (1994) *Science and Technology of Rubber*, 2nd edn, Academic Press, New York.
21. Mark, J.E. and Lal, J. (eds.) (1982) *Elastomers and Rubber Elasticity*, American Chemical Society, Washington, DC.
22. Labana, S.S. and Dickie, R.A. (eds.) (1984) *Characterization of Highly Cross-Linked Polymers*, American Chemical Society, Washington, DC.
23. Lal, J. and Mark, J.E. (eds.) (1986) *Advances in Elastomers and Rubber Elasticity*, Plenum Press, New York.
24. Singler, R.E. and Byrne, C.A. (eds.) (1987) *Elastomers and Rubber Technology*, US Government Printing Office, Washington, DC.
25. Kramer, O. (ed.) (1988) *Biological and Synthetic Polymer Networks*, Elsevier, London.
26. Baumgartner, A. and Picot, C.E. (1989) *Springer Proc. in Phys.*, Vol. 42.
27. Eichinger, B.E. (ed.) (1993) *Makromol. Chem., Macromol. Symp.*, Vol. 76.
28. Queslel, J.P. and Mark, J.E. (1986) in *Encyclopedia of Polymer Science and Engineering*, 2nd edn (ed. R.A. Meyers), Wiley-Interscience, New York, p. 365.
29. Queslel, J.P. and Mark, J.E. (1991) in *Encyclopedia of Physical Science and Technology*, 2nd edn (ed. R.A. Meyers), Academic Press, New York.
30. Mark, J.E. (ed.) (1996) *Physical Properties of Polymers Handbook*, American Institute of Physics Press, Woodbury, NY.
31. Labana, S.S. (ed.) (1977) *Chemistry and Properties of Crosslinked Polymers*, Academic Press, New York.

32. Leung, Y.-K. and Eichinger, B.E. (1984) *J. Chem. Phys.*, **80**, 3877.
33. Leung, Y.-K. and Eichinger, B.E. (1984) *J. Chem. Phys.*, **80**, 3885.
34. Dušek, K. (1986) *Adv. Polym. Sci.*, **78**, 1.
35. Stepto, R.F.T. (1986) in *Advances in Elastomers and Rubber Elasticity* (eds J. Lal and J.E. Mark) Plenum Press, New York.
36. Miller, D.R. and Macosko, C.W. (1987) *J. Polym. Sci., Polym. Phys. Ed.*, **25**, 2441.
37. Lee, K.J. and Eichinger, B.E. (1989) *Macromolecules*, **22**, 1441.
38. Lee, K.J. and Eichinger, B.E. (1990) *Polymer*, **31**, 406.
39. Lee, K.J. and Eichinger, B.E. (1990) *Polymer*, **31**, 414.
40. Eichinger, B.E. and Akgiray, O. (1994) in *Computer Simulation of Polymers* (ed. E.A. Colbourne), Longman, White Plains, NY, p. 263.
41. Mark, J.E. (1981) *J. Chem. Educ.*, **58**, 898.
42. Legge, N.R., Holden, G. and Schroeder, H.E. (eds.) (1988) *Thermoplastic Elastomers. A Comprehensive Review*, Oxford University Press, New York.
43. Taylor, G.K., Castner, E.S. and Galiatsatos, V. (1995) *J. Chem. Soc., Faraday Trans.*, **91**, 2655.
44. Mark, J.E. and Sullivan, J.L. (1977) *J. Chem. Phys.*, **66**, 1006.
45. Kosfeld, R., Hess, M. and Hansen, D. (1980) *Polym. Bulletin*, **3**, 603.
46. Mark, J.E. (1979) *Makromol. Chemie, Suppl.*, **2**, 87.
47. Meyers, K.O., Bye, M.L. and Merrill, E.W. (1980) *Macromolecules*, **13**, 1045.
48. Gottlieb, M., Macosko, C.W., Benjamin, G.S., Meyers, K.O. and Merrill, E.W. (1981) *Macromolecules*, **14**, 1039.
49. Meyers, K.O. and Merrill, E.W. (1982) in *Elastomers and Rubber Elasticity*, Vol. 193 (eds J.E. Mark and J. Lal), American Chemical Society, Washington, DC, p. 329.
50. Mark, J.E. (1985) *Acc. Chem. Res.*, **18**, 202.
51. Entelis, S.G., Evreinov, V.V. and Gorshkov, A.V. (1986) *Adv. Polym. Sci.*, **76**, 129.
52. Lanyo, L.C. and Kelley, F.N. (1987) *Rubber Chem. Technol.*, **60**, 78.
53. Smith, T.L., Haidar, B. and Hedrick, J.L. (1990) *Rubber Chem. Technol.*, **63**, 256.
54. Clarson, S.J., Wang, Z. and Mark, J.E. (1990) *Eur. Polym. J.*, **26**, 621.
55. Silva, L.K., Mark, J.E. and Boerio, F.J. (1991) *Makromol. Chemie*, **192**, 499.
56. Hanyu, A. and Stein, R.S. (1991) *Macromol. Symp.*, **45**, 189.
57. Roland, C.M. and Buckley, G.S. (1991) *Rubber Chem. Technol.*, **64**, 74.
58. Schimmel, K.-H. and Heinrich, G. (1991) *Coll. Polym. Sci.*, **269**, 1003.
59. Andrady, A.L., Llorente, M.A. and Mark, J.E. (1992) *Polym. Bulletin*, **28**, 103.
60. Oikawa, H. (1992) *Polymer*, **33**, 1116.
61. Hamurcu, E.E. and Baysal, B.M. (1993) *Polymer*, **34**, 5163.
62. Patel, S.K., Malone, S., Cohen, C., Gillmor, J.R. and Colby, R.H. (1992) *Macromolecules*, **25**, 5241.
63. Fischer, M. (1992) *Adv. Polym. Sci.*, **100**, 313.
64. Subramanian, P.R. and Galiatsatos, V. (1993) *Macromol. Symp.*, **76**, 233.
65. Mark, J.E. (1993) *New J. Chem.*, **17**, 703.
66. Bontems, S.L., Stein, J. and Zumbrum, M.A. (1993) *J. Polym. Sci., Polym. Chem. Ed.*, **31**, 2697.
67. Sharaf, M.A. and Mark, J.E. (1994) *Polymer*, **35**, 740.
68. Sharaf, M.A., Mark, J.E. and Al-Ghazal, A.A.-R. (1994) *J. Appl. Polym. Sci. Symp.*, **55**, 139.
69. Mark, J.E. (1994) *J. Inorg. Organomet. Polym.*, **4**, 31.
70. Gent, A.N., Liu, G.L. and Mazurek, M. (1994) *J. Polym. Sci., Polym. Phys. Ed.*, **32**, 271.
71. Shibayama, M., Takahashi, H., Yamaguchi, H., Sakurai, S. and Nomura, S. (1994) *Polymer*, **35**, 2944.
72. Mark, J.E. (1994) *Acc. Chem. Res.*, **27**, 271.
73. Trautenberg, H.L., Sommer, J.-U. and Goritz, D. (1995) *J. Chem. Soc., Faraday Trans.*, **91**, 2649.
74. Out, G.J.J., Turetskii, A.A., Snijder, M., Moller, M. and Papkov, V.S. (1995) *Polymer*, **36**, 3213.
75. Besbes, S., Bokobza, L., Monnerie, L., Bahar, I. and Erman, B. (1995) *Macromolecules*, **28**, 231.
76. Disch, S., Finkelmann, H., Ringsdorf, H. and Schuhmacher, P. (1995) *Macromolecules*, **28**, 2424.

77. Zentel, R. and Brehmer, M. (1995) *CHEMTECH*, **25**(5), 41.
78. Clarson, S.J. and Mark, J.E. (1996) in *Polymeric Materials Encyclopedia: Synthesis, Properties, and Applications*, (ed. J.C. Salamone), CRC Press, Boca Raton.
79. Erman, B. and Mark, J.E. (1988) *J. Chem. Phys.*, **89**, 3314.
80. Kloczkowski, A., Mark, J.E. and Erman, B. (1991) *Macromolecules*, **24**, 3266.
81. Sakrak, G., Bahar, I. and Erman, B. (1994) *Macromol. Theory Simul.*, **3**, 151.
82. Bahar, I., Erman, B., Bokobza, L. and Monnerie, L. (1995) *Macromolecules*, **28**, 225.
83. Mark, J.E. (1996) *Macromol. Symp.*, **101**, 423.
84. Mark, J.E., Eisenberg, A., Graessley, W.W., Mandelkern, L., Samulski, E.T., Koenig, J.L. and Wignall, G.D. (1993) *Physical Properties of Polymers*, 2nd edn, American Chemical Society, Washington, DC.
85. Mark, J.E. (1989) *J. Appl. Polym. Sci., Symp.*, **44**, 209.
86. de Gennes, P.G. (1979) *Scaling Concepts in Polymer Physics*, Cornell University Press, Ithaca, New York.
87. Garrido, L., Mark, J.E., Clarson, S.J. and Semlyen, J.A. (1985) *Polym. Comm.*, **26**, 53.
88. Iwata, K. and Ohtsuki, T. (1993) *J. Polym. Sci., Polym. Phys. Ed.*, **31**, 441.
89. Flory, P.J. (1976) *Proc. R. Soc. London, A*, **351**, 351.
90. Mark, J.E. (1985) *Brit. Polym. J.*, **17**, 144.
91. Gent, A.N. and Kuan, T.H. (1973) *J. Polym. Sci., Polym. Phys. Ed.*, **11**, 1723.
92. Gent, A.N. and Kuan, T.H. (1974) *J. Polym. Sci., Polym. Phys. Ed.*, **12**, 633.
93. Mohsin, M.A., Berry, J.P. and Treloar, L.R.G. (1986) *Brit. Polym. J.*, **18**, 145.
94. Mohsin, M.A. and Treloar, L.R.G. (1987) *Polymer*, **28**, 1893.
95. Xu, P. and Mark, J.E. (1990) *Rubber Chem. Technol.*, **63**, 276.
96. Xu, P. and Mark, J.E. (1991) *J. Polym. Sci., Polym. Phys. Ed.*, **29**, 355.
97. Xu, P. and Mark, J.E. (1991) *Makromol. Chemie*, **192**, 567.
98. Wang, S., Xu, P. and Mark, J.E. (1991) *Rubber Chem. Technol.*, **64**, 746.
99. Xu, P. and Mark, J.E. (1992) *Polymer*, **33**, 1843.
100. Mark, J.E., Wang, S., Xu, P. and Wen, J. (1992) in *Submicron Multiphase Materials*, Vol. 274 (eds R.H. Baney, L.R. Gilliom, S.-I. Hirano and H.K. Schmidt), Materials Research Society, Pittsburgh, PA, p. 77.
101. Wen, J. and Mark, J.E. (1994) *Polym. J.*, **26**, 151.
102. Mazan, J., Leclerc, B., Galandrin, N. and Couarraze, G. (1995) *Eur. Polym. J.*, **31**, 803.
103. Tanaka, T. (1978) *Phys. Rev. Lett.*, **40**, 820.
104. Tanaka, T. (1981) *Sci. Am.*, **244**, 124.
105. Ilavsky, M. (1982) *Macromolecules*, **15**, 782.
106. Ilavsky, M., Hrouz, J., Stejskal, J. and Bouchal, K. (1984) *Macromolecules*, **17**, 2868.
107. Tanaka, T., Sun, S.-T., Hirokawa, Y., Katayama, S., Kucera, J., Hirose, Y. and Amiya, T. (1987) *Nature*, **325**, 798.
108. Ilavsky, M. and Bouchal, K. (1988) in *Biological and Synthetic Polymer Networks* (ed. O. Kramer), Elsevier, London, p. 435.
109. Suzuki, A. and Tanaka, T. (1990) *Nature*, **346**, 345.
110. Kokufata, E., Zhang, Y.-Q. and Tanaka, T. (1991) *Nature*, **351**, 302.
111. Osada, Y. (1991) *Adv. Mater.*, **3**, 107.
112. Tanaka, T. (1992) in *Polyelectrolyte Gels*, Vol. 480 (eds R.S. Harland and R.K. Prud'homme), American Chemical Society, Washington, DC, p. 1.
113. Li, Y. and Tanaka, T. (1992) *Annu. Rev. Mater. Sci.*, **22**, 243.
114. Annaka, M. and Tanaka, T. (1992) *Nature*, **355**, 430.
115. Zhang, Y.-Q., Tanaka, T. and Shibayama, M. (1992) *Nature*, **360**, 142.
116. Osada, Y., Okuzaki, H. and Hori, H. (1992) *Nature*, **355**, 242.
117. Matsuo, E.S. and Tanaka, T. (1992) *Nature*, **358**, 482.
118. Katayama, S. (1992) *J. Phys. Chem.*, **96**, 5209.
119. Dušek, K. (ed.) (1993) *Responsive Gels, I*, Springer-Verlag, Berlin.
120. Dušek, K. (ed.) (1993) *Responsive Gels, II*, Springer-Verlag, Berlin.
121. Tsuchiya, M., Tsuji, K., Maki, K. and Tanaka, T. (1994) *J. Phys. Chem.*, **98**, 6187.
122. Hirai, T., Nemoto, H., Hirai, M. and Hayashi, S. (1994) *J. Appl. Polym. Sci.*, **53**, 79.
123. Osada, T., Okuzaki, H. and Gong, J.P. (1994) *Trends Polym. Sci.*, **2**, 61.
124. Tanaka, T. (1995) *Macromolecules*, **28**, 1704.
125. Tanaka, Y., Kagami, Y., Matsuda, A. and Osada, Y. (1995) *Macromolecules*, **28**, 2574.

126. Kuroki, Y. and Sekimoto, K. (1995) *J. Chem. Phys.*, **102**, 8626.
127. Yoshida, R., Ichijo, H., Hakuta, T. and Yamaguchi, T. (1995) *Macromol. Rapid Commun.*, **16**, 305.
128. Osada, Y. and Matsuda, A. (1995) *Nature*, **376**, 219.
129. Galiatsatos, V. and Mark, J.E. (1987) *Polym. Bulletin*, **17**, 197.
130. Galiatsatos, V. and Mark, J.E. (1987) *Macromolecules*, **20**, 2631.
131. Galiatsatos, V. and Mark, J.E. (1990) in *Advances in Silicon-Based Polymer Science. A Comprehensive Resource* (eds J.M. Zeigler and F.W.G. Fearon), American Chemical Society, Washington, DC.
132. Monnerie, L. (1983) *Faraday Symp. Chem. Soc.*, **18**, 1.
133. Queslel, J.P., Erman, B. and Monnerie, L. (1985) *Macromolecules*, **18**, 1991.
134. Spiess, H.W. (1992) *Angew. Makromol. Chemie*, **202/203**, 331.
135. Jezzard, P., Wiggins, C.J., Carpenter, T.A., Hall, L.D., Jackson, P., Clayden, N.J. and Walton, N.J. (1992) *Adv. Mater.*, **4**, 82.
136. Noda, I., Dowrey, A.E. and Marcott, C. (1987) in *Fourier Transform Infrared Characterization of Polymers* (ed. H. Ishida), Plenum Press, New York.
137. Noda, I. (1989) *J. Am. Chem. Soc.*, **111**, 8116.
138. Noda, I., Dowrey, A.E. and Marcott, C. (1993) *Macromol. Symp.*, **72**, 121.
139. Noda, I., Dowrey, A.E. and Marcott, C. (1993) *Polym. News*, **18**, 167.
140. Wignall, G.D. (1993) in *Physical Properties of Polymers*, 2nd edn (eds J.E. Mark, A. Eisenberg, W.W. Graessley, L. Mandelkern, E.T. Samulski, J.L. Koenig and G.D. Wignall), American Chemical Society, Washington, DC, p. 313.
141. Higgins, J.S. and Benoit, H. (1994) *Neutron Scattering from Polymers*, Clarendon Press, Oxford.
142. Wignall, G.D. (1996) in *Physical Properties of Polymers Handbook* (ed. J.E. Mark), American Institute of Physics Press, Woodbury, NY, p. 299.
143. Landry, M.R., Coltrain, B.K., Landry, C.J.T. and O'Reilly, J.M. (1995) *J. Polym. Sci., Polym. Phys. Ed.*, **33**, 637.
144. McCarthy, D.W., Mark, J.E. and Schaefer, D.W. (1997) *J. Polym. Sci., Polym. Phys. Ed.* (submitted).
145. McCarthy, D.W., Mark, J.E., Clarson, S.J. and Schaefer, D.W. (1997) *J. Polym. Sci., Polym. Phys. Ed.* (submitted).
146. Flory, P.J. and Yoon, D.Y. (1974) *J. Chem. Phys.*, **61**, 5358.
147. Yoon, D.Y. and Flory, P.J. (1974) *J. Chem. Phys.*, **61**, 5366.
148. Fixman, M. and Alben, R. (1973) *J. Chem. Phys.*, **58**, 1553.
149. Shen, M. (1978) in *Science and Technology of Rubber* (ed. F.R. Eirich), Academic Press, New York, p. 155.
150. Ronca, G. and Allegra, G. (1975) *J. Chem. Phys.*, **63**, 4990.
151. Flory, P.J. (1977) *J. Chem. Phys.*, **66**, 5720.
152. Flory, P.J. (1977) in *Contemporary Topics in Polymer Chemistry* (eds E.M. Pearce and J.R. Schaefgen), Plenum Press, New York, Vol. 2.
153. Flory, P.J. (1979) *Polymer*, **20**, 1317.
154. Flory, P.J. (1985) *Polym. J.*, **17**, 1.
155. Flory, P.J. and Erman, B. (1982) *Macromolecules*, **15**, 800.
156. Oeser, R., Ewen, B., Richter, D. and Farago, B. (1988) *Phys. Rev. Lett.*, **60**, 1041.
157. Ewen, B. and Richter, D. (1989) in *Molecular Basis of Polymer Networks*, Vol. 42 (eds A. Baumgartner and C.E. Picot), p. 82.
158. Ewen, B. and Richter, D. (1992) in *Elastomeric Polymer Networks* (eds J.E. Mark and B. Erman), Prentice Hall, Englewood Cliffs, NJ.
159. Stepto, R.F.T. and Taylor, D.J.R. (1995) *J. Chem. Soc. Faraday Trans.*, **91**, 2639.
160. Guth, E. and James, H.M. (1941) *Ind. Eng. Chem.*, **33**, 624.
161. James, H.M. and Guth, E. (1942) *Ind. Eng. Chem.*, **34**, 1365.
162. James, H.M. (1947) *J. Chem. Phys.*, **15**, 651.
163. James, H.M. and Guth, E. (1947) *J. Chem. Phys.*, **15**, 669.
164. James, H.M. and Guth, E. (1953) *J. Chem. Phys.*, **21**, 1039.
165. Duiser, J.A. and Staverman, A.J. (1965) in *Physics of Non-Crystalline Solids* (ed. J.A. Prins), North-Holland, Amsterdam.
166. Eichinger, B.E. (1972) *Macromolecules*, **5**, 496.

167. Graessley, W.W. (1975) *Macromolecules*, **8**, 186.
168. Graessley, W.W. (1975) *Macromolecules*, **8**, 865.
169. Pearson, D.S. (1977) *Macromolecules*, **10**, 696.
170. Kloczkowski, A., Mark, J.E. and Erman, B. (1989) *Macromolecules*, **22**, 1423.
171. Erman, B., Kloczkowski, A. and Mark, J.E. (1989) *Macromolecules*, **22**, 1432.
172. Mark, J.E. (1993) in *Physical Properties of Polymers*, 2nd edn (eds J.E. Mark, A. Eisenberg, W.W. Graessley, L. Mandelkern, E.T. Samulski, J.L. Koenig and G.D. Wignall), American Chemical Society, Washington, DC, p. 3.
173. Erman, B. and Flory, P.J. (1982) *Macromolecules*, **15**, 806.
174. Erman, B. and Flory, P.J. (1983) *Macromolecules*, **16**, 1601.
175. Erman, B. and Flory, P.J. (1983) *Macromolecules*, **16**, 1607.
176. Erman, B. and Monnerie, L. (1985) *Macromolecules*, **18**, 1985.
177. Erman, B. (1981) *J. Polym. Sci., Polym. Phys. Ed.*, **19**, 829.
178. Erman, B. and Monnerie, L. (1989) *Macromolecules*, **22**, 3342.
179. Erman, B. and Monnerie, L. (1992) *Macromolecules*, **25**, 4456.
180. Fontaine, F., Noel, C., Monnerie, L. and Erman, B. (1989) *Macromolecules*, **22**, 3352.
181. Kloczkowski, A., Mark, J.E. and Erman, B. (1995) *Macromolecules*, **28**, 5089.
182. Kloczkowski, A., Mark, J.E. and Erman, B. (1995) *Comput. Polym. Sci.*, **5**, 37.
183. Freed, K.F. (1971) *J. Chem. Phys.*, **55**, 5588.
184. Edwards, S.F. (1968) *J. Phys. A, Gen Phys.*, **1**, 15.
185. Edwards, S.F. and Freed, K.F. (1970) *J. Phys. C*, **3**, 739, 750, 760.
186. Edwards, S.F. (1971) in *Polymer Networks: Structural and Mechanical Properties* (eds A.J. Chompff and S. Newman), Plenum, New York.
187. Edwards, S.F. (1971) *Brit. Polym. J.*, **3**, 140.
188. Deam, R.T. and Edwards, S.F. (1976) *Philos. Trans. R. Soc. London, Ser. A*, **280**, 317.
189. Edwards, S.F. and Vilgis, T.A. (1986) *Polymer*, **27**, 483.
190. Queslel, J.P. and Mark, J.E. (1987) in *Encyclopedia of Physical Science and Technology* (ed. R.A. Meyers), Academic Press, New York.
191. Flory, P.J. (1969) *Statistical Mechanics of Chain Molecules*, Interscience, New York.
192. Nagai, K. (1964) *J. Chem. Phys.*, **40**, 2818.
193. Liberman, M.H., Abe, Y. and Flory, P.J. (1972) *Macromolecules*, **5**, 550.
194. Ingwall, R.T., Czurylo, E.A. and Flory, P.J. (1973) *Biopolymers*, **12**, 1137.
195. Liberman, M.H., DeBolt, L.C., and Flory, P.J. (1974) *J. Polym. Sci., Polym. Phys. Ed.*, **12**, 187.
196. Erman, B. and Bahar, I. (1988) *Macromolecules*, **21**, 452.
197. Flory, P.J. and Abe, Y. (1969) *Macromolecules*, **2**, 335.
198. Abe, Y. and Flory, P.J. (1970) *J. Chem. Phys.*, **52**, 2814.
199. DeBolt, L.C. and Mark, J.E. (1987) *Polymer*, **28**, 416.
200. Mark, J.E. and Curro, J.G. (1983) *J. Chem. Phys.*, **79**, 5705.
201. Mark, J.E. and Curro, J.G. (1984) *J. Chem. Phys.*, **80**, 5262.
202. Mark, J.E. and Curro, J.G. (1984) *J. Chem. Phys.*, **81**, 6408.
203. Mark, J.E. and Curro, J.G. (1985) *J. Polym. Sci., Polym. Phys. Ed.*, **23**, 2629.
204. Curro, J.G. and Mark, J.E. (1985) *J. Chem. Phys.*, **82**, 3820.
205. Curro, J.G., Schweizer, K.S., Adolf, D. and Mark, J.E. (1986) *Macromolecules*, **19**, 1739.
206. DeBolt, L.C. and Mark, J.E. (1988) *J. Polym. Sci., Polym. Phys. Ed.*, **26**, 865.
207. Curro, J.G. and Mark, J.E. (1984) *J. Chem. Phys.*, **80**, 4521.
208. Cifra, P. and Bleha, T. (1995) *J. Chem. Soc. Faraday Trans.*, **91**, 2465.
209. Ogden, R.W. (1982) in *Mechanics of Solids*, Vol. 28 (eds H.G. Hopkins and M.J. Sewell), p. 499.
210. Ogden, R.W. (1986) *Rubber Chem. Technol.*, **59**, 361.
211. Ogden, R.W. (1987) *Polymer*, **28**, 379.
212. Pak, H. and Flory, P.J. (1979) *J. Polym. Sci., Polym. Phys. Ed.*, **17**, 1845.
213. Brotzman, R.W. and Mark, J.E. (1986) *Macromolecules*, **19**, 667.
214. Gottlieb, M. and Gaylord, R.J. (1983) *Polymer*, **24**, 1644.
215. Brereton, M.G. and Klein, P.G. (1988) *Polymer*, **29**, 970.
216. Mooney, M. (1940) *J. Appl. Phys.*, **11**, 582.
217. Mooney, M. (1948) *J. Appl. Phys.*, **19**, 434.
218. Rivlin, R.S. (1948) *Phil. Trans. R. Soc., Part A*, **240**, 459.

219. Rivlin, R.S. (1948) *Phil. Trans. R. Soc., Part A*, **240**, 491.
220. Rivlin, R.S. (1948) *Phil. Trans. R. Soc., Part A*, **240**, 509.
221. Rivlin, R.S. (1948) *Phil. Trans. R. Soc., Part A*, **241**, 379.
222. Mark, J.E. (1979) *Polym. Eng. Sci.*, **19**, 409.
223. Tan, Z., Jaeger, R. and Vancso, G.J. (1994) *Polymer*, **35**, 3230.
224. Gottlieb, M. and Gaylord, R.J. (1984) *Macromolecules*, **17**, 2024.
225. Neuberger, N.A. and Eichinger, B.E. (1988) *Macromolecules*, **21**, 3060.
226. McKenna, G.B. and Crissman, J.M. (1993) in *Macromolecules 92* (ed. J. Kahovec), VSP, Utrecht, The Netherlands, p. 67.
227. Horkay, F. and McKenna, G.B. (1996) in *Physical Properties of Polymers Handbook* (ed. J.E. Mark), American Institute of Physics Press, Woodbury, NY, p. 379.
228. Dubault, A., Deloche, B. and Herz, J. (1987) *Macromolecules*, **20**, 2096.
229. Erman, B. and Mark, J.E. (1989) *Macromolecules*, **22**, 480.
230. Erman, B. (1987) *Macromolecules*, **20**, 1917.
231. Kraus, G. (ed.) (1965) *Reinforcement of Elastomers*, Interscience, New York.
232. Kraus, G. (1971) *Adv. Polym. Sci.*, **8**, 155.
233. Boonstra, B.B. (1979) *Polymer*, **20**, 691.
234. Rigbi, Z. (1980) *Adv. Polym. Sci.*, **36**, 21.
235. Donnet, J.-B. and Vidal, A. (1986) *Adv. Polym. Sci.*, **76**, 103.
236. Edwards, D.C. (1990) *J. Mater. Sci.*, **25**, 4175.
237. Guth, E. and Gold, O. (1938) *Phys. Rev.*, **53**, 322.
238. Smallwood, H.M. (1944) *J. Appl. Phys.*, **15**, 758.
239. Guth, E. (1945) *J. Appl. Phys.*, **16**, 20.
240. Witten, T.A., Rubinstein, M. and Colby, R.H. (1993) *J. Phys. II France*, **3**, 367.
241. Kloczkowski, A., Sharaf, M.A. and Mark, J.E. (1993) *Comput. Polym. Sci.*, **3**, 39.
242. Sharaf, M.A., Kloczkowski, A. and Mark, J.E. (1994) *Comput. Polym. Sci.*, **4**, 29.
243. Kloczkowski, A., Sharaf, M.A. and Mark, J.E. (1994) *Chem. Eng. Sci.*, **49**, 2889.
244. Yuan, Q.W., Kloczkowski, A., Mark, J.E. and Sharaf, M.A. (1996) *J. Polym. Sci., Polym. Phys. Ed.*, **34**, 1674.
245. Mark, J.E. (1996) *J. Comput.-Aided Mats. Design*, **3**, 311.

8 Viscoelastic behaviour of network materials

M. ILAVSKY

8.1 Introduction

The deformation behaviour of polymers, including polymer networks, is considerably dependent on temperature, rate of deformation and time or frequency. The knowledge of mechanical properties under a broad range of external parameters not only determines the fields of application, but also characterizes the physical structure of polymer materials. The fixation of chains in chemical junctions leads to specific features in the physical behaviour of network materials, for example large recoverable deformations, better thermal stability and insolubility.

If a polymer molecule is subjected to an external force, it assumes new (the most probable under the given external conditions) conformations by diffusive motion, the rate of which depends on the effective viscosity of the medium. Such time-dependent motions lead to a time-dependent (visco-elastic) mechanical behaviour. A typical dependence of the real component J' of the complex compliance J^* and of the loss tangent $\tan \delta$, on temperature T or angular rate of deformation ω, can be seen schematically in Figure 8.1.

In the low-temperature region (high frequencies), the polymer is in the glassy state (region A), where only local segmental motions of short parts of the main chain (usually vibrations) or of the side chains (hindered rotations) take place. This motion gives rise to the low-temperature secondary transitions (β, γ, ...); the temperatures (T_β, T_γ, ...) and the frequency positions of these transitions are given primarily by the chemical and geometrical structure of the monomeric unit (i.e. submolecular structure). With increasing temperature, the main transition region (sometimes called also the α transition) is reached (region B), in which all mechanical functions are markedly frequency- or temperature-dependent. The beginning of this transition is given by the glass-transition temperature T_g which is predominantly determined by the submolecular structure. To a lesser extent, the value of molecular weight and junction-point density (i.e. molecular structure), together with crystallinity and phase separation in block copolymers (i.e. supermolecular structure) also affect T_g. In this region, the motion of the main chain is subsequently released by the cooperative motion of chain segments; while, over short times, the conformations of short segments change, over long times, the conformations of long segments of the main-chain change.

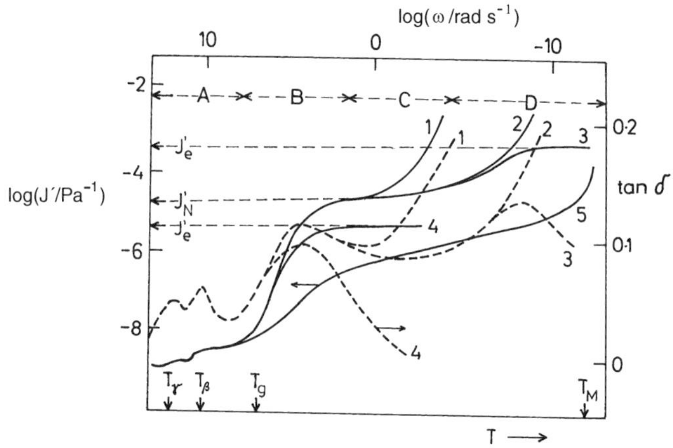

Figure 8.1 Schematic dependence of the storage compliance J' and loss tangent $\tan\delta$ on temperature T or angular rate of deformation ω.

At still higher temperatures, the so-called rubberlike (plateau) region (region C) is reached; in this region only small dependences of mechanical functions on temperature and frequency (or time) are observed. The values of mechanical functions are determined by intermolecular physical interactions (entanglements) and, with chemical networks, also by the junction-point density. In the case of a non-network polymer, the values and distribution of molecular weights determine the mechanical behaviour in the plateau region. From the value of the compliance in the plateau region, J'_N (usually determined as the inflection point in the dependence of J' on T or ω), the concentration of entanglements can be determined. For non-network polymer, as the entanglement constraints have only a temporary character, a further rise in temperature causes a transition into the range of a polymer melt (region D), where the motion of chains is completely released. In this region the importance of the flow deformation component (characterized by the steady-flow viscosity η) is dominant. In the case of semicrystalline polymer, the flow region is reached above the melting temperature T_M.

The introduction of chemical junctions into a polymer material stops the flow and leads to the final, equilibrium value of the storage compliance J'_e, which equals the equilibrium compliance J_e (or equilibrium modulus $G_e = 1/J_e$). For loose networks, with $J'_e \gg J'_N$, mechanical equilibrium is experimentally difficult to attain and extrapolation procedures are usually used. In such networks, two peaks in the frequency (or temperature) dependence of the loss tangent $\tan\delta$ or imaginary component of complex compliance J'' are usually found; these transitions correspond to the fast (the glass process) and the slow (high-temperature, secondary relaxation process) relaxations. Increasing junction-point concentration diminishes the height of the low-frequency peak and shifts it towards the higher-frequency one. At $J'_e \approx J'_N$, the two peaks usually overlap and a further increase in the junction-point density causes a decrease in the width and height of the J'' peak. Mechanical behaviour, similar to that of covalent networks, is also shown by systems in which junction points are formed by parts of chains in the glassy state (block copolymers, e.g., styrene-butadiene-styrene) or by parts of polymer chains in the crystalline state.

In addition to the investigation of ready-to-use polymers, mechanical measurements can also be used in the investigation of structure formation and in monitoring network-forming (polymerization) reactions. Whereas in linear polymerization, the steady-flow viscosity η increases with reaction time and reaches a final value, in a network-forming reaction, the viscosity is infinite at the gel point (GP) and, later on, the value of equilibrium modulus G_e increases with reaction time. At a constant reaction temperature, the increase in η and G_e is due to the increase in the conversion of the network-forming reaction. A large change in mechanical properties (e.g., G_e can change from 0 to 10^7 Pa) can be utilized effectively in controlling material production (e.g. by the RIM process (Chapter 5)). In recent years, it has also been demonstrated that, at the GP, a specific (power-law) viscoelastic behaviour is observed.

8.2 Measurements of viscoelastic properties

If a polymer sample is subjected to stress, both its shape and volume change. Hence, from the physical viewpoint, there exist two basic types of deformation – shear (corresponding to a change in shape) and compression (corresponding to a change in volume); by using them, any other type of deformation (e.g. simple extension) can be constructed. Hence, both basic types, which are time- and temperature-dependent, fully characterize mechanical behaviour of the polymer [1, 2].

In the case of a homogeneous isotropic material, the relationship between the shear stress σ and shear strain γ is determined by the complete loading history (Boltzmann superposition principle). For creep (an increase in strain for a given stress history), we have the relationship [2–4]

$$\gamma(t) = \int_0^t J(t - x)(\mathrm{d}\sigma(x)/\mathrm{d}x)\,\mathrm{d}x \tag{8.1}$$

in which $J(t)$ is the shear-compliance function and t is time. Thus, the total strain at time t, $\gamma(t)$, is obtained by the summation of infinitesimal load contributions $\mathrm{d}\sigma(x_i)$ at times x_i; every load gives rise to an independent contribution $J(t - x_i)\,\mathrm{d}\sigma(x_i)$, and the viscoelastic creep behaviour is fully determined by the compliance $J(t)$. Stress relaxation, a change in the force with time for a given deformation history, can be described by the similar relationship

$$\sigma(t) = \int_0^t G(t-x)(\mathrm{d}\gamma(x)/\mathrm{d}x)\,\mathrm{d}x \tag{8.2}$$

in which $G(t)$ is the shear-modulus function.

If a sample is subjected to a constant stress σ_0 at time $x = 0$, we have from equation (8.1)

$$\gamma(t) = \sigma_0 J(t) \tag{8.3}$$

If a constant strain γ_0 is imposed on a sample at time $x = 0$, then equation (8.4) is obtained from equation (8.2)

$$\sigma(t) = \gamma_0 G(t) \tag{8.4}$$

Equation (8.4) describes the stress relaxation of linear viscoelastic material (generalized Hooke's law).

In some cases, an experiment is used in which strain γ increases at a rate $\dot{\gamma} = \text{const.}$ ($\gamma(x) = \dot{\gamma}(x)$.) The shear modulus $G(t)$ can then be determined from equation (8.2) in the form

$$G(t) = (1/\dot{\gamma})\,\mathrm{d}\sigma(t)/\mathrm{d}t \tag{8.5}$$

In dynamic experiments, the cyclic shear strain γ^* produces cyclic stress σ^* with the same angular frequency ω, which, however, is phase-shifted by an angle δ. Hence the relationships

$$\gamma^* = \gamma_0 \exp(i\omega t) \qquad \text{and} \qquad \sigma^* = \sigma_0 \exp[i(\omega t + \delta)] \tag{8.6}$$

are valid. The complex shear-modulus G^* can be defined by

$$G^* = G' + iG'' = \sigma^*/\gamma^* = (\sigma_0/\gamma_0)\exp(i\delta) \tag{8.7}$$

From equation (8.7), the real (storage) component is $G' = (\sigma_0/\gamma_0)\cos\delta$, the imaginary (loss) component is $G'' = (\sigma_0/\gamma_0)\sin\delta$, and the loss tangent is $\tan\delta = G''/G'$. The complex compliance J^* is defined by

$$J^* = J' - iJ'' = \gamma^*/\sigma^* = 1/G^* \tag{8.8}$$

From equation (8.8), the storage component is $J' = G'/(G'^2 + G''^2)$, the loss component is $J'' = G''/(G'^2 + G''^2)$, and $\tan\delta = J''/J' = G''/G'$.

Each of the functions $G(t)$, $J(t)$, $G^*(\omega)$ and $J^*(\omega)$ fully characterizes the linear-viscoelastic shear-deformation behaviour. Exact interrelations of these functions can be found [2–5]. Basically, all the other functions can be calculated from the knowledge of one function, even though in some cases this is a difficult problem. (The interrelations are given by integral equations.) In transformations of static to dynamic functions, the time t is replaced by $1/\omega$ and, by combining static and dynamic methods of measurements, the time interval from 10^{-3} to 10^6 s can be covered.

Viscoelastic functions for bulk compression can be defined similarly to the shear functions [2]; a set of shear and compression functions thus fully characterizes the linear-viscoelastic, mechanical behaviour.

An example of an experiment, in which the change both in the shape and in the volume of the sample is reflected, is simple elongation (see Chapter 7). For Young's elasticity modulus $E(t)$, i.e., $E(t) = \sigma_t(t)/\varepsilon_t$, where σ_t and ε_t are, respectively, tensile stress and tensile deformation; $\varepsilon_t = (l - l_0)/l_0$, where l and l_0 are, respectively, the deformed and initial sample lengths, we have

$$1/E(t) = 1/3G(t) + 1/9K(t) \tag{8.9}$$

where $K(t)$ is the bulk modulus ($K = -P/(\Delta V/V)$, P is the pressure, and $-\Delta V/V$ is the relative decrease in the volume). The Young's modulus can also be expressed in the form

$$E = 2G(1 + \mu) \tag{8.10}$$

where μ is Poisson's ratio ($\mu = [1 - (1/V)(\partial V/\partial \varepsilon_t)]/2$). The ratio μ is a measure of lateral contraction due to longitudinal deformation, i.e., $\mu = \mathrm{d}\ln(w/w_0)/\mathrm{d}\ln(l/l_0)$, where w and w_0, respectively, are the deformed and initial sample widths. If there is no change in volume with deformation, $\mu = 1/2$ and $E = 3G$; this behaviour is usually observed for networks in the rubbery state.

8.3 Mechanical models, viscoelastic spectra

The time dependences of mechanical functions can be represented by models which consist of elastic springs and viscous dashpots [2, 4]. The mechanical behaviour of springs is described by Hooke's law, $\sigma_H = G\gamma_H$, and the behaviour of the dashpots satisfies Newton's law, $\sigma_N = \eta(\mathrm{d}\gamma_N/\mathrm{d}t)$, in which η is the viscosity. The simplest models are those suggested by Maxwell (spring and dashpot in series) and Voigt (spring and dashpot in parallel). Whereas Maxwell's model is suited for the description of stress relaxation, the Voigt's model can be used in the description of creep.

To describe experimental deformation behaviour more adequately, one needs a number of Maxwell models connected in parallel, or a number of Voigt models in series. In such cases, stress relaxation at constant γ_0 can be expressed by

$$\sigma(t) = \gamma_0 \sum_i G_i \mathrm{e}^{-t/\tau_i} = \gamma_0 G(t) \tag{8.11}$$

where the constants G_i and τ_i are related to the ith member of the series. Similarly, for creep at constant σ_0 the relationships

$$\gamma(t) = \sigma_0 \sum_i \frac{1}{G_i}(1 - \mathrm{e}^{-t/\tau_i}) = \sigma_0 \sum_i J_i(1 - \mathrm{e}^{-t/\tau_i}) = \sigma_0 J(t) \tag{8.12}$$

are valid, where J_i is the compliance of the ith member.

Equations (8.11) and (8.12) were generalized further for the continuous case, by introducing viscoelastic spectra – the relaxation $H(\tau)$ and retardation $L(\tau)$ spectra. The modulus is defined by

$$G(t) = \int_{-\infty}^{\infty} H(\tau)\,\mathrm{e}^{-t/\tau}\,\mathrm{d}(\ln\tau) + G_\mathrm{e} \tag{8.13}$$

where G_e is the equilibrium modulus for $t \rightarrow \infty$ ($G_\mathrm{e} \neq 0$ in networks). Similarly, from the retardation spectrum we have the relationship

$$J(t) = J_\mathrm{g} + \int_{-\infty}^{\infty} L(\tau)(1 - \mathrm{e}^{-t/\tau})\,\mathrm{d}(\ln\tau) + t/\eta \tag{8.14}$$

in which J_g is compliance for $t \rightarrow 0$ (glassy state) and η is the steady-flow viscosity (for networks $\eta \rightarrow \infty$). There also exist relationships between complex functions G^* and J^* and the relaxation and retardation spectra [2–4]. The dynamic compliances are given by

$$J'(\omega) = J_\mathrm{g} + \int_{-\infty}^{\infty} [L(\tau)/(1 + \omega^2\tau^2)]\,\mathrm{d}(\ln\tau) \tag{8.15}$$

and

$$J''(\omega) = \int_{-\infty}^{\infty} [L(\tau)\omega\tau/(1 + \omega^2\tau^2)]\,\mathrm{d}(\ln\tau) + 1/\omega\eta \tag{8.16}$$

When setting $\omega = 0$ in equation (8.15), we obtain, for the equilibrium compliance J_e of a viscoelastic solid (network) and steady-state (recoverable) compliance J_e^0 of a viscoelastic liquid, the relationships

$$J_\mathrm{e} = J_\mathrm{g} + \int_{-\infty}^{\infty} L(\tau)\,\mathrm{d}(\ln\tau) \tag{8.17}$$

and

$$J_\mathrm{e}^0 = J_\mathrm{g} + \int_{-\infty}^{\infty} L(\tau)\,\mathrm{d}(\ln\tau) \tag{8.18}$$

Calculations of spectra and individual mechanical functions may be performed by using the Laplace or Fourier transformation [3, 4]. However, the experimental functions are complicated and cannot be expressed in the analytical form. In recent years, a variety of numerical methods for the determination of $H(\tau)$ or $L(\tau)$ spectra have been used [5–9] and it was shown that this is an ill-conditioned problem. For this reason, approximate (graphical) methods have also been derived. For instance, the retardation spectrum can be expressed, in a first approximation, through the loss compliance J'', by using the expression [3]

$$L(\tau) = (2/\pi)J''(\omega)|_{\tau=1/\omega} \tag{8.19}$$

8.4 Superposition hypothesis of viscoelastic behaviour

The time (or frequency) dependences of viscoelastic functions depend considerably on external parameters (temperature, pressure, degree of swelling). A change in external parameters leaves the forms of the functions roughly unaffected, but displaces them along the time (frequency) axis. Thus, the dependences of the functions obtained for various constant values of external parameters within a limited time interval can be superimposed, giving the dependence of the superimposed (master) function over a broad time interval at a constant value of the external parameter. Using the superposition, we also obtain the dependence of the horizontal shift factor on the external parameter. Mechanical functions are usually corrected by a small vertical shift before superposition [2].

If the external parameters are denoted by ϕ_i, then the modulus is a function of time and the parameters, $G(t, \phi_i)$. Measurement yields a set of curves $G(t, \phi_i)$, where ϕ_i are various values of a single parameter (e.g., temperature (Figure 8.2)). Superposition of curves of the moduli measured at ϕ_i to a constant reference value of ϕ_0 can be carried out by using the transformation

$$G_{\mathrm{p}}(t, \phi_0) = b(\phi_i, \phi_0)G[t/a(\phi_i, \phi_0), \phi_i] \tag{8.20}$$

in which $G_{\mathrm{p}}(t, \phi_0)$ is the reduced (master) curve for the modulus superimposed onto the value of ϕ_0, $b(\phi_i, \phi_0)$ is the vertical shift factor and $a(\phi_i, \phi_0)$ is the horizontal shift factor. The function $G_{\mathrm{p}}(t, \phi_0)$ is plotted against the reduced time $t/a(\phi_i, \phi_0)$ (Figure 8.2). Equations similar to equation (8.20) are also valid for components of the dynamic modulus, $G^*(\omega)$ (with $t = 1/\omega$).

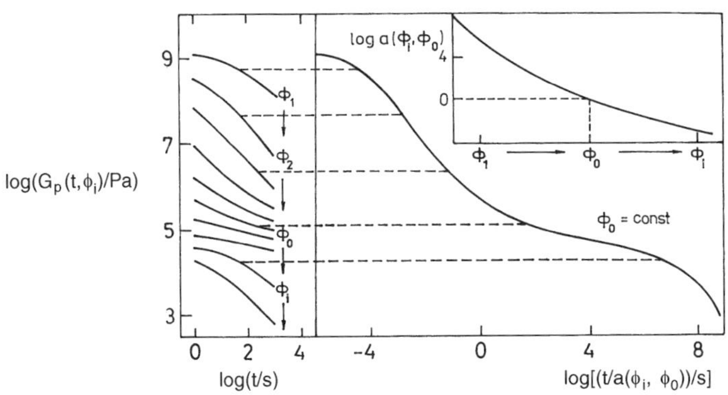

Figure 8.2 Application of superposition hypothesis using time t and an external parameter ϕ for shear modulus G. The horizontal shift factor is $\log a(\phi_i, \phi_0)$.

Superposition of the compliance curves, $J(t, \phi_i)$, can be performed by the transformation

$$J_p(t, \phi_0) = b^{-1}(\phi_i, \phi_0)J[t/a(\phi_i, \phi_0), \phi_i] \tag{8.21}$$

in which $J_p(t, \phi_0)$ is the master compliance curve superimposed onto the value of ϕ_0. Relations similar to equation (8.21) also hold for components of the complex compliance $J^*(\omega)$ (with $t = 1/\omega$).

Time–temperature superposition is the best procedure known for defining master curves of viscoelastic behaviour, and is also that most frequently used. With increasing temperature, viscoelastic functions are shifted towards shorter times or higher frequencies; a change in temperature by $\sim$10 °C displaces curves by approximately one order of magnitude on the time axis. It has been found experimentally that the temperature dependence of the horizontal shift factor $a(T, T_0)$ of amorphous polymers in the main-transition region satisfies the so-called Williams–Landel–Ferry (WLF) equation in the form [2, 10]

$$\log a(T, T_0) = -C_1^0(T - T_0)/(C_2^0 + T - T_0) \tag{8.22}$$

where C_1^0 and C_2^0 are constants which depend on the choice of temperature T_0 and on the structure of the polymer. In the glassy state the shift factor satisfies an Arrhenius dependence, rather than equation (8.22)

$$\log a(T, T_0) = (\Delta H/R)(1/T_0 - 1/T) \tag{8.23}$$

where R is the gas constant and ΔH is the activation energy.

Increasing the concentration of a low-molecular-weight compound (diluent) in the bulk polymer manifests itself as an equivalent rise in temperature, i.e., it shifts mechanical functions to shorter times usually without essential changes in their forms. This fact allows a time–(degree-of-swelling) superposition to be used in data treatment. Also, increasing the external pressure affects the time dependences of mechanical functions, similar to decreases in temperature. Here too, the superposition method can be applied [2]. The swelling and pressure dependences of the horizontal shift factor can be described by means of an equation similar to equation (8.22).

The universal application of the WLF equation suggests that the temperature, swelling and pressure dependences of relaxation times (and, thus, the shift factor a) may be determined to a first approximation by a single (structural) parameter. In agreement with the semi-empirical free-volume concept of the glass transition, it was assumed that the total polymer volume V^t consists of the occupied volume V^0 and the free volume V^f at a given external parameter ϕ. The value of the fractional free volume $f_\phi = V^f/V^t$ can then be related to a relaxation time τ_ϕ by [11, 12]

$$\tau_\phi = A \exp(B/f_\phi) \tag{8.24}$$

where A, B are constants. The horizontal shift factor is given by

$$\log a(\phi, \phi_0) = \log(\tau_\phi/\tau_{\phi_0}) = (B/2.303)(1/f_\phi - 1/f_{\phi_0}) \tag{8.25}$$

To a first approximation, the temperature, swelling and pressure dependences of the horizontal shift factor a can be derived from equation (8.25), assuming that for f_ϕ

$$f_{T,v_2,p} = f_{T_0,v_{20},p_0} + \alpha_f(T - T_0) - \beta'_f(v_2 - v_{20}) - \beta_f(p - p_0) \tag{8.26}$$

holds, where α_f, β'_f and β_f, respectively, are the temperature, swelling and pressure coefficients of the change in the fractional free volume, v_2 is the volume fraction of the polymer in the swollen state ($v_2 = V_d/V_{sw}$, where V_d and V_{sw} are dry and swollen volumes, respectively), and p is a pressure. Modifications of the free-volume theory for the dependence of f on physical aging [13, 14], or on the shape, flexibility and polarity of a monomeric unit, can also be found in the literature [15–19].

The WLF equation can also be derived from the thermodynamic theory of Gibbs and DiMarzio [20, 21], based on a cooperative motion of polymer-chain segments in the main-transition region. According to Adam and Gibbs [22], such motion leads to the conformational rearrangement of segments and, for any rearrangement, there is, of necessity, the cooperation of many elementary segmental motions. At a given temperature, there exists a certain (minimal) amount of elementary motion which allows cooperative, conformational rearrangements. With decreasing temperature, the degree of cooperation increases, i.e. the minimal amount of elementary motions that must occur simultaneously increases. The theory is based on the molar conformational entropy expressed as a function of temperature, $S_c(T)$, in the form

$$S_c = s_c^* N_A / n^* \tag{8.27}$$

where n^* is the minimal number of elementary motions at temperature T (n^* represents the minimal size of the cooperatively rearranging region), s_c^* is the conformational entropy of the minimal region and N_A is the Avogadro constant. The probability P that a change in conformational arrangement can occur at T is given by

$$P = A\exp(-n^* E_A/kT) = A\exp(-E_A s_c^*/RTS_c) = 1/\tau \tag{8.28}$$

because the relaxation time is inversely proportional to P. E_A is the height of the potential-energy barrier hindering the elementary motion and k is the Boltzmann constant.

The conformational entropy plays a particularly important role in the temperature dependence of the relaxation time τ, according to equation (8.28). The theory by Gibbs and DiMarzio [21] postulates, for any glass-forming liquid, the existence of certain temperature T_2 at which the conformational entropy decreases to zero (cooperatively rearranging regions grow to infinite size). Below T_2, of course, S_c remains equal to zero; above T_2 it increases with

T. In order to express the temperature dependence of S_c, one can look at the temperature dependence of ΔC_p – the discontinuity in isobaric heat capacity at T_g. Since, substantially, ΔC_p does not change with T, two approximations may be used [22, 23],

$$\Delta C_p = \text{constant} \qquad \text{or} \qquad \Delta C_p = \text{constant}/T \tag{8.29}$$

From equation (8.29), these can be obtained by integration

$$S_c = \Delta C_p \ln(T/T_2) \qquad \text{or} \qquad S_c = (\Delta C_p T_g/T_2)(1 - T_2/T) \tag{8.30}$$

The substitution of the second expression of equation (8.28) into equation (8.26) leads to a relation for the shift factor $a(T, T_0)$ which is formally identical with that in equation (8.22). Thus, the free-volume approach to superposition and that based on the cooperative character of conformational rearrangements are formally equivalent [23]. When the conformational entropy S_c is expressed as a function of the external pressure [24] p, or of the volume fraction of polymer v_2 [25], then the pressure and swelling dependences of the horizontal shift factor can be described, just as in free-volume theory. More extensive discussions on cooperative behaviour at the glass transition can be found in the literature [26–29].

8.5 Molecular theory of viscoelastic behaviour

8.5.1 Linear systems

The theoretical determination of the form and position of mechanical functions or spectra in dependence on time and temperature using the known polymer structure still remains an open problem. Molecular theories relate the time dependence of these functions to diffusive changes in the chain conformations subjected to external stress. Usually, the chain is modelled by an equivalent, freely-jointed chain composed of statistical segments, the end-to-end chain distance of which is described in terms of Gaussian statistics; this model quite adequately represents any sufficiently long, real chain [30–32]. The conformations of short segments change over short times and that of long segments, over long times.

The most frequently used, Rouse theory [30], predicts, for the time dependence of the shear modulus $G(t)$, the equation

$$G(t) = nkT \sum_{i=1}^{N} \exp(-t/\tau_i) \tag{8.31}$$

where n is the number of molecules per unit volume and N is the number of the submolecules in a molecule. The ith relaxation time is given by

$$\tau_i = \frac{P\xi_0 \langle r_0^2 \rangle}{24N^2 \sin^2(\pi i/2N)} \tag{8.32}$$

where P is the degree of polymerization, ξ_0 is the monomeric friction coefficient and $\langle r_0^2 \rangle$ is the mean-square end-to-end distance of the molecule.

In the region $i < N/5$ (longer times), equation (8.32) can be approximated by

$$\tau_i = P\xi_0 \langle r_0^2 \rangle / 6\pi^2 kTi^2 \tag{8.33}$$

From the comparison of equations (8.13) and (8.32), it is possible to calculate the relaxation spectrum

$$H(\tau) = (n/2\pi)(P\xi_0 \langle r_0^2 \rangle kT/6)^{1/2} \tau^{-1/2} \tag{8.34}$$

In the region where this equation is fulfilled, the retardation spectrum $L(\tau) = 1/\pi^2 H(\tau) \sim \tau^{1/2}$, the modulus $G(t) \sim t^{-1/2}$ and storage modulus $G'(\omega) \sim \omega^{1/2}$. Such dependences are usually experimentally observed at the end of the main-transition region, close to the rubbery plateau. Usually, equation (8.34) is used for determinations of the monomeric friction coefficient ξ_0, which is predominantly dependent on the structure and polarity of the monomeric unit [2].

The above-mentioned theory describes the conformational motion of one isolated molecule and, in accord with equation (8.33), predicts that the longest relaxation time τ_1 (motion of whole molecule) is $\sim M^2$ (where M is the molecular weight and $\langle r_0^2 \rangle \sim P$). As the viscosity, $\eta = G\tau$ and $G \sim 1/M$ (in equation (8.31), $n = \rho/M$, where ρ is the density), this model predicts that $\eta \sim M$. The linear increase in η with increasing M is in accord with experiment for low molecular weights [2, 34]; in this region, the steady-state compliance $J_e^0 = \tau/\eta$ is proportional to M [35].

It is known from experiment that, for polymers of high M, the dependence of viscosity η on M is stronger, $\eta \sim M^{3.3-3.6}$, and the compliance J_e^0 is essentially independent of M. The change in exponent from 1 to $\sim$3.4 takes place at $M = M_e$, and M_e is usually interpreted as molecular weight of the chain between physical junctions or entanglements. (The term entanglement serves as a measure of the effect of topological constraints of other chains on the selected chain [35–39].) For $M > 2M_e$, a physical network structure is formed and one can expect that mobility of segments with $M > M_e$ is slowed down. It was assumed [2] that for such segments, the effective monomeric frictional coefficient $\xi_{0,ef}$ is higher than ξ_0 ($\xi_{0,ef} = Q_e \xi_0$). If one requires $\eta \sim M^{3.4}$, then $Q_e \sim (M/2M_e)^{2.4}$ and this decrease in internal mobility, permitting motion of segments with $M > M_e$, gives rise to the plateau region. The width of the plateau region Δ can be used for determination of M_e; $\Delta = \log Q_e = 2.4 \times \log(M/2M_e)$, where Δ is the difference between $\log t$ values at the points where the slope of the curve $\log G$ versus $\log t$ equals $-1/2$ (the end of the main transition region and the beginning of the flow region).

For a description of the dynamics of a chain in an entanglement network, the tube model was suggested [40–42]. A molecule reptates in the tube formed by the strands of other chains and the central line L of the tube tracks the current chain conformation. The short-range segmental motion within the tube can be described by Rouse dynamics (for $t \leq \tau_1$, $M < M_e$). Over long times, the chain can escape from the tube by a snake-like motion (curvilinear diffusion along its own contour). The important model parameters are the central line and the tube diameter a. De Gennes [41–43] and Doi and Edwards [44–46] have found that $\eta \sim M^3$ if the value of L is constant. During the reptational motion, the length L and diameter a can generally change. If the change in L and a is taken into account, then an exponent higher than 3 is obtained for the dependence of η on M, showing better agreement with experiment [47–49]. A viscosity exponent varying between 3.2 and 3.52 was found also by Rubinstein in a computer simulation of an entangled melt, using the model of a self-avoiding-walk chain in a tube [50]. Reviews of the dynamics of melts can be found in the literature [51, 52].

8.5.2 Network systems

The anchoring of chains by junctions leads to an equilibrium contribution to the viscoelastic functions and makes the network 'remember' its states during network formation. It is known that the internal energy of a network is a function of temperature because the chain dimensions change as a result of setting free previously hindered rotations [53, 54]. This fact was taken into account by introducing the reference state of the network [55–57], in which $\langle r_0^2 \rangle$ corresponds to a free chain. On average, no force acts upon the chain ends in this state. A similar situation occurs during the swelling of a polymer network. In a polymer solution, the solvent changes the concentration and dimensions of the polymer chains. For a network, changes in concentration and chain dimensions due to the presence of the solvent must also be taken into consideration (strain effect due to the solvent).

Modification of the Rouse–Mooney (R-M) molecular theory of the viscoelastic behaviour of networks (in which chains are fixed by junctions which cannot move) for the case where the external force is applied to the network in an isotropic state different from that in the network-formation state (usually considered as the reference state) has led to the expression for the relaxation time τ_i [58] (similar to equation (8.33))

$$\tau_i = P\xi_0 \langle r_0^2 \rangle \langle \alpha_{i,s}^2 \rangle / 6\pi^2 kT(i-1)^2 \tag{8.35}$$

where the dilatation factor $\langle \alpha_{i,s}^2 \rangle = \langle r_{i,s}^2 \rangle / \langle r_0^2 \rangle$ and $\langle r_{i,s}^2 \rangle$ is the mean square end-to-end distance of the chain in the isotropic swollen state. It follows from equation (8.35) that for $i = 1$, $\tau_1 \to \infty$. (This is the consequence of anchoring of the chain ends by junctions.) For the shear modulus $G(t)$ it was found [58] that

$$G(t) = \nu_{i,s} kT \langle \alpha_{i,s}^2 \rangle \left(1 + \sum_{i=2}^{N} e^{-t/\tau_i}\right) \tag{8.36}$$

where $\nu_{i,\mathrm{s}}$ is the number of elastically active network chains (EANCs) in unit isotropically swollen volume (similar to equation (8.31)). The equilibrium modulus ($G_\mathrm{e} = \nu_{i,\mathrm{s}} kT \langle \alpha_{i,\mathrm{s}}^2 \rangle$) coincides with the relationship derived in the kinetic theory of rubber elasticity for the affine model [59–61] (front factor $A = 1$, see Chapter 7). From equation (8.36), the relaxation spectrum was derived in the form [58]

$$H(\tau) = (\nu_{i,\mathrm{s}} \langle \alpha_{i,\mathrm{s}}^2 \rangle^{3/2} / 2\pi)(P\xi_0 \langle r_0^2 \rangle kT/6)^{1/2} \tau^{-1/2} \tag{8.37}$$

Usually, it is assumed that the reference state of network is identical with the network-formation state [56, 57]. If only the volume-effect of solvent is taken into account (disregarding possible specific effects), then the dilation factor becomes [58]

$$\langle \alpha_{i,\mathrm{s}}^2 \rangle_{T^0, v_2^0} = (\langle r_{i,\mathrm{s}}^2 \rangle / \langle r_0^2 \rangle)_{T^0, v^0} = 1 \tag{8.38}$$

where T^0 and v_2^0, respectively, are the temperature and the volume fraction of polymer at network formation. If the same network is measured at temperature $T \neq T^0$ and $v_2 \neq v_2^0$, then we have

$$\langle \alpha_{i,\mathrm{s}}^2 \rangle_{T,v} = \frac{\langle \alpha_{i,\mathrm{d}}^2 \rangle_T}{\langle \alpha_{i,\mathrm{d}}^2 \rangle_{T^0}} \left(\frac{v_2^0}{v_2} \right)^{2/3} = \phi_1(T, T_0) \phi_2(v_2, v_2^0) \tag{8.39}$$

where $\langle \alpha_{i,\mathrm{d}}^2 \rangle = \langle r_{i,\mathrm{d}}^2 \rangle / \langle r_0^2 \rangle$ is the dilation factor of the dry state. The function ϕ_1, which is the ratio of the end-to-end distance of the chain in the isotropic dry state $\langle r_{i,\mathrm{d}}^2 \rangle$ at temperatures T and T_0, describes the temperature dependence of $\langle r_0^2 \rangle$; usually the parameter $\delta = \mathrm{d} \ln \langle r_0^2 \rangle / \mathrm{d}T$ is a constant which characterizes the temperature dependence of the unperturbed chain dimensions [62].

From equation (8.37), it is possible to obtain the value of the monomeric friction coefficient:

$$\log \xi_0 = 2 \log H + \log \tau + \log \left(\frac{6}{kT} \right) + \log \frac{4\pi^2 [f(T - T^0)]^3}{P \langle r_0^2 \rangle_{T^0} \langle \nu_{i,\mathrm{s}}^2 \rangle_{T^0} \langle \alpha_{i,\mathrm{s}}^2 \rangle_{T^0}} \tag{8.40}$$

where the function $f(T - T^0) = \exp\{(\delta/2)(T - T^0)\}$. This expression is in agreement with that for linear systems [2], except for the factor $\log\{[f(T - T^0)]^3 / \langle \alpha_{i,\mathrm{s}}^2 \rangle_{T_0}^3\}$, which characterizes the temperature and volume dependences of the unperturbed-chain dimensions. If the network is formed in the absence of diluent ($\langle \alpha_{i,\mathrm{s}}^2 \rangle_{T^0} = 1$), the correction to ξ_0 is given by $3 \log[f(T - T^0)]$. For instance, in the case of natural rubber [54] ($\delta = 0.4 \times 10^{-3}\,\mathrm{K}^{-1}$), when the temperature of measurement T is lower by $\sim$150 K than network formation temperature T^0, the correction is $\approx$10% of ξ_0. If the network was formed in the presence of solvent (v_2^0) and measured in the dry state ($v_2 = 1$) at temperature T^0, then $\langle \alpha_{i,\mathrm{s}}^2 \rangle_{T^0} = (v_2^0)^{2/3}$ and the correction can attain much higher values. (If $v_2^0 = 0.5$, the correction is $\approx$60% of ξ_0.)

An interesting modification of the reduced variables for time–temperature superposition follows from equation (8.36). Whereas the commonly used vertical shift factor (equation (8.31)) is $b(T, T^0) = T^0 \rho_{T_0}/T\rho_T$, the modified one is [58]

$$b(T, T^0) = (T^0/T)(\rho_{T^0}/\rho_T)^{1/3}\langle r_0^2\rangle_{T^0}/\langle r_0^2\rangle_T \tag{8.41}$$

Tensile creep measurements performed with poly(butyl methacrylate) networks in the main-transition and rubberlike regions have shown [63] that use of the modified factor $b(T, T^0)$ gives better continuous superimposed curves of reduced compliances than use of the classical one. The dynamic–mechanical measurements of TEPDM (terpolymer of ethylene, propylene and diene monomer) networks [64–66] or epoxy networks [67–69] have led to the same conclusions.

Even larger modifications of reduced variables follow from equations (8.36) and (8.39) for the time–(degree-of-swelling) superposition. If v_{20} is the reference volume-fraction of polymer in the swollen state to which we superimpose, then the vertical shift-factor is

$$b(v_2, v_{20}) = (v_{20}/v_2)^{1/3} \tag{8.42}$$

instead of that commonly used, $b(v_2, v_{20}) = v_{20}/v_2$. It also follows from equation (8.35) that the horizontal shift factor,

$$a(v_2, v_{20}) = \tau_{v_2}/\tau_{v_{20}} = [(\xi_0)_{v_2}/(\xi_0)_{v_{20}}](v_{20}/v_2)^{2/3} \tag{8.43}$$

besides the relative change in the friction factor ξ_0 with degree of swelling, also contains the factor $(v_{20}/v_2)^{2/3}$, which represents the relative change in the mean-square end-to-end distance of the chain with swelling. It has been proved experimentally that the use of equation (8.42) leads to an improvement of the time–(degree-of-swelling) superposition of the compliance data on variously swollen poly(2-hydroxyethyl methacrylate) networks in diethylene glycol [70, 71]. However, the correction still remained inadequate for systems of low junction-point density.

Equation (8.36) also allows one to analyze quantitatively the effects of network-formation conditions upon viscoelastic behaviour. Two new superpositions have been proposed [58]:

1. Time–(temperature of network formation) for networks prepared at different temperatures but the same dilution and measured at the same temperature.
2. Time–(concentration of polymer at network formation) (v_2^0) for networks prepared at the same temperature but different dilutions and measured at the same temperature.

For instance, if two networks were prepared at the same temperature but at different concentrations of the polymer at network formation, v_2^1 and v_2^2, and are measured at the network-formation temperature and constant volume fraction v_2 then equation (8.36) gives the vertical shift factor

$$b(v_2^1, v_2^2) = (v_2^1/v_2^2)^{2/3} \tag{8.44}$$

The creep behaviour of poly(2-hydroxyethyl methacrylate) networks formed at various water contents and measured in the dry state [72] proved, to a first approximation, the correctness of equation (8.44).

Another characteristic parameter of the viscoelastic behaviour of polymer networks is the ratio of the maximum of the loss compliance $J''_\mathrm{m}(\omega)$ (or of the retardation spectrum spectra $L_\mathrm{m}(\tau)$) to the equilibrium compliance J_e. The unmodified R-M theory has predicted [73] that the ratio $J''_\mathrm{m}/J_\mathrm{e} = 0.362$ is constant, independent of the number of EANCs. A modification of the R-M theory for the case that submolecules in a chain are not monodisperse but show various length distributions, or network junctions are not fixed, leads to the ratio $J''_\mathrm{m}/J_\mathrm{e}$ in the range from 0.29 to 0.42. (As $L_\mathrm{m}/J_\mathrm{e} \approx (2/\pi)J''_\mathrm{m}/J_\mathrm{e}$, the ratio $L_\mathrm{m}/J_\mathrm{e}$ lies in the range 0.18–0.26.) Introduction of the reference state of a network led [2] to equation (8.36), from which it follows that every relaxation time τ_i is associated with the contribution $\nu_{i,\mathrm{s}}kT\langle\alpha_{i,\mathrm{s}}^2\rangle$. Due to this fact, we can write, for the components of G' and G'', the relationships

$$G' = \nu_{i,\mathrm{s}}kT\langle\alpha_{i,\mathrm{s}}^2\rangle\left(1 + \sum_{n=2}^{N}\frac{\omega^2\tau_i^2}{1+\omega^2\tau_i^2}\right) \tag{8.45}$$

and

$$G'' = \nu_{i,\mathrm{s}}kT\langle\alpha_{i,\mathrm{s}}^2\rangle\left(1 + \sum_{n=2}^{N}\frac{\omega\tau_i}{1+\omega^2\tau_i^2}\right) \tag{8.46}$$

from which $J'' = G''/(G'^2 + G''^2)$ can be calculated. Although J'' and, in particular, the value and position of the peak J''_m depend on the reference state, the ratio $J''_\mathrm{m}/J_\mathrm{e}$ is $\langle\alpha_{i,\mathrm{s}}^2\rangle$-independent and has the value 0.362 ($L_\mathrm{m}/J_\mathrm{e} = 0.23$), the same as in the unmodified R-M theory. Values of $L_\mathrm{m}/J_\mathrm{e}$ in the range 0.16–0.22 were found experimentally for various network systems [2, 72–76].

So far, we have discussed the dynamics of networks without entanglements. This approach can describe the viscoelastic behaviour of systems of high junction-point density (with $J_\mathrm{e} < J_\mathrm{N}$) or the fast (glass) processes in loose networks [2, 77]. Molecular interpretations of the slow processes in loose networks are often attributed to the relaxation of dangling ends in the presence of topological constraints [77–89]. Usually, the tube model is used, as in the case of linear melts, and the diffusion mechanism is a reptation (snake-like) motion of the dangling chains. According to de Gennes [87, 88], the reptation time t_P for dangling chain with one end fixed to reach equilibrium depends exponentially on its length (degree of polymerization P)

$$t_P = \tau_1 P^{3/2}\exp(\alpha P) \tag{8.47}$$

where τ_1 corresponds roughly to the maximum Rouse relaxation time and α is a material constant. Using this equation, Curro *et al.* [82], after summing over contributions to the modulus $G(t)$ from a distribution of branch lengths derived for a random crosslinking process derived, for the time dependence of the modulus $G(t)$, an empirical Chasset–Thirion (C-T) type of equation [90]:

$$G(t) = G_e[1 + (t/t_0)^{-m}] = G_e[1 + (t/\tau_{av})^{-\nu_e/\alpha\rho_m}] \tag{8.48}$$

where t_0 and m are constants; the average relaxation time $\tau_{av}(= t_0) \approx \tau_1(\nu_e)^{-\alpha\rho_m/\nu_e}$ ($\nu_e = G_e/(\langle\alpha_{i,s}^2\rangle RT)$ is the concentration of EANCs) and $m = \nu_e/\alpha\rho_m$ (ρ_m is the monomer density). With increasing network density ν_e, the value of the exponent m increases and $\tau_{av} \to \tau_1$; the increase in m with ν_e is consistent with the data on loose natural-rubber networks [90].

The C-T equation (8.48) gives an excellent description of stress–relaxation data for many loose network systems [89, 91–97], deep in the rubbery state (i.e. at long times, when all other relaxation processes have relaxed away). It was shown that on introducing equation (8.48) into equation (8.13), the relaxation spectrum $H(\tau)$ takes the power-law form [92]

$$H(\tau) = (2.303 G_e t_0^m/\Gamma(m))\tau^{-m} \tag{8.49}$$

where $\Gamma(m)$ is the gamma function. As the time dependence of $G(t)$ is small, the creep compliance can be given by

$$J(t) = 1/G(t) = J_e/[1 + (t/t_0)^{-m}] \tag{8.50}$$

This equation fitted well, long-term creep data from natural-rubber networks [91]. The inverse Laplace transform of equation (8.50) cannot be made, but it is possible to use a precise numerical method [98] for calculation of the retardation spectrum $L(\tau)$. It was found [72] that $L(\tau)$ shows a maximum at time $\tau_m = t_0$ and that this maximum is symmetrical. In such a way, reptation theory explains the slow-relaxation maximum in the frequency dependence of the loss compliance J'' (see equation (8.19)) and its shift to faster processes with increasing junction-point density.

An alternative explanation of the slow relaxation process was offered by Heinrich *et al.* [99–101]. These authors associated the slow relaxation peak in retardation spectrum with the relaxation of the constraint release parameter β, which characterized the deformation dependence of lateral tube dimensions around the chain. The slow relaxation process was introduced by a fractional-exponential time-dependence of β [101]:

$$\beta^2(t) = \beta_e^2 + (1 - \beta_e^2)\,e^{-(t/t_c)^{\varphi}} \tag{8.51}$$

where β_e characterizes the equilibrium state, φ is a constant and t_c is the characteristic time of the chain-constraint release process, which depends on the number of chemical and trapped-entanglement junctions. For the time-dependence of the modulus $G(t)$, the stretched exponential form was found to be

$$G(t) = G_e(1 + k\,e^{-(t/t_c)^{\varphi}}) \tag{8.52}$$

where k is a constant and $t_c = \tau_1(M_c/M_e)^{\varphi} \approx \tau_1 \nu_e^{-\varphi}$ is the characteristic relaxation time. The theory yields the following expression for the equilibrium, total shear-modulus of the network (cf. Chapter 7):

$$G_e = G_c + G_N = [(f_e - 2)/f_e]\nu_e RT + G_N^0 \beta_e^2 \tag{8.53}$$

where G_c is the modulus associated with chemical junction points and G_N represents the trapped entanglement contribution, f_e is the number-average junction-point functionality and G_N^0 is the plateau modulus.

A similar fractional-exponential-law equation to equation (8.52) has been predicted by Gaylord *et al.* [102] for stress relaxation originating in the hindered diffusion of dangling ends in presence of entanglements within the continuous-time random-walk model. This means that constraint-release effects of the tube dimensions may hardly be distinguished from the effects of dangling chains on the relaxation properties of networks. It was also shown [102] that equation (8.52) is practically indistinguishable from the power-law form of equation (8.48). Qualitatively similar results have been reached from computer simulations of the dynamics of networks [103–105]. Motions with times longer than those directly attributable to dangling ends were found in simulations [106].

Instead of the parameter β_e^2, the trapping factor T_e is often used in the literature ($\beta_e^2 \approx T_e$) [107–114]. This factor characterizes the number of trapped entanglements, with the limiting values of zero at the gel point and unity for a perfect network without defects. The factor T_e is proportional to the probability of contacts between pairs of segments located in EANCs and can be calculated from theories of network formation (i.e. theory of branching processes [112–114]). The use of equation (8.51) for the time-dependence of T_e explains the slow relaxation process in this case. From equation (8.51), it further follows that the long-time process vanishes in perfect networks ($T_e \rightarrow 1$).

8.6 Viscoelastic properties in the critical (gel) state: self-similarity

A polymerization with at least one of the reactants (monomers) having a functionality greater than two is characterized by disproportionate increases in the number- and weight-average molecular weights, M_n and M_w. As time proceeds, the steady-state viscosity η increases [115–117] and, at a threshold time t_c, the viscosity and M_w diverge. At time t_c, the critical conversion α_c at the gel point (GP) is reached [118–119]. In the pre-gel state, the system is soluble (of liquid-like structure) and, after passing the GP, the insoluble gel fraction w_g, characterized by a finite equilibrium modulus G_e, is formed in the reacting system (giving a solid-like structure). The material at the GP, termed the critical gel (CG), is in an intermediate state between solid and liquid.

It is interesting to note that the GP can generally be reached in two ways:

1. Increasing the conversion of a reacting system at constant composition. CG properties are usually determined during the reaction [120–128] although an attempt to stop the reaction close to the GP by quenching the sample was also made [129].
2. Using the critical-molar-ratio (CMR) method [130–133] in systems where the network is formed from end-linked reactants with various reactive groups (such as polyurethanes or epoxides). This method consists of finding the initial CMR of reactive groups, r^c (i.e. finding the critical composition), at which the GP occurs just when the conversion of minority groups approaches 100%.

Schematic changes in structural parameters during network formation from triol and diisocyanate are shown [133] in Figure 8.3. Whereas a change in conversion at constant composition generally leads to reacting structures, a change in the composition of a system at complete reaction gives structurally stable samples and CG properties can be better examined experimentally. It should be mentioned that the classical GP determination is based on solubility experiments.

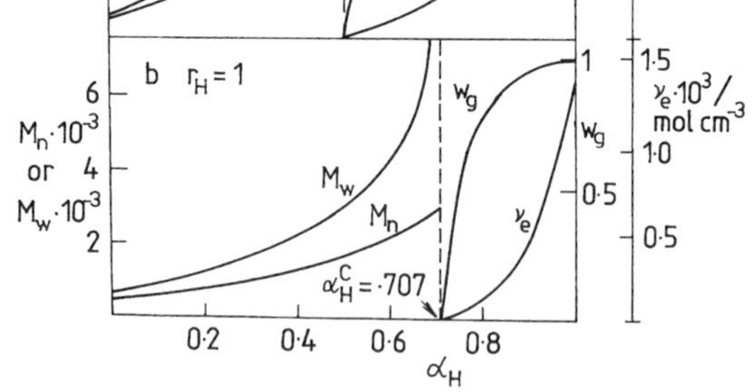

Figure 8.3 Schematic changes in structural parameters during network formation using a monodisperse triol (weight-average functionality $f_{w,H} = 3$ and $M_{n,H} = 750$) and diisocyanate ($f_{w,I} = 3$ and $M_{n,I} = 250$). Number, M_n, and weight, M_w, average molecular weights, weight fraction of gel, w_g, and concentration of EANCs, ν_e, versus α_H, the conversion of hydroxy groups. α_I is the conversion of isocyanate groups, α_H^c is critical GP conversion and r_H is the initial molar ratio of hydroxy to isocyanate groups (r_H^c is the critical value for the GP). (For more details see [33].)

8.6.1 *Dynamics at the gel point*

The very complex dynamic–mechanical behaviour of the branched structures at the GP is due to the high polydispersity as one can expect that polydispersity will imply a broad distribution of relaxation times. In recent years, many dynamic–mechanical experiments have been made in the vicinity of and at the GP [120–133]. The first proposal was that the crossover of the dynamic storage (G') and loss (G'') shear moduli is characteristic [120] of the GP. (The loss tangent $\tan\delta = G''/G' = 1$, and $G'' > G'$ for the pre-gel and $G'' < G'$ for the post-gel state.) Later on, experiments revealed that, at the GP, values of $\tan\delta > 1$ can also be found and that only a frequency-independent $\tan\delta$ was characteristic of the CG. Finally, it was concluded that the CG is characterized by the power-law, static, shear relaxation modulus [123, 134–136]

$$G(t) = St^{-n} \tag{8.54}$$

where S (Pa s^n) is the gel strength ($S = G_0\tau_0^n$, where modulus G_0 and relaxation time τ_0 are specific materials parameters), and n is a critical exponent with $0.2 < n < 0.8$.

From equation (8.54), after Fourier transformation, the storage $G'(\omega)$ and loss $G''(\omega)$ shear moduli at the GP are given by

$$G'(\omega) = G''(\omega)/\tan\delta = \Gamma(1-n)\cos\delta S\omega^n \tag{8.55}$$

and the phase angle $\delta = n\pi/2$ is independent of frequency. Equation (8.55) shows that a power-law behaviour is also found in dynamic experiments. In recent years, the validity of equation (8.55) has been proved experimentally for many chemically and physically gelling systems [133–139].

The power-law behaviour is due to the mechanical self-similarity which is a consequence of the structural self-similarity developed during the network-formation process over length scales from chains up to sample dimensions. The properties of branched structures near the gel point can be modelled either by the classical, mean-field theory [118, 119, 140–144] or by percolation theory [88, 145–150]. It was demonstrated by many researchers that percolation theory is quite successful in predicting static properties in the vicinity of the sol–gel transition and can also be used for modelling dynamic properties (e.g. the determination of the exponent in equation (8.55)) of branched structures. Several versions of percolation theory can be found in the literature, such as random bond-percolation [145, 146], random site-percolation [147], correlated percolation [148, 149] or chain percolation [150].

Usually, it is assumed that all molecules which constitute the sol near the gel point have the same fractal dimensionality d_f, i.e. the mass of a cluster of scales with the radius R as $M \sim R^{d_f}$. The distribution of the cluster masses is self-similar and this similarity implies that the number of clusters of mass M is [151]

$$N(M) \sim M^{\tau} \tag{8.56}$$

where $\tau - 1 = d/d_f$, d being the spatial dimensionality. (This equation is known as a hyperscaling relation [145].) Percolation theory predicts $d_f = 5/2$ and for $d = 3$ it follows that the polydispersity exponent $\tau = 2.2$.

For a polydisperse system of fractal dimensionality d_f, when excluded-volume interactions are included and entanglements are ignored (Rouse modes dominate the dynamics), percolation theory gives [145, 152, 153]

$$n = d_f(\tau - 1)/(d_f + 2) = d/(d_f + 2) \tag{8.57}$$

Therefore, it follows from equation (8.57) that $n = 2/3$ ($d = 3$, $d_f = 2.5$ and $\tau = 2.2$). From investigations of the dynamic–mechanical behaviour of various systems, values of n close to 0.65 were found by several authors [139, 154, 155]. On the other hand, different values of n ranging from 0.2 to 0.8 have also been reported for various systems, especially for non-stoichiometrically balanced network-forming reactions of a four-functional end-linker with bifunctional poly(dimethylsiloxane)s [134–136]. Therefore, a modification of the percolation approach was suggested in which excluded-volume interactions were not considered, because one can expect that the situation in the vicinity of the gel point is similar to that in a melt where the excluded-volume effect plays a negligible role.

The progressive screening of the excluded-volume effect leads to a shrinkage of the sizes of chains (increase in d_f); this was proved experimentally by scattering experiments [158]. Therefore, equation (8.57) was modified to [153]

$$n = d/(\overline{d_f} + 2) = [d(d + 2 - 2d_f)]/[2(d + 2 - d_f)] \tag{8.58}$$

where $\overline{d_f}$ is the fractal dimension of the system when the excluded-volume effect is fully screened and $\overline{d_f} = 2d_f/(d + 2 - 2d_f)$ (see [152]). For $d = 3$, it follows from equations (8.57) and (8.58) that, for the unscreened case, n changes from 1 to 2/3 as d_f changes from 1 to 2.5, whereas, for the screened case, n changes from 9/8 to 0 over the same d_f interval. The experimental values of n, ranging from 0.2 to 0.8, can then be explained by changes in the extent of the excluded-volume screening effect, which depends on the particular chemical system.

8.6.2 *Experimental results*

Experimental investigation of the dynamic–mechanical, viscoelastic behaviour at the gel-point threshold is usually carried out with a Rheometrics apparatus and small-strain, oscillatory shear measurements are performed using parallel-plate geometry [133–139]. Data using a series of frequency sweeps ($\omega \approx 0.1$–$100\,\mathrm{rad\,s^{-1}}$) at various strains between $\gamma = 0.2$ and 0.001 are usually collected.

An example of the changes in the real, G', and imaginary, G'', components of the complex shear modulus and of the loss tangent, $\tan\delta (= G''/G')$, during a curing reaction is shown in Figure 8.4. The reacting polyurethane-forming

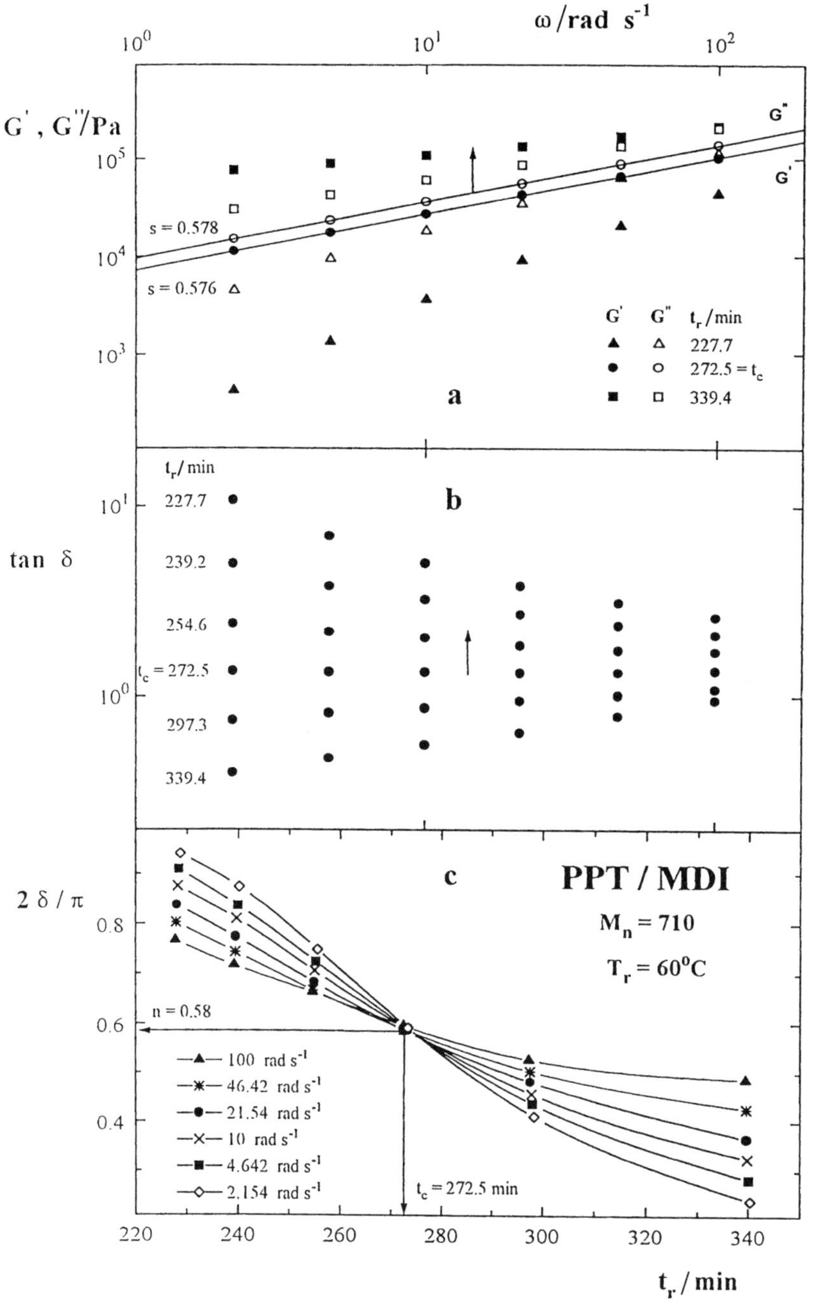

Figure 8.4 Dependences of (a) dynamic moduli G' and G'', and (b) loss tangent $\tan\delta$ on frequency ω. (c) Dependence of the ratio $2\delta/\pi$ on reaction time t_r. t_c denotes the critical GP time; curing temperature $T_r = 60\,°C$ [133].

system consists of a poly(oxypropylene) triol (PPT, $M_n = 710$) and 4,4′-diphenylmethane diisocyanate (MDI); the curing temperature T_r was 60 °C and the stoichiometric, initial ratio of hydroxy (OH) and isocyanate (NCO) groups, $r_H = [OH_0]/[NCO]_0 = 1$, was used [133]. Generally, a stoichiometric mixture shows the earliest structure build-up and gel formation. Typical evolutions of the phase angle δ at six frequencies as functions of the reaction time t_r are shown in Figure 8.4(c) for reaction times near the GP. The critical GP time occurs at 272.5 min; this relatively long time corresponds to a slow reaction and one can expect small changes in the dynamic functions due to the changes in conversion during the short (60 s) frequency sweeps.

Figure 8.4(b) shows interpolated values of the loss tangent tan δ (interpolation gives only small corrections) as functions of frequency ω at six reaction times t_r, around the GP. As can be expected, tan δ decreases as ω increases in the liquid state ($t_r < t_c$) and increases with ω in the solid state ($t_r > t_c$). It seems that for $t_r = t_c$, tan δ is independent of ω in accord with equation (8.55). The critical exponent (the exponent at the GP), $n = 0.58$ for this system. The frequency dependences of the dynamic moduli G' and G'', shown in Figure 8.4(a), also demonstrate the power-law behaviour for the CG at t_c (the frequency dependences of $\log G'$ and $\log G''$ on $\log \omega$ are parallel) and the critical exponents (slopes of the $\log G'$ and $\log G''$ versus $\log \omega$ plots) agree with those calculated from the phase angle. Although, before the GP, the storage modulus G' rapidly decreases with decreasing ω, after the GP it shows a rubbery plateau. The storage and loss moduli G' and G'' of a CG sample can be used for the calculation of the gel strength S in equation (8.54) ($S = 7500$ Pa s^n). As was shown earlier [122, 133], CG rheological parameters are not universal and critical exponents n and the gel strengths S depend on the chemical compositions of systems; usually a lower n-value is associated with a higher S-value and vice versa.

The sol–gel transition is due to the divergence of the weight-average cluster mass and various physical quantities follow the standard critical formula [145] $X \sim \varepsilon^{\delta}$, where δ is the critical exponent and $\varepsilon = |t_r - t_c|$ is the reduced time, which indicates the deviation from the GP. Prior to the GP, the steady-flow viscosity diverges as $\eta \sim \varepsilon^{-1.4}$ (in the Rouse limit, when hydrodynamic interactions are screened) and the steady-state, creep compliance diverges as $J_e^0 \sim \varepsilon^{-2.8}$ (see [159]). Just at the GP, the system has an infinite viscosity and a steady-state compliance (zero elastic modulus $\approx 1/J_e^0$). The values of G' and G'' at constant ω and curing temperature T_r can be changed by changing the extent of reaction (i.e. ε). The scaling relations for the frequency dependences of viscoelastic functions were combined with the scaling relations for the dependences of these functions on the extent of reaction and a time – cure superposition principle was suggested [159, 160]. This principle is, to some extent, analogous to the time–temperature superposition, which is based on the assumption that all viscoelastic modes are governed by the same temperature-dependent friction coefficient

ξ_0. Time–cure superposition does not result from a change in friction but from changes in the sizes of clusters near the GP (changes in connectivities). An increase in t_r results in a change in cluster size; whereas clusters become larger for $t_r < t_c$ with increasing t_r, they are smaller for $t_r > t_c$, as t_r increases. As expected, smaller clusters relax more quickly than larger ones and this change in time-scale can be absorbed by renormalizing the frequency scale. The approach predicts that the dependence of $G^*(\omega\tau_z)J_e^0$ on the dimensionless $\omega\tau_z$ gives a universal (master) curve near the GP (τ_z is a characteristic time that scales like $\eta J_e^0 \sim \varepsilon^{-4.2}$). Such master curves were obtained for epoxy as well as for silica gel-point thresholds [159, 160].

For CG structures, unexpected dynamic-mechanical results were obtained by the critical molar ratio (CMR) method for polyurethane-forming [133] and epoxy-forming [161] systems. As mentioned earlier, the CMR method consists of finding the critical initial molar ratio of reactive groups (i.e. finding the critical composition) at which the GP occurs just when the conversion of minority groups approaches 100% (see Figure 8.3). For the polyurethane-forming systems from two poly(oxypropylene) triols (PPTs), $M_n = 710$ and $M_n = 2630$, and a diisocyanate (MDI), the CG structure was obtained at the critical molar ratio $r_H^c = ([OH_0]/[NCO_0])_c$ given by (cf. equation (2.2))

$$r_H^c = (\alpha_I^c)^2(f_{w,H} - 1)(f_{w,I} - 1) \tag{8.59}$$

where α_I^c is the conversion of NCO groups and $f_{w,H}$ and $f_{w,I}$ are the reduced second moments of the functionality distributions of hydroxy and isocyanate reactants, respectively. The samples with $r_H > r_H^c$ are soluble and the systems with $r_H < r_H^c$ contain both sol and gel. The mechanical measurements over broad frequency and temperature ranges, from glassy region to rubbery (or flow) region, were carried out with the stable samples [133]. The values of CMRs, $r_H^c = 1.72$ ($M_n = 710$) and $r_H^c = 1.55$ ($M_n = 2630$), were determined from extraction experiments.

The dependences of the loss tangent, $\tan\delta_p$, on the reduced frequency, ωa_T (after $\omega - T$ superposition), at reference temperature $T_0 = 30\,°C$ are shown in Figure 8.5. The shape and position of the main-transition region (maximum in $\tan\delta_p$) is only slightly dependent on the ratio r_H. As expected, the frequency of the $\tan\delta_p$ maxima of samples with the longer triol are some three orders of magnitude higher than those of samples prepared from the short triol. This shift is due to higher contents of the polar urethane groups in systems with the short triol. Marked changes in the shapes of the frequency dependences of $\tan\delta_p$ with the ratio r_H can be observed in the rubbery and flow regions. For all network samples, with $r_H < r_H^c$, $\tan\delta_p$ decreases with decreasing ωa_{T_0} (solid-like behaviour). For soluble samples, with $r_H > r_H^c$, $\tan\delta_p$ increases with decreasing ωa_{T_0} (liquid-like behaviour). From Figure 8.5 it follows further that the end of the main-transition region, characterized by the frequency of the minimum in $\tan\delta_p$, marks a short-time limit of the power-law region. (The behaviour becomes

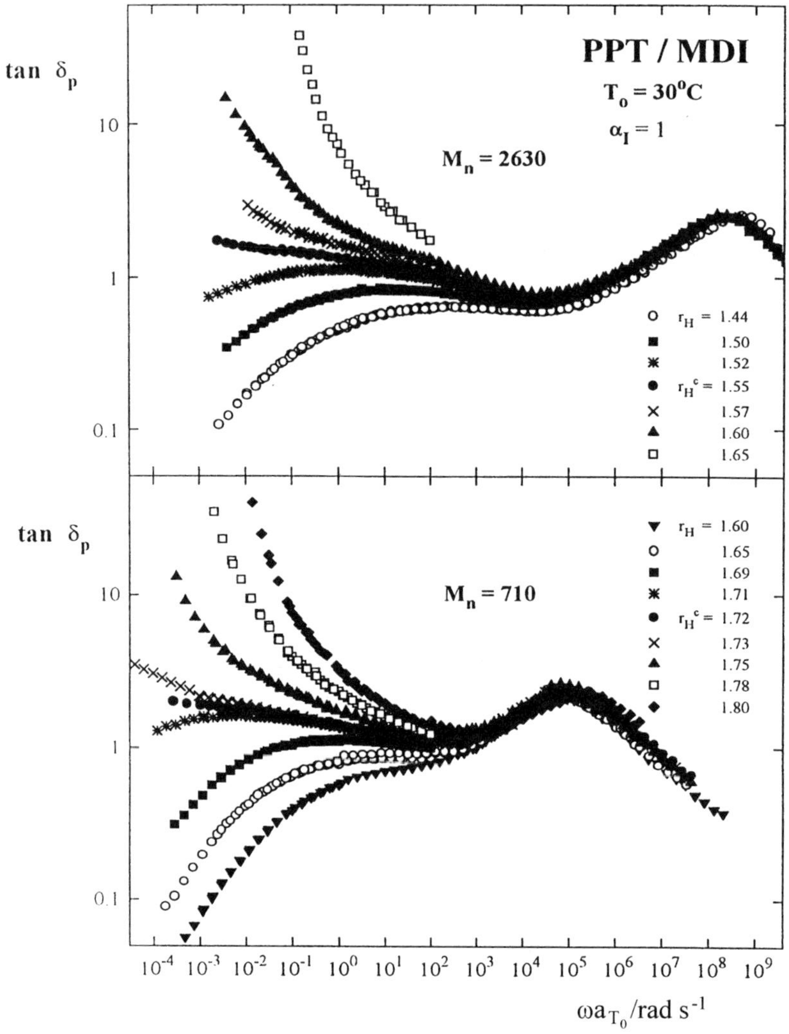

Figure 8.5 Dependences of the $\omega - T$ superimposed loss tangent $\tan \delta_p$ on the reduced frequency ωa_{T_0} for reference temperature $T_0 = 30\,^\circ\mathrm{C}$. Samples from long ($M_n = 2630$) or short ($M_n = 710$) triols reacting with MDI were prepared at the indicated r_H values; α_I denotes the conversion of isocyanate groups [133].

dominated by the local motions of monomeric units and parts of chains.) However, a long-time limit of power-law behaviour is neither expected, nor observed.

For both the samples with CG structures prepared at r_H^c, linear dependences of the superimposed moduli $\log G_p'$ and $\log G_p''$ on $\log \omega a_{T_0}$ were found in the low-frequency region. Surprisingly, the slopes s_1 found for G_p'

(0.7 and 0.63 for the short and long triol, respectively) differ from the slopes s_2 found for the G''_p (0.66 and 0.59 for the short and long triol, respectively). Due to this fact, $\tan\delta_p$ is slightly dependent on ωa_{T_0} for both CG structures (Figure 8.5). The proportionality

$$\tan\delta_p \sim (\omega a_{T_0})^m \tag{8.60}$$

was found, where $m = s_2 - s_1 = -0.04$ for both the samples prepared at r_H^c. Different power-law exponents, s_1 and s_2, were found also by Durand *et al.* [124] for a polyurethane-forming system using triol and hexamethylene diisocyanate with an excess of OH groups, and by Cuvelier and Launay [162] for biopolymer gelling systems.

For all the stable CMR samples shown in Figure 8.5, the frequency–temperature superposition could be performed easily and the temperature dependences of the horizontal shift factors satisfied the WLF equation over the whole temperature range [133]. The dependences of the free-volume parameters (fractional free volume f and thermal expansion coefficient α_f) on r_H were continuous, with no abrupt changes at the critical ratio r_H^c. Therefore, the GP has no specific effect on the temperature dependences of the relaxation times.

Interesting behaviour was found in the frequency dependences of the superimposed storage and loss compliances, J'_p and J''_p, in the rubbery and flow regions (Figure 8.6). For the CG sample with $r_H^c = 1.55$ and 1.72, linear dependences of $\log J'_p$ and $\log J''_p$ on $\log\omega a_{T_0}$ can be seen, and the coresponding slopes s'_1 and s'_2 differ (again, $s'_2 - s'_1 = -0.04$). For networks prepared with $r_H < r_H^c$, limiting equilibrium compliances $J'_e = J_e$ can be determined by extrapolation of the data to $\omega a_{T_0} \to 0$. On the other hand, for soluble samples with $r_H > r_H^c$, limiting steady-state compliances $J'^0_e = J^0_e \neq 0$ can be found. Both limiting compliances J'^0_e and J'_e decrease with increasing deviation from the CG structures. The CG samples prepared at r_H^c show the highest J'_p values and linear dependences of $\log J'_p$ and $\log J''_p$ on $\log\omega a_{T_0}$ (i.e. extrapolation to $\omega a_{T_0} \to 0$ leads to infinite values of J'^0_e and J'_e). It was suggested that highest J'_p values at given ωa_{T_0} are typical of the limiting critical dynamic mechanical behaviour of CG structures prepared by the CMR method. Similar results were obtained for flexible epoxy systems based on the diglycidyl ether of Bisphenol A (DGEBA) and a poly(oxypropylene) di- or triamine (Jeffamine) [161]. The exponent m of equation (8.60) had the value of –0.05.

8.7 Examples of viscoelastic properties of network materials

As previously mentioned, at high frequencies (short times) the viscoelastic properties of networks do not differ much from those of non-network polymers. In the low-frequency range, however, the network structure and

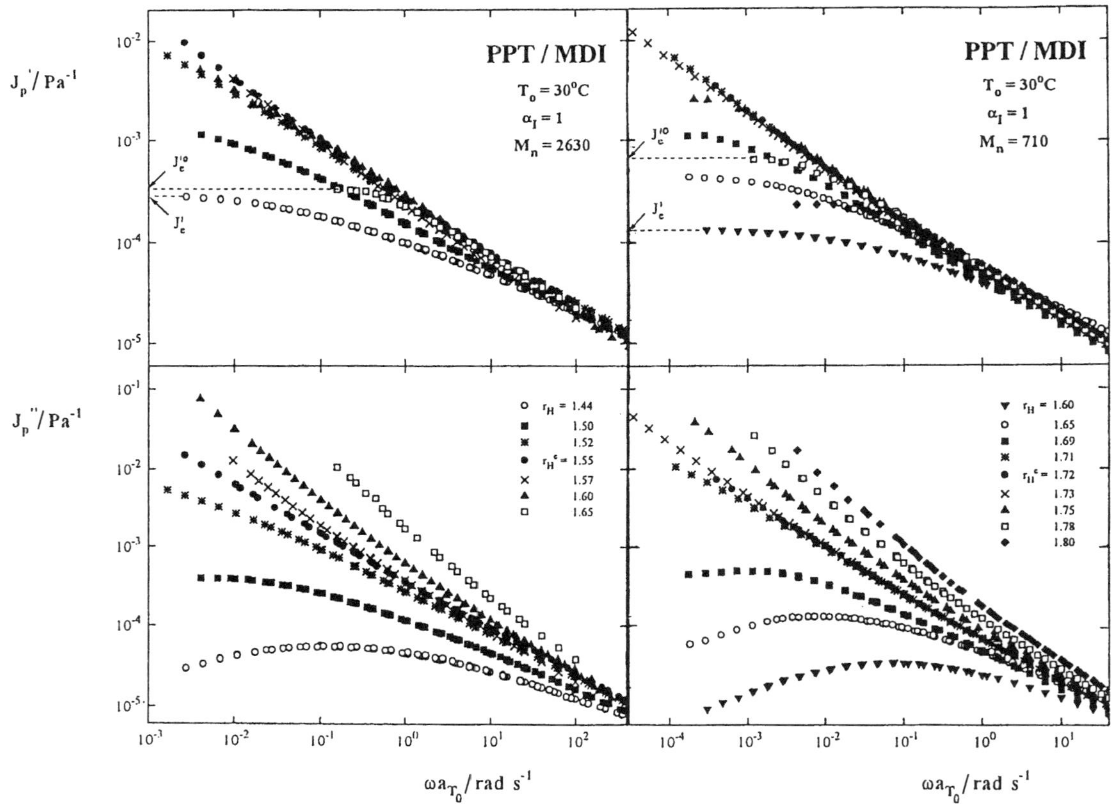

Figure 8.6 Dependences of the superimposed storage J'_p and loss J''_p compliances on the reduced frequency ωa_{T_0} for $T_0 = 30\,^{\circ}\mathrm{C}$ for the samples in Figure 8.5. $J_e^{\prime 0}$ and J'_e are steady-state and equilibrium storage compliances, respectively [133].

its cooperativity, caused by the presence of chemical and physical (entanglement) junctions, become dominant. The problem of cooperativity in viscoelastic behaviour will be more complex if network defects (such as dangling chains) are formed during network formation. The dynamics of such networks include the dynamics of defects (i.e. the relaxation of pendant chains) and the influence of defects on the dynamics of the EANCs. Therefore, one can expect that the shapes of the dependences of viscoelastic functions on frequency (or temperature) at the ends of the main-transition and rubbery regions will be affected not only by the overall junction-point density, but also by the detailed structure of the network – the network topology.

8.7.1 Networks prepared by random chain copolymerization

A typical effect of junction points on the shape and position of retardation spectra (the dependence of $L(\tau)$ on retardation time τ) is shown in Figure 8.7 for networks prepared by chain copolymerization of butyl methacrylate monomer with various amounts of polyfunctional (divinyl) reactant – ethylene glycol dimethacrylate [63]; the spectra were calculated from the superimposed creep compliances $J_p(t)$. It was found that application of the classical time–temperature superposition to measured creep data did not give a continuous master curve in the neighbourhood of the rubberlike region; superposition over the whole time regime required introduction of the change in the unperturbed chain dimensions, $\langle r_0^2 \rangle$, with temperature in accord with equation (8.41). With increasing content of the polyfunctional reactant, the main-transition part of the spectrum is shifted by approximately one decade to longer times, without other changes in shape. Although the

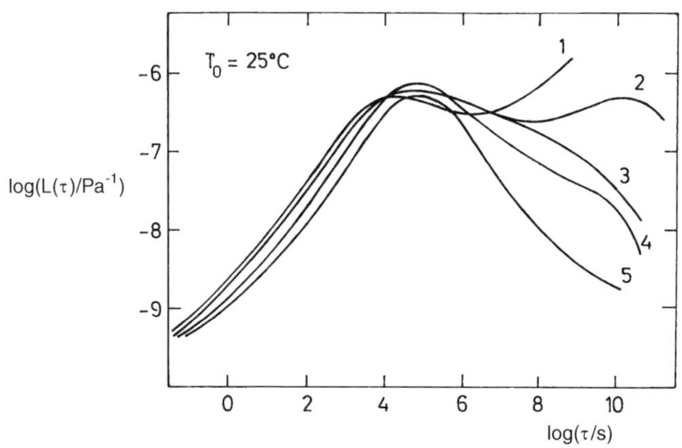

Figure 8.7 Dependence of the retardation spectra $L(\tau)$ on time τ for poly(butyl methacrylate) networks shifted to $T_0 = 25\,°C$. Concentrations of polyfunctional reactant, ethylene glycol dimethacrylate, $c \times 10^4$ mol cm^{-3}: (1) 0; (2) 0.09; (3) 0.44; (4) 0.84; (5) 1.65 [63].

linear sample exhibits flow behaviour, the system with the lowest amount of polyfunctional reactant shows the expected two maxima. Increasing polyfunctional-reactant concentration diminishes the height of the secondary relaxation peak (located at the higher retardation times) and shifts it to shorter times. Generally, an increase in the content of polyfunctional reactant diminishes the occurrence of long-retardation times in the spectrum and decreases the broadness of $L(\tau)$ (making the spectrum more symmetric).

The effects on viscoelastic behaviour of the amount of diluent (water) present in a network-forming copolymerization is shown in Figure 8.8 for poly(2-hydroxyethyl methacrylate) networks [72]. The networks were

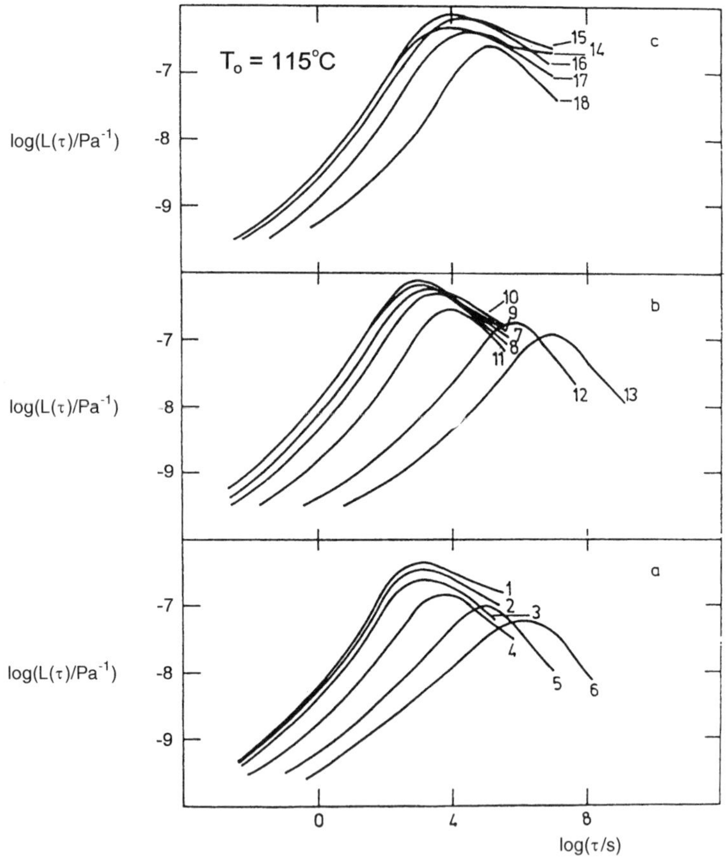

Figure 8.8 Dependences of the retardation spectra $L(\tau)$ on time τ for poly(2-hydroxyethyl methacrylate) networks prepared at various dilutions and polyfunctional reactant concentrations shifted to $T_0 = 115\,°C$. v_2^0 is the volume fraction of polymer at network formation in water and c is the concentration of polyfunctional reactant, ethylene dimethacrylate: (a) $v_2^0 = 1$; (b) $v_2^0 = 0.6$; (c) $v_2^0 = 0.3$. $c \times 10^4\,\text{mol cm}^{-3}$; (1) 0.27; (2) 0.63; (3) 1.34; (4) 2.60; (5) 3.93; (6) 5.15; (7) 0.09; (8) 0.30; (9) 0.62; (10) 1.24; (11) 2.63; (12) 5.16; (13) 6.80; (14) 0.08; (15) 0.27; (16) 0.65; (17) 1.36; (18) 2.80 [72].

prepared at various dilutions (the volume fractions of polymer at network formation, $v_2^0 = 1$, 0.6 and 0.3) and various concentrations of polyfunctional reactant (ethyleneglycol dimethacrylate, $c = 0.08\text{–}6.8 \times 10^{-4}\,\text{mol}\,\text{cm}^{-3}$) and were dried and measured in the dry state. With decreasing v_2^0, the maxima of the retardation spectra L_m increase, and the positions of the spectra are shifted towards shorter retardation times τ. The results are in quantitative agreement with the modified R-M theory, in which the deformation effect of the diluent is taken into account. At a constant reference temperature of 115 °C, the retardation times of the maxima τ_m increase with increasing EANC concentration ν_e. However, it was found that the reduced value τ_m^r (from which the deformational effects of the diluent and frictional coefficient ξ_0 are subtracted) given by (see equation (8.35))

$$\log \tau_\text{m}^\text{r} = \log \tau_\text{m} - \log \xi_0 - (2/3) \log v_2^0 = \log K - 2 \log \nu_\text{e} \tag{8.61}$$

decreases with ν_e at a rate which quantitatively agrees with equation (8.61) (Figure 8.9). This means that the correction factor for the content of diluent $(2 \log v_2^0/3)$ is, at least qualitatively, correct.

Havránek and Heinrich have collected measurements of creep compliance in the main-transition and rubberlike regions [101] of different types of cross-linked, network systems – natural rubber (NR) crosslinked simultaneously by sulfur (S) and dicumyl peroxide (DCP), ethylene-propylene rubber (EPR) crosslinked by sulfur and DCP – and the non-stoichiometric polyurethane networks, prepared from PPTs and MDI, mentioned in Section 8.6.2. The data were converted to retardation spectra $L(\tau)$, which were further shifted to the glass-transition temperatures T_gs of the individual networks. They concluded that for low values of ν_e, two maxima in $L(\tau)$ exist. Whereas the shape and position, $\tau_{\text{m}1}$, of the short-time (glassy process)

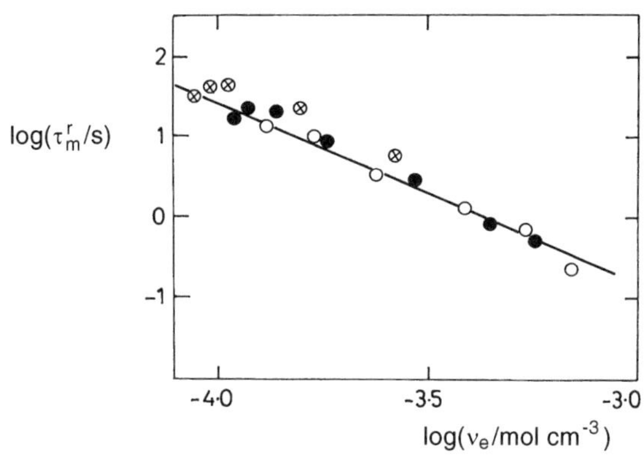

Figure 8.9 Dependence of the reduced value of the retardation-time maximum τ_m^r on junction-point concentration ν_e for the poly(2-hydroxyethyl methacrylate) networks shown in Figure 8.8: (○) samples 1–6; (●) samples 7–13; (⊗) samples 14–18; (——) theoretical slope, $s = -2$ (equation (8.61)) [72].

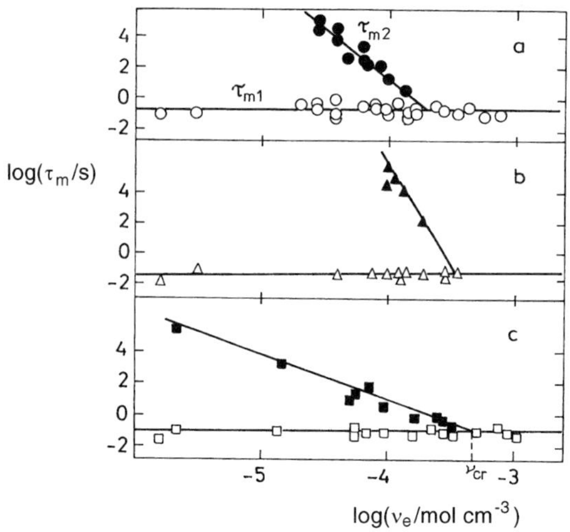

Figure 8.10 Dependence of the short-time maximum τ_{m1} (glassy process), and long-time maximum τ_{m2} (low relaxation process) on the junction-point concentration ν_e: (a) natural-rubber networks; (b) ethylene-propylene-rubber networks; (c) polyurethane networks from PPTs and MDI [101].

maximum is approximately independent of ν_e, for the long-time maximum (slow relaxation process), the shape and position, τ_{m2}, strongly depend on ν_e (Figure 8.10). Two relaxation processes overlap at critical EANC concentration ν_{cr} and, in the range $\nu_e < \nu_{cr}$, they found the dependence

$$\tau_{m2} \sim \nu_e^{-\varphi} \tag{8.62}$$

with $\varphi \approx 6.7$ for NR, $\varphi \approx 12$ for EPR and $\varphi \approx 3$ for PPT/MDI networks. These results, in particular the slow relaxation process, were semiquantitatively interpreted [101] within the scope of the earlier-mentioned theory (see equation (8.52)), in which an exponential-time dependence of the constraint-release parameter β (characterizing tube lateral dimensions) was introduced.

8.7.2 Networks prepared by endlinking

The viscoelastic behaviour of networks prepared from telechelic prepolymers by endlinking has been much studied in the literature [163–181]. Due to a comparatively simple build-up of network structure, a detailed knowledge of starting reactants and reaction mechanisms, these systems are used in testing theories of formation and mechanical behaviour [169, 174–178]. Application of the theory of branching processes (TBP) to such networks has led to a detailed description of their structures [178, 179], which has made possible a discussion of the relations between the structural parameters and mechanical and other properties. Mostly, polyurethane networks from polyols and polyisocyanates and epoxide networks based on polyamines and polyepoxides have been used in the literature. By changing the initial ratio of reactive groups or the molecular weights of components, one can obtain networks having broad variations in structure (broad ranges of junction-point density and network topology and lengths and numbers of dangling chains). Whereas networks prepared with stoichiometric ratios of reactive groups have simple EANC structures, off-stoichiometric networks have complex EANC structures, with various dangling chains (DCs) attached to backbones (BCs) of various lengths [76, 86, 178–182].

An example of the effect of the detailed structure of EANCs on the viscoelastic behaviour of networks can be demonstrated by the viscoelastic creep behaviour of polyurethane networks based on the PPTs and MDI described in Figure 8.5. In this case, networks with initial molar ratios $r_H = 0.6–1.5$ were prepared [77, 184]; reactions were carried out up to the highest conversion of minority groups. With networks with an excess of MDI ($r_H < 1$), the formation of allophanate groups was avoided by the addition of cyclohexanol to the reacting mixtures [187]. From master compliance curves, obtained using t–T superposition, the retardation spectra $L(\tau)$ were calculated and used to determine the half-width, $h = \log \tau_2 - \log \tau_1$, where τ_2 and τ_1 are the retardation times for which $L(\tau_1) = L(\tau_2) = L_m/2$. In Figure 8.11, the dependence of h is plotted against r_H. As expected, the narrowest spectra were found for $r_H = 1$. With increasing deviations from stoichiometry, the values of h increase. Networks from the longer triol have higher values of h over the whole r_H range, compared with those made from the shorter PPT.

The increase in h was correlated with the increase in number-average molecular weight of the EANCs, $(M_{EC})_n$, BCs, $(M_{BC})_n$ or DCs, $(M_{DC})_n$ (all these parameters were calculated from the TBP [186]). Figure 8.12

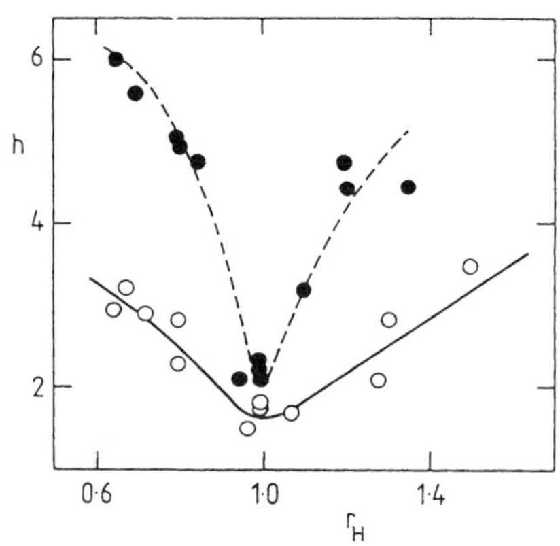

Figure 8.11 Dependence of the half-width of retardation spectrum h on the initial molar ratio of the reactive groups r_H for PPT + MDI networks: (○) networks from the shorter PPT ($M_n = 710$); (●) networks from the longer PPT ($M_n = 2630$) [186].

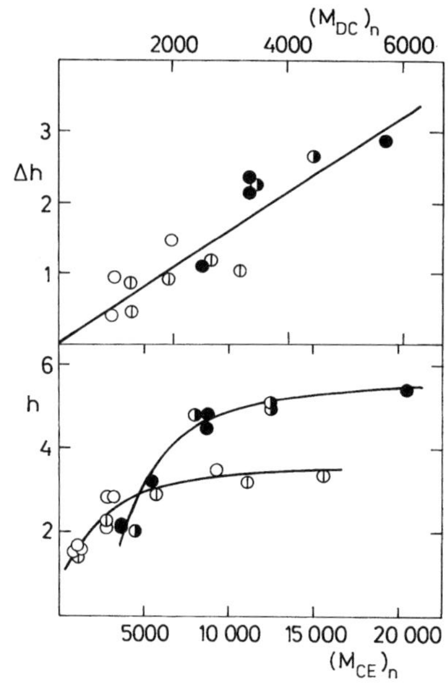

Figure 8.12 Dependence of the half-width of the retardation spectrum h on the number-average molecular weight of EANCs $(M_{CE})_n$ and the dependence of the difference in the half-widths $\Delta h = h - h_x$ (h_x corresponds to the value of h of the perfect network with $r_H = 1$) on the number-average molecular weight of dangling chains $(M_{DC})_n$. Shorter PPT ($M_n = 710$): (⦶) $r_H < 1$; (○) $r_H \geq 1$. Longer PPT ($M_n = 2630$): (◑) $r_H < 1$; (●) $r_H \geq 1$ [186].

shows that, for both network series, the increase in h with $(M_{EC})_n$ is independent of whether the networks were prepared with an excess of OH or NCO groups. On the other hand, the values of h of networks from the longer triol are higher than those from the shorter triol. This means that the half-width is affected not only by total concentration of EANCs but also by imperfections in the EANC structure. The same conclusion follows from the plot of h versus $(M_{BC})_n$. The best fit for all networks was found when the difference $\Delta h = h - h_x$ (where h_x corresponds to the h value of a perfect network with $r_H = 1$) was plotted against $(M_{DC})_n$. It may be concluded, therefore, that the increase in the spectrum width in the rubberlike region with increasing deviation from stoichiometry is caused mainly by the increasing lengths of dangling chains.

The same polyurethane networks from PPTs and MDI were also studied by dynamic–mechanical measurements [77]. Detailed analysis of the shape of the master complex-compliance curve $J_p^*(\omega)$ on reduced frequency ωa_{T_0} showed that the experimental curves could be decomposed into two relaxation processes J_i^* of the Cole–Cole form

$$J_i^* = \frac{\Delta J_i}{1 + (i\omega\tau_i)^{a_i}} \qquad i = 1, 2 \tag{8.63}$$

where ΔJ_i is the relaxation strength, τ_i is the relaxation time ($\tau_i = \omega_{m,i}^{-1}$, with $\omega_{m,i}$ the frequency of the maximum in J_i'') and a_i is a measure of the broadness of the relaxation process ($a_i \approx 0.7$ and 0.4 were found for the high- and the low-frequency processes, respectively). The relaxation strength ΔJ_1 of the faster of the two processes (corresponds to the glass-transition process) was practically independent of r_H, whereas the relaxation time τ_1 increased with decreasing r_H (due to an increasing content of MDI, i.e. polarity of the system). However, the ΔJ_2 and τ_2 values of the low-frequency process considerably increased with the difference of r_H from 1. It is interesting to mention that for a 'perfect' network, with $r_H = 1$, both processes practically overlapped, in accord with equation (8.51), as the trapping factor $T_e \approx \beta_e^2$ of a stoichiometric network is unity. As in the case of the half-width of the retardation spectrum h, it was shown [188] that ΔJ_2 and τ_2 of the low-frequency process correlate, to a first approximation, with the number-average molecular weight of the dangling chains $(M_{DC})_n$.

Interesting viscoelastic behaviour was observed in the case of epoxy–amine networks. Such systems are of great importance, both practical and theoretical [179–181]. With flexible epoxide networks made from the diglycidyl ether of Bisphenol A (DGEBA) and a poly(oxypropylene) diamine or triamine (Jeffamine), dynamic mechanical behaviour similar to that of polyurethane systems was found [76]. Frequency–temperature superposition could be made easily and the temperature dependences of the horizontal shift factors could be described by the WLF equation (equation (8.22)), with values of the free-volume constants similar to those usually found for other systems. For off-stoichiometric networks, prepared with various ratios of amine (NH_2) and epoxide (E) groups, $r_A = 2[NH_2]/[E] = 1$ to 3.5, the half-width h of the retardation spectrum increased with increasing deviation from stoichiometry; in this case, it could not be determined which structural parameter (the lengths of EANCs, BCs or DCs) had the strongest effect on the shape of the spectrum [76].

Hard epoxy networks, based on aromatic diglycidylamines, N,N'-diglycidylaniline (DGA) or N,N,N',N'-tetraglycidyl-4,4′-diaminodiphenylmethane (TGDDM), and reacted with 4,4′-diaminodiphenylmethane (DDM), showed a viscoelastic behaviour different from that of the flexible networks [68, 69, 189]. Extensive cyclization and a positive substitution effect in DGA (increase in reactivity of the second epoxy group after the first one has reacted (see Chapter 3)) are dominant for the reactions of DGA and TGDDM with amines. These special features make network formation using DGA and TGDDM more complicated than that using DGEBA. Examples of the temperature dependences of the horizontal shift factors $\log a_{T_\alpha}$ (T_α is the temperature of the maximum in the temperature dependence of the loss modulus G'' measured at frequency $f = \omega/2\pi = 1$ Hz), for TGDDM/DDM networks [69] prepared with excess amine groups ($r_A \geq 1$), are shown in Figure 8.13. In the glassy region, these dependences follow the Arrhenius

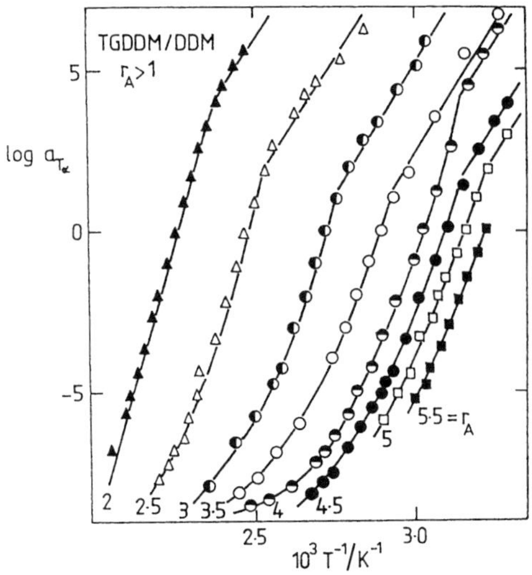

Figure 8.13 Dependence of the horizontal shift factor $\log a_{T_\alpha}$ (T_α is the temperature of the maximum in the dependence of G'' on T measured at $f = \omega/2\pi = 1$ Hz for individual samples) on $1/T$ of epoxide networks from tetraepoxide (TGDDM) and diamine (DDM). r_A denotes the initial molar ratio of amine to epoxide groups [69].

equation (equation (8.23)) with an activation energy $\Delta H = 290\,\text{kJ}\,\text{mol}^{-1}$, which is roughly independent of the composition. It was also found that, at higher temperatures (in the main-transition region), the dependences of $\log a_{T_\alpha}$ on $1/T$ of networks with $r_A < 2.5$ can be described by the Arrhenius equation. However, the dependences of $\log a_{T_\alpha}$ on $1/T$ for networks with $r_A \geq 2.5$ are adequately described by the WLF equation. It was suggested that the change in the character of the dependences is associated with very short EANCs and also with the structure of the pendant chains. Similar behaviour was found for DGA–DDM and DGEBA–1,3-phenylenediamine systems [68, 190]. In all cases, it was also found that the change in the internal energy of the network has affected the vertical shift factor b_{T_α} and equation (8.41) had to be applied to the data.

The hard epoxy networks [69] also showed unexpected shapes of the frequency dependences of superimposed compliance curves (Figure 8.14). For TGDDM/DDM samples with excess of amine groups, a large shift of the loss compliance J''_p to higher frequencies (about 27 decades) was found, but the half-width h of the J''_p maximum (which should be the same as from $L(\tau)$, see equation (8.19)) decreases with increasing r_A (decreasing concentration of EANCs, ν_e). The unusual temperature dependence of the horizontal shift factor $\log a_{T_\alpha}$, together with the unexpected decrease in h with decreasing ν_e, was interpreted as resulting from the increasing stiffness of the EANCs with increasing ν_e; the increasing stiffness probably leads to the broadening of the free-volume distribution function [69]. With increasing

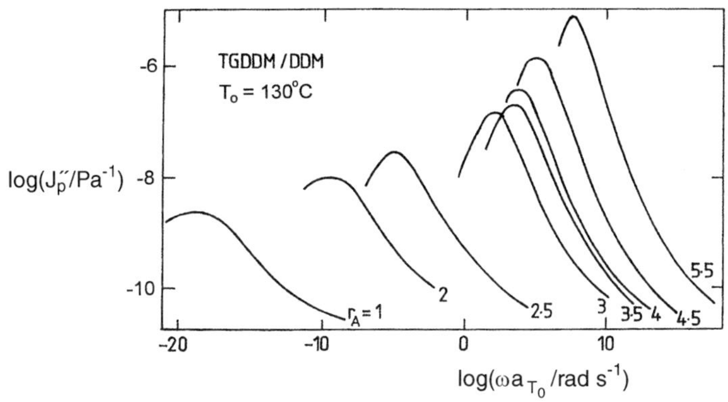

Figure 8.14 Dependence of the superimposed loss compliance J''_p on reduced frequency ωa_{T_0} for the TGDDM/DDM epoxy networks of Figure 8.13, $T_0 = 130°C$ [69].

r_A, the maximum of the loss compliance J''_m (and thus the maximum of $L(\tau)$, $L_m = (2/\pi)J''_m$) increased. The ratio $J''_m/J_e \approx 0.18$ suggested a broad distribution of submolecules in chains.

8.7.3 *Effects of chemical (topological) clusters on viscoelastic behaviour*

Many network systems are formed from reactants of various chain structures (e.g. soft and hard, long and short, etc.) By network-forming reactions, these reactants can form branched structures which may be composed of different types of unit, forming soft regions and hard regions, i.e., chemical clusters [169, 191, 192]. A typical example of such chemical, topological clusters occurs in the formation of polyurethane networks [191] from three or more reactants (e.g., from long (flexible) diol, (stiff) diisocyanate and short (stiff) triol). As the relative reactivity of functional groups may not be equal and can change during the reaction (substitution effect), polymerization leads to the formation of hard (diisocyanate plus short triol) and soft (diisocyanate plus long diol) clusters of various sizes. The structures of these topological clusters are determined by the rules of chemical bond formation between the starting reactants and not by the physical interactions between them. The extent of cluster formation depends not only on the functionalities, the reactivities and initial molar ratios of reactants but also on the method of preparation, e.g., whether the network was formed in a one-stage or two-stage process. In a one-stage process, all components are mixed together, whereas in a two-stage process the reaction of macrodiol with an excess of diisocyanate (usually $r_H = 0.5$) occurs first, followed by reaction of the product with the short triol as the second step.

The effects of number and size of chemical clusters on dynamic–mechanical behaviour was studied [193] using polyurethane networks based on poly(oxypropylene) diol (PPD, $M_n = 2100$), MDI and trimethylolpropane (TMP). Networks were prepared at various ratios of triol hydroxy to isocyanate groups $r_{HT} = [OH_{TMP}]/([NCO]/2) = 1$ to 1.5, using the one- and two-stage processes. In the one-stage process, the sizes of the clusters are controlled by the difference in the reactivities of the OH groups of TMP (primary) and PPD (secondary). This is not the case in the two-stage process. Small-angle X-ray scattering (SAXS) experiments have indeed revealed that the one-stage networks have larger numbers of smaller clusters than the two-stage ones. It was further found by SAXS experiments that the sizes of the inhomogeneities increase with increasing r_{HT}.

Frequency–temperature superposition of both components of the complex compliance J_p^* could be applied to all networks and the shift factor $\log a_{T_0}$ satisfied the WLF equation [193]. The concentration of MDI (polarity of the systems) was approximately the same in all networks. Hence, their polarities were approximately the same and the temperature dependence of $\log a_{T_0}$ was independent of r_{HT} and method of network preparation. However, the two-stage networks exhibited more complex frequency dependences of J'_p and J''_p at low ωa_{T_0} (Figure 8.15). All superimposed curves could be

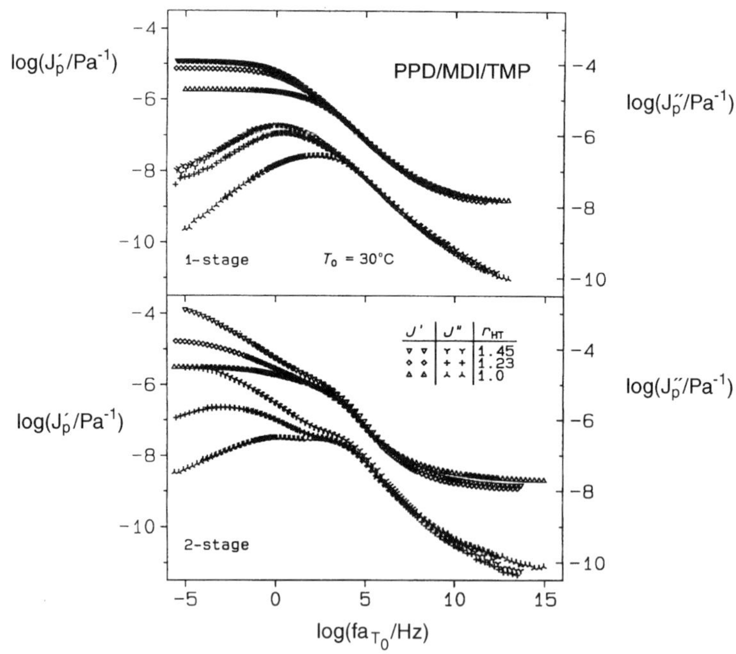

Figure 8.15 Dependence of the superimposed storage J'_p and loss J''_p compliances on the reduced frequency fa_{T_0} for three-component (PPD–MDI–TMP) polyurethane systems, shifted to $T_0 = 30\,°C$. Networks prepared by the one- and two-stage processes. r_{HT} denotes the initial ratio, $[OH_{TMP}]_0/([NCO]/2)$ [193].

decomposed into two relaxation processes of the Cole–Cole form, in accord with equation (8.61). It was found that for the fast (glass) process, the values of ΔJ_1 and τ_1 values were roughly independent of r_{HT} and the network-formation process. On the other hand, ΔJ_2 and τ_2 of the slow process increased with r_{HT} and the two-stage networks showed much higher values of ΔJ_2 and τ_2 than the one-stage networks. Obviously, the larger numbers of smaller clusters in the one-stage networks reduces the motion of branched EANCs at the lower frequencies (deep in the rubbery region), more than the smaller number of larger clusters present in the two-stage materials.

8.8 Concluding remarks

The subject of the viscoelastic behaviour of covalent networks is very broad and cannot possibly be covered in a single chapter. Indeed, this chapter has dealt mainly with results and systems, with which the author and his co-workers have been involved for the past 25 years. There exist, of course, other important areas, such as the viscoelastic behaviours of interpenetrating polymer networks (see Chapter 6), liquid-crystalline materials and reinforced elastomers, that have not been covered by this review. Nevertheless, it is hoped that the chapter has highlighted the fascination and subtleties of the problems discussed and will encourage others to carry out research on the viscoelasticity of network materials.

Acknowledgements

The author thanks the Grant Agency of the Czech Republic and Grant Agency of the Academy of Sciences of the Czech Republic for financial support, given under Grant Nos 203/95/1318 and A4050511.

Glossary of symbols and abbreviations

ΔC_p	discontinuity in the isobaric heat-capacity at T_g
E, $E(t)$	Young's modulus
E_A	height of potential-energy barrier for elementary conformational motion
G	shear modulus
$G(t)$	shear modulus; shear-modulus function in stress–relaxation
$G_P(t, \phi_0)$	reduced master curve for shear modulus superimposed onto a reference value ϕ_0 of an external parameter (e.g. temperature)
G^*	complex shear-modulus
G'	storage shear-modulus; real component of G^*

G''	loss shear-modulus; imaginary component of G^*
G_c	shear modulus of a network due to chemical junction-points
G_e	equilibrium (infinite-time or zero-frequency) shear-modulus ($= J_e^{-1}$)
G_N	shear modulus associated with trapped entanglements
G_N^0	(rubbery) plateau shear-modulus
$H(t)$	relaxation spectrum
ΔH	activation energy
$J(t)$	shear-compliance function in creep
$J_P(t, \phi_0)$	reduced master curve for compliance superimposed onto a reference value ϕ_0 of an external parameter (e.g. temperature)
J^*	complex shear-compliance
J'	storage shear-compliance; real part of J^*
J''	loss shear-compliance; imaginary component of J^*
J_e	equilibrium (infinite-time or zero-frequency) shear-compliance
J_g	glassy-state (zero-time or infinite-frequency) shear-compliance
J'_N	storage compliance in the rubber-like (plateau) region of dynamic–mechanical behaviour
J'_P	storage compliance in the rubbery and flow (plateau) regions
J''_P	loss compliance in the rubbery and flow (plateau) regions
J''_m	maximum of loss shear-compliance
J_e^0	steady-state, recoverable compliance of a viscoelastic liquid
J'_e	equilibrium storage shear-compliance
ΔJ_i	relaxation strength in dynamic–mechanical behaviour
$K(t)$	bulk modulus
L	molecular-tube length in an entangled polymer system
$L(t)$	retardation spectrum
$L_m(\tau)$	maximum in the retardation spectrum
M_e	molecular weight of the chain between physical junctions or entanglements
N	number of submolecules in a molecule
P	degree of polymerization
S	gel (mechanical) strength
S_c	molar conformational entropy per elementary chain-conformational motion at temperature T
S_c^*	molar conformational entropy per conformational-rearranging region at temperature T
T^0	temperature of network formation
T_g	glass-transition temperature; main (α)-transition temperature
T_m	melting temperature
T_0	reference temperature for defining the horizontal shift factor, $a(T, T_0)$
T_r	curing (reaction) temperature
T_2	minimum temperature for conformational rearrangements

$T_\beta, T_\gamma, \ldots$	secondary $(\beta, \gamma, \ldots)$-transition temperatures
a	molecular-tube diameter in an entangled polymer system
$a(T, T_0)$; a_{T_0}	horizontal shift-factor when temperature T is the external parameter
$a(\phi_i, \phi_0)$	horizontal shift-factor for measurements of G and J at ϕ_i, used to evaluate the master curves, $G_P(t, \phi_0)$ and $J_P(t, \phi_0)$
$b(\phi_i, \phi_0)$	vertical shift-factor for measurements of G and J at ϕ_i, used to evaluate the master curves, $G_P(t, \phi_0)$ and $J_P(t, \phi_0)$
d	spatial dimensionality
d_f	fractal dimensionality of critically branched molecules
$\bar{d}_f$	d_f in the absence of excluded volume
f	frequency
$f_{w,H}$; $f_{w,I}$	reduced second moments of the functionality distributions of hydroxy and isocyanate reactants (weight-average functionalities)
f_ϕ	fractional free-volume at a given external parameter ϕ
n	number of molecules per unit volume
n^*	number of elementary conformational motions in a conformationally rearranging region at temperature T
r^c	CMR of reactive groups in a polymerization
r_A	$(= 2[NH_2]_0/[E]_0)$ initial, molar ratio of amine (NH_2) to epoxide (E) groups on the basis of the stoichiometry of the amine-epoxide reaction
r_H	$(= [OH]_0/[NCO]_0)$ initial, molar ratio of hydroxy to isocyanate groups
r_H^c	critical value of r_H, giving the GP at complete reaction of the minority group
r_{HT}	$(= [OH_{TMP}]_0/([NCO]_0/2))$ initial, molar ratio of the hydroxy groups of TMP to the isocyanate groups required for the chain-extension of TMP
$\langle r_{i,s}^2 \rangle$	mean-square end-to-end distance of an expanded chain in an isotropically swollen network
$\langle r_o^2 \rangle$	unperturbed mean-square end-to-end distance of a free polymer chain, defining the reference state of chains in a network
t	current time in viscoelastic behaviour
t_P	relaxation time of a dangling chain in a network
t_c	characteristic time of the chain-constraint-release process in a tube model; gel time
t_r	curing (reaction) time
v_2	volume fraction of polymer in a swollen state
v_2^0	volume fraction of polymer at network formation
v_{20}	reference volume-fraction of polymer in a network for the superposition of dynamic–mechanical properties

w_g	weight fraction of gel (gel fraction)
x	past time in viscoelastic behaviour
α_c	critical converson of reactive groups at the GP
α_f	temperature coefficient of the fractional change in free volume; thermal expansion coefficient
$\langle \alpha_{i,s}^2 \rangle$	$(= \langle r_{i,s}^2 \rangle / \langle r_o^2 \rangle)$ dilation factor for the expansion of $\langle r_o^2 \rangle$ by the isotropic swelling of a network
$\langle \alpha_{i,d}^2 \rangle$	$(= \langle r_{i,d}^2 \rangle / \langle r_o^2 \rangle)$ dilation factor for the expansion or contraction of $\langle r_o^2 \rangle$ caused by drying a network
α_I^c	conversion of isocyanate groups at the GP
β	constraint-release parameter in the tube model
β_f	pressure coefficient of the change in fractional free-volume
β_f'	swelling coefficient of the change in fractional free-volume
δ	$(= \mathrm{d}\ln\langle r_o^2 \rangle / \mathrm{d}T)$ temperature coefficient of unperturbed-chain dimensions
$\tan\delta$	loss tangent
ε_t	(uniaxial) tensile deformation
γ	shear strain
γ^*	cyclic shear-strain
γ_H	strain in a Hookean spring
γ_N	strain in Newtonian, viscous flow
γ_0	initial, constant, applied strain
η	steady-flow (shear) viscosity; Newtonian viscosity
μ	Poisson's ratio
ν_e	concentration of EANCs
$\nu_{i,s}$	number of EANCs per unit isotropically-swollen volume
ξ_0	monomeric friction coefficient
$\xi_{0,ef}$	effective monomeric friction coefficient
ρ	density
ρ_m	monomer density
σ	shear stress
σ^*	cyclic stress
σ_H	stress in a Hookean spring
σ_N	stress in Newtonian, viscous flow
σ_t	tensile stress
σ_0	initial, constant, applied stress
τ	relaxation- or retardation-time constant
τ_{av}	average relaxation time
τ_i	ith relaxation time
τ_m	retardation time at the maximum in a retardation spectrum
τ_φ	relaxation time, related to f_φ
τ_m^r	reduced retardation time at the maximum in a retardation spectrum
ϕ_i	(general) external parameter, affecting viscoelastic behaviour

ω	angular frequency
Γ	gamma function
BC	backbone (chain)
CG	critical gel (state)
CMR	critical molar ratio of reactive groups
DC	dangling chain in a network
DCP	dicumyl peroxide
DDM	4,4′-diaminodiphenylmethane
DGA	*N*,*N*′-diglycidylamine
DGEBA	diglycidyl ether of Bisphenol A
EANC	elastically active network-chain
EPR	ethylene-propylene rubber (PEDPM)
GP	gel point
MDI	4,4′-diphenylmethane diisocyanate
NR	natural rubber
PEDPM	poly(ethylene-*co*-propylene-*co*(diene monomer))
PPD	poly(oxypropylene) diol
PPT	poly(oxypropylene) triol
SAXS	small-angle X-ray scattering
TBP	theory of branching processes
TGDDM	*N*,*N*,*N*′,*N*′-tetraglycidyl-4,4′-diaminodiphenylmethane
TMP	trimethylol propane

References

1. Leaderman, H. (1958) in *Rheology*, Vol. 2 (ed. F.R. Eirich), Academic Press, New York.
2. Ferry, J.D. (1980) *Viscoelastic Properties of Polymers*, 3rd edn, Wiley, New York.
3. Gross, B. (1953) *Mathematical Structure of Theories of Viscoelasticity*, Herman, Paris.
4. Bland, D.R. (1960) *The Theory of Linear Viscoelasticity*, Pergamon Press, Oxford.
5. Tschoegl, N.W. (1971) *Rheol. Acta*, **10**, 595.
6. Wiff, D.R. (1978) *J. Rheol.*, **22**, 589.
7. Plazek, D.J., Raghupathi, N. and Orbon, S.J. (1979) *J. Rheol.*, **23**, 477.
8. Whaba, G. (1978) in *Solution Methods for Integral Equations* (ed. M.A. Goldberg), Plenum Press, New York.
9. Honerkamp, J. and Wiese, J. (1989) *Macromolecules*, **22**, 4372.
10. Williams, M.L., Landel, R.F. and Ferry, J.D. (1955) *J. Am. Chem. Soc.*, **77**, 3701.
11. Doolittle, A.K. (1951) *J. Appl. Phys.*, **22**, 1471.
12. Boyer, R.F. and Simha, R. (1973) *J. Polym. Sci.*, **B11**, 33.
13. Cohen, M.H. and Grest, G.S. (1979) *Phys. Rev.*, **B20**, 1077.
14. Grest, G.S. and Cohen, M.H. (1981) *Adv. Chem. Phys.*, **48**, 455.
15. Fujita, H. and Kishimoto, H. (1958) *J. Colloid Sci.*, **13**, 418.
16. Ilavský, M. and Hasa, J. (1968) *Collect. Czech. Chem. Commun.*, **33**, 2142.
17. Ilavský, M. and Kolařík, J. (1969) *Collect. Czech. Chem. Commun.*, **34**, 2473.
18. Ilavský, M., Hrouz, J. and Havlíček, I. (1979) *J. Macromol. Sci.*, **B16**, 259.
19. Conghlin, C.S., Mauritz, K.A. and Storey, R.F. (1991) *Macromolecules*, **24**, 1526.
20. Gibbs, J.H. and DiMarzio, A.E. (1958) *J. Chem. Phys.*, **28**, 373.
21. DiMarzio, E.A. and Gibbs, J.H. (1963) *J. Polym. Sci.*, **1A**, 1417.
22. Adam, A. and Gibbs, J.H. (1965) *J. Chem. Phys.*, **43**, 139.

23. Havlíček, I., Vojta, V., Ilavský, M. and Hrouz, J. (1980) *Macromolecules*, **13**, 357.
24. Havlíček, I., Ilavský, M. and Hrouz, J. (1982) *J. Macromol. Sci. Phys.*, **B21**, 425.
25. Havlíček, I., Ilavský, M. and Hrouz, J. (1978) *J. Polym. Sci., Polym. Phys. Ed.*, **16**, 653.
26. Donth, E.J. (1992) *Relaxation Thermodynamics in Polymers. Glass Transition*, Akademie-Verlag, Berlin.
27. Jäckle, J. (1986) *Rep. Prog. Phys.*, **49**, 171.
28. Fredrickson, G.H. (1988) *Annu. Rev. Phys. Chem.*, **39**, 149.
29. Böhmer, R. and Angell, C.A. (1994) in *Disorder Effects on Relaxational Processes* (eds R. Richert and A. Blumen), Springer-Verlag, Berlin.
30. Rouse, P.E. (1953) *J. Chem. Phys.*, **21**, 1272.
31. Bueche, F. (1954) *J. Chem. Phys.*, **22**, 603.
32. Zimm, B.H. (1956) *J. Chem. Phys.*, **24**, 269.
33. Mooney, M. (1959) *J. Polym. Sci.*, **34**, 599.
34. Berry, G.C. and Fox, T.G. (1968) *Adv. Polym. Sci.*, **5**, 261.
35. Nemoto, N., Odani, H. and Kurata, M. (1972) *Macromolecules*, **5**, 531.
36. Bueche, F. (1952) *J. Chem. Phys.*, **20**, 1959.
37. Graessley, W.W. (1974) *Adv. Polym. Sci.*, **16**, 1.
38. Graessley, W.W. (1979) *Adv. Polym. Phys.*, **47**, 1942.
39. Bueche, F. (1962) *Physical Properties of Polymers*, Interscience, New York.
40. Edwards, S.F. (1967) *Proc. Phys. Soc.*, **92**, 9.
41. De Gennes, P.G. (1971) *J. Chem. Phys.*, **55**, 572.
42. De Gennes, P.G. (1974) *J. Phys. Lett. France*, **35**, L-133.
43. De Gennes, P.G. (1980) *J. Chem. Phys.*, **72**, 4756.
44. Doi, M. and Edwards, S.F. (1978) *J. Chem. Soc., Faraday Trans. 2*, **74**, 1789.
45. Doi, M. and Edwards, S.F. (1978) *J. Chem. Soc., Faraday Trans. 2*, **74**, 1818.
46. Doi, M. and Edwards, S.F. (1979) *J. Chem. Soc., Faraday Trans. 2*, **75**, 38.
47. Des Cloizeaux, J. (1984) *J. Phys. Lett. France*, **45**, L-17.
48. Scher, H. and Schesinger, M.F. (1986) *J. Chem. Phys.*, **84**, 5922.
49. Schweizer, K.S. (1989) *J. Chem. Phys.*, **91**, 5802, 5822.
50. Rubinstein, M. (1987) *Phys. Rev. Lett.*, **59**, 1946.
51. Lodge, T.P., Rotstein, N.A. and Prager, S. (1990) *Advanced Chem. Physics*, Vol. 79 (eds I. Prigogine and S.A. Rice), Wiley, New York.
52. Kremer, K. and Grest, G.S. (1995) in *Monte Carlo and Molecular Dynamics Simulations in Polymer Science* (ed. K. Binder), Oxford University Press, New York.
53. Volkenstein, M.V. (1963) *Configurational Statistics of Polymer Chains*, Interscience, New York.
54. Flory, P.J. (1969) *Statistical Mechanics of Chain Molecules*, Interscience, New York.
55. Flory, P.J. (1956) *J. Am. Chem. Soc.*, **78**, 5222.
56. Dušek, K. and Prins, W. (1969) *Adv. Polym. Sci.*, **6**, 1.
57. Flory, P.J., Hoeve, C.A.J. and Ciferri, A. (1959) *J. Polym. Sci.*, **44**, 337.
58. Ilavský, M., Hasa, J. and Havlíček, I. (1972) *J. Polym. Sci. A-2*, **10**, 1775.
59. Flory, P.J. (1976) *Proc. R. Soc. Lond.*, A**351**, 351.
60. Queslel, J.P. and Mark, J.E. (1984) *Adv. Polym. Sci.*, **65**, 135.
61. Mark, J.E. and Erman, B. This book, Chapter 7.
62. Ciferri, A. (1964) *J. Polym. Sci. A*, **2**, 3089.
63. Ilavský, M. and Hasa, J. (1973) *J. Polym. Sci. Polym. Phys. Ed.*, **11**, 539.
64. Scholtens, B.J.R. (1984) *J. Polym. Sci. Polym. Phys. Ed.*, **22**, 317.
65. Scholtens, B.J.R. (1985) *Br. Polym. J.*, **17**, 134.
66. Scholtens, B.J.R. (1985) *J. Polym. Sci., Polym. Phys. Ed.*, **23**, 1325.
67. Ilavský, M., Hrouz, J. and Havlíček, I. (1987) in *Crosslinked Epoxies*, Walter de Gruyter, Berlin.
68. Ilavský, M., Zelenka, J., Špaček, V. and Dušek, K. (1992) *Polym. Networks Blends*, **2**, 95.
69. Ilavský, M., Dušek, K., Zelenka, J. and Dobáš, I. (1994) *Polym. Networks Blends*, **4**, 187.
70. Hasa, J. and Ilavský, M. (1969) *Collect. Czech. Chem. Commun.*, **34**, 2189.
71. Ilavský, M. and Hasa, J. (1969) *Collect. Czech. Chem. Commun.*, **34**, 2199.
72. Ilavský, M., Hasa, J., Vojta V. and Janáček, J. (1976) *J. Polym. Sci. Polym. Phys. Ed.*, **14**, 285.
73. Straton, R.A. and Ferry, J.D. (1963) *J. Phys. Chem.*, **67**, 2781.

74. Hrouz, J. and Janáček, J. (1972) *J. Polym. Sci. A-2*, **10**, 1318.
75. Valentine, H.R., Ferry, J.D., Homma, T. and Ninomiya, K. (1968) *J. Polym. Sci. A-2*, **6**, 479.
76. Ilavský, M., Hrouz, J., Šomvársky, J. and Dušek, K. (1989) *Makromol. Chem. Macromol. Symp.*, **30**, 13.
77. Havránek, A., Ilavský, M., Nedbal, J., Böhm, M., v. Soden, W. and Stoll, B. (1987) *Colloid Polym. Sci.*, **265**, 8.
78. Curro, J.G. and Pinkus, P. (1983) *Macromolecules*, **16**, 559.
79. Neuburger, N.A. and Eichinger, B.E. (1985) *J. Chem. Phys.*, **83**, 884.
80. Gaylord, R.J., Joss, B., Bendler, J.T. and di Marzio, E.A. (1985) *Br. Polym. J.*, **17**, 126.
81. Bibbó, M.A. and Vallés, E.M. (1984) *Macromolecules*, **17**, 360.
82. Curro, J.G., Pearson, D.S. and Helfand, E. (1985) *Macromolecules*, **18**, 1157.
83. Plazek, D.J. (1966) *J. Polym. Sci. Polym. Phys. Ed.*, **4**, 745.
84. Gottlieb, M., Macosko, C.W., Benjamin, G.S., Meyers, K.O. and Merril, E.W. (1981) *Macromolecules*, **14**, 1039.
85. Ilavský, M., Ulmer, H.W., te Nijenhuis, K. and Mijs, W.J. (1994) *J. Non-Cryst. Solids*, 172–174, 935.
86. Bastide, J. and Picot, C.J. (1979) *J. Polym. Sci. Polym. Phys. Ed.*, **17**, 1441.
87. De Gennes, P.G. (1975) *J. Phys. Lett. France*, **36**, 1199.
88. De Gennes, P.G. (1979) *Scaling Concepts in Polymer Physics*, Cornell University Press, Ithaca, New York.
89. Pearson, D.S. and Helfand, E. (1984) *Macromolecules*, **17**, 888.
90. Chasset, R. and Thirion, P. (1965) in *Proceedings of the Conference on Physics of Non-Crystalline Solids* (ed. J.A. Prins), North-Holland, Amsterdam.
91. Dickie, R.A. and Ferry, J.D. (1966) *J. Phys. Chem.*, **70**, 2594.
92. Ilavský, M. and Prins, W. (1970) *Macromolecules*, **3**, 415.
93. Ilavský, M. and Prins, W. (1970) *Macromolecules*, **3**, 425.
94. Ilavský, M., Hasa, J. and Dušek, K. (1975) *J. Polym. Sci. Polym. Symp.*, **53**, 239.
95. Ulbrich, K., Dušek, K., Ilavský, M. and Kopeček, J. (1978) *Eur. Polym. J.*, **14**, 45.
96. Ilavský, M., Talašová E. and Dušek, K. (1980) *Eur. Polym. J.*, **16**, 191.
97. Maekawa, E., Mancke, R.G. and Ferry, J.D. (1965) *J. Phys. Chem.*, **69**, 2811.
98. Vojta, V. and Ilavský, M. (1973) *Collect. Czech. Chem. Commun.*, **38**, 861.
99. Heinrich, G., Straube, E. and Helmis, G. (1988) *Adv. Polym. Sci.*, **85**, 33.
100. Heinrich, G. and Straube, E. (1987) *Polym. Bull.*, **17**, 255.
101. Havránek, A. and Heinrich, G. (1988) *Acta Polym.*, **10**, 564.
102. Gaylord, R.J., Joss, B., Bendler, J.T. and Di Marzio, E.A. (1985) *Br. Polym. J.*, **17**, 126.
103. Edwards, S.F. and Vilgis, T.A. (1988) *Rep. Prog. Phys.*, **59**, 243.
104. Duering, E.R., Kremer, R. and Grest, G.S. (1993) *Macromolecules*, **26**, 3241.
105. Duering, E.R., Kremer, R. and Grest, G.S. (1994) *J. Chem. Phys.*, **101**, 8169.
106. Neuburger, N.A. and Eichinger, B.E. (1985) *J. Chem. Phys.*, **83**, 884.
107. Langley, N.R. (1968) *Macromolecules*, **1**, 348.
108. Pearson, D.S. and Graessley, W.W. (1978) *Macromolecules*, **11**, 528.
109. Pearson, D.S. and Graessley, W.W. (1979) *Macromolecules*, **12**, 123.
110. Miller, D.R. and Macosko, C.W. (1987) *J. Polym. Sci., Polym. Phys. Ed.*, **25**, 2441.
111. Miller, D.R. and Macosko, C.W. (1988) *J. Polym. Sci., Polym. Phys. Ed.*, **26**, 1.
112. Dušek, K. and Ilavský, M. (1980) *Colloid Polym. Sci.*, **258**, 605.
113. Ilavský, M., Bogdanova, L.M. and Dušek, K. (1984) *J. Polym. Sci., Polym. Phys. Ed.*, **22**, 265.
114. Ilavský, M. and Dušek, K. (1983) *Polymer*, **24**, 981.
115. Mussati, F.G. and Macosko, C.W. (1973) *Polym. Eng. Sci.*, **13**, 236.
116. Macosko, C.W. (1985) *Br. Polym. J.*, **17**, 239.
117. Adam, M., Delsanti, M. and Durand, D. (1985) *Macromolecules*, **18**, 2285.
118. Flory, P.J. (1953) *Principles of Polymer Chemistry*, Cornell University, Ithaca, New York.
119. Zimm, B. and Stockmayer, W.H. (1949) *J. Chem. Phys.*, **17**, 1301.
120. Tung, C.Y.M. and Dynes, P.J. (1982) *J. Appl. Polym. Sci.*, **27**, 569.
121. Vilgis, T.A. and Winter, H.H. (1988) *Colloid Polym. Sci.*, **266**, 494.
122. Scanlan, J.C. and Winter, H.H. (1991) *Macromolecules*, **24**, 47.
123. Matějka, L. (1991) *Polym. Bull.*, **26**, 109.

124. Durand, D., Delsanti, M., Adam, M. and Luck, L.M. (1987) *Europhys. Lett.*, **3**, 97.
125. Venkataraman, S.K., Coyne, L., Chambon, F., Gottlieb, M. and Winter, H.H. (1989) *Polymer*, **30**, 2222.
126. Muller, R., Gerard, E., Dugand, P., Rempp, P. and Gnanou, Y. (1991) *Macromolecules*, **24**, 1321.
127. Ishida, H. and Smith, M.E. (1991) *Rheol. Acta*, **30**, 184.
128. Scanlan, J.C. and Winter, H.H. (1991) *Makromol. Chem., Macromol. Symp.*, **45**, 11.
129. Chambon, F. and Winter, H.H. (1985) *Polym. Bull.*, **13**, 499.
130. Matějka, L. and Dušek, K. (1980) *Polym. Bull.*, **3**, 489.
131. Budinski-Simendiċ, J., Petroviċ, Z.S., Ilavský, M. and Dušek, K. (1993) *Polymer*, **34**, 5157.
132. Ilavský, M., Šomvársky, J., Bouchal, K. and Dušek, K. (1993) *Polym. Gels Networks*, **1**, 159.
133. Ilavský, M., Bubeníková, Z., Bouchal, K. and Fähnrich, J. (1996) *Polymer*, **17**, 3851.
134. Winter, H.H. and Chambon, F. (1986) *J. Rheol.*, **30**, 367.
135. Chambon, F. and Winter, H.H. (1987) *J. Rheol.*, **31**, 683.
136. De Rosa, M.E., Mours, M. and Winter, H.H. (1997) *Polym. Gels Networks*, **5**, 69.
137. te Nijenhuis, K. and Winter, H.H. (1989) *Macromolecules*, **22**, 411.
138. Michon, C., Cuvalier, G. and Lannay, B. (1993) *Rheol. Acta*, **32**, 94.
139. Lairez, D., Adam, M., Emery, J.R. and Durand, D. (1992) *Macromolecules*, **25**, 286.
140. Stockmayer, W.H. (1943) *J. Chem. Phys.*, **11**, 45; (1944) **12**, 125.
141. Gordon, M. (1962) *Proc. R. Soc. London A*, **268**, 240.
142. Gordon, M. and Ross-Murphy, S.B. (1975) *Pure Appl. Chem.*, **43**, 1.
143. Kajiwara, K. and Gordon, M. (1975) *J. Chem. Phys.*, **59**, 3628.
144. Dušek, K. (1989) in *Telechelic Polymers. Synthesis and Applications* (ed. E. Goethals), CRC Press, Boca Raton.
145. Stauffer, D. (1985) *Introduction to Percolation Theory*, Taylor and Francis, London.
146. Essam, J.W. (1980) *Rep. Prog. Phys.*, **43**, 833.
147. Fisher, M.E. and Essam, J.W. (1961) *J. Math. Phys.*, **2**, 609.
148. Herrman, D.W. and Staufer, D. (1980) *Z. Phys.*, **B37**, 69.
149. Coniglio, A. (1976) *Phys. Rev.*, **B13**, 2194.
150. Ord, G. and Whittington, S.G. (1982) *J. Phys.*, **A15**, 29.
151. Martin, J.E. and Wilcoxon, J.P. (1988) *Phys. Rev. Lett.*, **61**, 373.
152. Muthukumar, M. (1985) *J. Chem. Phys.*, **83**, 3162.
153. Muthukumar, M. (1989) *Macromolecules*, **22**, 4656.
154. Durand, D., Delsanti, M., Adam, M. and Luck, J.M. (1987) *Europhys. Lett.*, **3**, 297.
155. Muller, R., Gerard, E., Dugand, P., Rempp, P. and Gnanou, J. (1991) *Macromolecules*, **24**, 1321.
156. Antonietti, M., Fölsch, K.J., Sillescu, H. and Pakula, T. (1989) *Macromolecules*, **22**, 2812.
157. De Rosa, M.E., Mours, M. and Winter, H.H. (1997) *Polym. Gels Networks*, **5**, 69.
158. Antonietti, M., Ehlich, D. and Lindmer, P. (1989) *Macromolecules*, **22**, 2802.
159. Martin, J.E., Adolf, D. and Odineh, J. (1990) *Makromol. Chem., Macromol. Symp.*, **40**, 1.
160. Adolf, D. and Martin, J.E. (1990) *Macromolecules*, **23**, 3700.
161. Ilavský, M., Bubeníková, Z., Bouchal, K. and Nedbal J. (1998) in *The Wiley Polymer Networks Group Review* (eds W.J. Mijs and K. te Nijenhuis), Wiley.
162. Cuvelier, G. and Launay, B. (1990) *Makromol. Chem., Macromol. Symp.*, **40**, 23.
163. Critchfield, F.E. and Gerkin, R.M. (1976) *J. Elast. Plast.*, **8**, 396.
164. Feger, C. and MacKnight, W.J. (1985) *Macromolecules*, **18**, 280.
165. Smith, T.L. (1982) *ACS Symp. Ser.*, **193**, 419.
166. Vallés, E.M. and Macosko, C.W. (1979) *Macromolecules*, **12**, 673.
167. Grillet, A.C., Galy, J., Gérard, J.F. and Pascault, J.P. (1991) *Polymer*, **32**, 1885.
168. Vakil, U.M. and Martin, G.C. (1992) *J. Appl. Polym. Sci.*, **46**, 2089.
169. K. Dušek (1989) in *Telechelic Polymers: Synthesis and Applications* (ed. E.J. Goethals), CRC Press, Boca Raton.
170. Ilavský, M., Ulmer, H.W., te Nijenhuis, K. and Mijs, W.J. (1994) *Polym. Gels Networks*, **2**, 73.
171. Ilavský, M. and Hrouz, J. (1991) *J. Non-Cryst. Solids*, **131/133**, 906.
172. Masiulanis, B., Hrouz, J., Baldrian, J., Ilavský, M. and Dušek, K. (1987) *J. Appl. Polym. Sci.*, **34**, 1941.

173. Nedbal, J., Fähnrich, J., Valentová, H., Ilavský, M. and Bubeníková, Z. (1996) *Polym. Bull.*, **36**, 233.
174. Ilavský, M., Bogdanova, L. and Dušek, K. (1984) *J. Polym. Sci., Polym. Phys. Ed.*, **22**, 265.
175. Mark, J.E. (1979) *Makromol. Chem., Suppl.*, **2**, 87.
176. Mark, J.E. and Sung, P.H. (1980) *Eur. Polym. J.*, **16**, 1223.
177. Lee, K.J. and Eichinger, B.E. (1990) *Polymer*, **31**, 414.
178. Ilavský, M. and Dušek, K. (1986) *Macromolecules*, **19**, 2139.
179. Dušek, K. (1986) *Adv. Polym. Sci.*, **78**, 1.
180. Morgan, R.J. (1985) *Adv. Polym. Sci.*, **72**, 1.
181. Kong, E.S.-W. (1986) *Adv. Polym. Sci.*, **80**, 125.
182. Ilavský, M., Šomvársky, J., Bouchal, K. and Dušek, K. (1993) *Polym. Gels Networks*, **1**, 159.
183. Ilavský, M. (1995) in *Keynote Lectures in Selected Topics of Polymer Science* (ed. E. Riande), CSIC, Madrid.
184. Havránek, A., Nedbal, J., Berčík, C., Ilavský, M. and Dušek, K. (1980) *Polym. Bull.*, **3**, 497.
185. Ilavský, M. and Dušek, K. (1982) *Polym. Bull.*, **8**, 359.
186. Krakovský, I., Havránek, A., Ilavský, M. and Dušek, K. (1988) *Coll. Polym. Sci.*, **266**, 324.
187. Ilavský, M. and Dušek, K. (1983) *Polymer*, **24**, 981.
188. Dušek, K. and Ilavský, M. (1989) *Progr. Colloid Polym. Sci.*, **80**, 26.
189. Zelenka, J., Ilavský, M., Dobáš, I. and Dušek, K. (1991) *Polym. Networks Blends*, **1**, 98.
190. Kitch, M. and Suzuki, K. (1976) *Kobunshi Ronbunshu*, **33**, 19.
191. Dušek, K. and Ilavský, M. (1979) *Polym. Eng. Sci.*, **19**, 246.
192. Dušek, K. and Šomvársky, J. (1995) *Faraday Disc., Chem. Soc.*, **101**, 147.
193. Ilavský, M., Šomvársky, J., Pleštil, J. and Nedbal, J. (1996) in *Proceedings of the 5th Latin American and 3rd Ibero American Polymer Symposium*, Mar del Plata, 1996.

9 Formation, structure and properties of physical networks

S.B. ROSS-MURPHY

9.1 Introduction

Historically, we can divide polymer networks into two main classes: covalent network materials (including bulk elastomers) and 'entanglement networks'. Almost all the other chapters in this book cover these two categories in some detail. The rheological properties, in particular, are described in detail in Chapters 7 and 8, so this section only introduces sufficient aspects to help clarify the succeeding material. For example, entanglement networks are formed by the topological interactions of polymer chains, either in the melt or in solution, when the product of concentration and molecular weight becomes greater than some critical molecular weight, M_e [1, 2]. In this case they behave as 'pseudogels' at frequencies higher (timescales shorter) than the lifetime of the topological entanglements. This depends for linear chains on M^3, where M is molecular weight (Section 8.5.1). Covalent network materials, however, are formed by a variety of routes including cross-linking high-molecular-weight linear chains, either chemically or by radiation (Chapter 3), by endlinking reactant chains with branching units (Chapters 2 and 3), or by the polymerization of oligomeric multifunctional precursors (Chapter 2). They are true macromolecules, with a molecular weight that is nominally infinite, and they therefore possess an infinite relaxation time (Section 8.5.2) and an equilibrium modulus [1].

The discrimination of covalent from entanglement networks can be made, for example, by the technique of mechanical spectroscopy. In this, a small-deformation oscillatory, shear-strain of frequency ω is applied to the material, and the real and imaginary parts, G' and G'', of the complex shear modulus are measured. For entanglement networks, as the frequency is decreased there is a 'cross-over' in G' and G''. At very low frequencies, in the 'terminal zone', they flow as high-viscosity liquids. By contrast, covalent networks show no such effect, and for a rubbery gel (i.e., one above its glass-transition temperature), G' and G'' are parallel, and essentially frequency insensitive [1, 2]. A further discrimination can be made by adding excess solvent to the bulk 'gel'. Entanglement network systems will dissolve to form a more dilute polymer solution, whereas the gel will swell, but not dissolve.

9.1.1 Characterization of gels in terms of gelation mechanism and structure

In what has subsequently proved to be one of the most quoted articles in gels and gelation, Paul Flory in 1974 in his introduction to a Faraday Discussion on gels and gelation [3] proposed a classification of gels based on the following:

I. Well-ordered lamellar structures, including gel mesophases;
II. Covalent polymeric networks, completely disordered;
III. Polymer networks formed through physical aggregation, predominantly disordered, but with regions of order;
IV. Particular, disordered structures.

In this chapter, these are called Flory I, II, III and IV gels, respectively. Clearly, Flory II gels and networks are the major province of this book, but the present chapter is, in principle, concerned with all the remaining three. In fact, this chapter concentrates mostly on Flory III systems, because the majority of published work has been carried out on these. Flory I gels cover a wide range of systems, including liquid-crystalline materials, but the properties tend not to be macromolecular, and are not discussed here.

It is important to appreciate that isolated point-like interactions, which are not intrinsically 'strong' – say $<10\,kT$ – do not tend to produce materials generally recognized to be gels. In making this assertion we are implicitly accepting the much quoted definition of Jordan-Lloyd: 'the colloidal condition, the gel, is one which is easier to recognize than to define' [4]. Rather, they tend to produce what are called 'weak gels' – actually a rather imprecise term used by different groups to describe a range of material properties. An example might be an ionomer system such as a polymer sulfated at each end, dispersed in a non-polar solvent. The following section discusses weak gels and strong gels in more detail. For the present, it is sufficient to say that such an ionomer system can be well described by modifications of, for example, Doi–Edwards entanglement tube-theories, described in Section 8.5.1, with added 'sticky' interactions [2]. Such a system clearly does not have an equilibrium modulus: it just has a rather longer relaxation time than the equivalent unsubstituted polymer at the same volume occupancy.

This chapter concentrates on systems where the relaxation time is extremely long and, within the usual experimental window, there is no indication of anything other than equilibrium modulus behaviour. In this respect, we echo the recent rheological definition of a gel by Almdal *et al.* [5]. This states that a gel is a soft, solid or solid-like material, which consists of two or more components, one of which is a liquid, present in substantial quantity. The gel shows a flat mechanical spectrum in an oscillatory-shear experiment. This requires that G' exhibits a plateau extending to times of the order of seconds, whereas G'' is considerably smaller.

9.1.2 *Physical junction zones in polymer gels and networks*

Flory III and IV gels consist of chains 'physically' rather than chemically anchored at junction zones of small but infinite energy, or of finite lifetime, or both. Such physical gels include those from both biological and synthetic polymers [6, 7]. The junction zones are not points on chains, as are covalent junctions, but involve more extended 'zones' or even particulate structures. In many cases, the nature of the junction zones is not known unambiguously, but can include Coulombic, dipole–dipole, hydrophobic and hydrogen-bonding interactions. Synthetic physical gels include ionomers and A-B-A type block copolymers. Biopolymer gels [7, 8] include cold-set gelatin and plant polysaccharide systems (Flory III) and heat-set globular protein gels (Flory IV).

A physical description of such systems becomes very complex, because the potential number of, say, junction zones per primary chain, the 'functionality' f and the extent (molecular weight) of the junction can be estimated, if at all, only indirectly. In many cases there is a subsequent lateral aggregation of chains, after the initial contact. These factors must influence the actual number of physical junctions and, consequently, the modulus of the final gel. Since this modulus will usually reflect both entropic and enthalpic contributions, an *a priori* description of the modulus of these materials is bound to involve a number of approximations. Nevertheless, the present chapter serves both as a guide to the different examples of physical gels, and also as an introduction to their theoretical description. The theory that has been developed naturally draws heavily on earlier approaches developed for covalent gels and networks. These are introduced, while pointing out their natural limitations.

9.2 Strong and weak physical gels

At this stage, a distinction can be made between so-called strong and weak physical gels [8, 9]. Furthermore, the latter can be further divided into rheological and thermodynamic weak gels. In the second case the energy per junction zone is only a few kT, so thermodynamic weak gels can be cleaved by thermal fluctuation, only to reform. By contrast, many thermoreversible, biopolymer strong gels can only be melted by passing through an order–disorder transition (with its associated midpoint transition temperature). Both strong and (rheological) weak gels respond as solids at small deformations. Whereas the strong gels, such as gelatin, rupture at larger deformation, and cannot reform unless reheated through the order–disorder temperature, the weak gels are really structured fluids, so they will flow, rather than fail, at large deformations. This distinction must be somewhat arbitrary since, if the non-covalent junction zones can be ruptured by mechanical perturbation, they must have bond energies of only a few kT units, and therefore may also melt on heating.

Ultimately, such discussion serves only to focus attention on the inadequacy of using the term 'weak gel' to describe a range of different phenomena. Rheologically, this means that although such a system lies within the linear (stress-proportional-to-strain) viscoelastic regime, the maximum strain for linear behaviour, γ_0, is very small. For strong gels, we can say that $\gamma_0 \approx 20\%$, whereas for weak gels it can be up to 1000 times smaller. Colloidal and particulate networks are often of this type.

It is clear that we need to be able to estimate the 'strength' of physical-gel interactions. A first assumption is that a single physical junction zone would have a free energy or lifetime which was of Boltzmann form. However, the actual physical junction zone then requires a degree of molecular cooperativity, unlike a junction point (Figure 9.1). We would then expect the net free energy of this junction zone to be greater than the sum of its component units. If we are to try to relate the modulus of the gel to macromolecular properties we must be able to factorize the contribution from the total number of units in junction zones, that is, to separate the effect of the network connectivity from the free energy of any individual network junction-zone. Further, even though this procedure can be carried out formally, to be confident of the conclusions we require structural detail, at the molecular and macromolecular distance scales (say 0.1–10 nm and 10–10^4 nm, respectively). This is because another major difference between networks with covalent and non-covalent junctions is that the former are effectively homogeneous (self-similar) over the entire distance scale from, say, 10 nm to 10^7 nm, whereas this is certainly not the case for the latter. In many cases there is a specific, and often intricate, hierarchy of structures.

Examples of weak physical gels include ionomer systems in non-polar solvents. For example, polymers sulfated at one end give star aggregates,

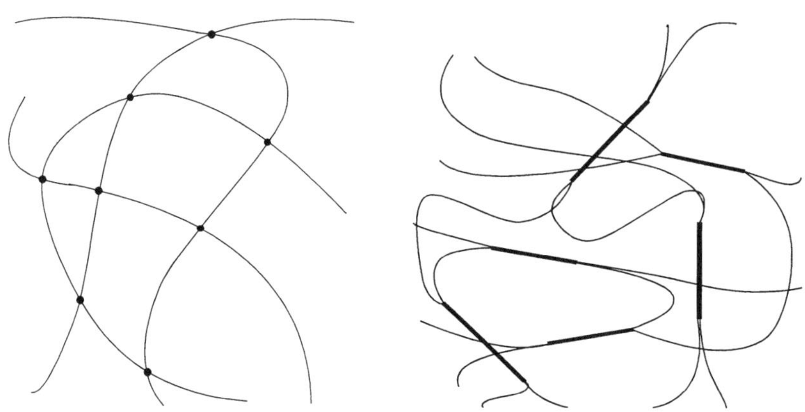

Figure 9.1 Diagrammatic representations of (left) a (chemical) gel network having junction points and (right) a physical gel network having junction zones.

whereas those sulfated at each end give weak gels. Another example is hydrogen-bond-modified polybutadiene, in which the polymer has been tagged chemically so that hydrogen-bonded groups are distributed either at single sites or blockwise along the chains [10]. The archetypal biopolymer weak gel is the microbial polysaccharide, xanthan, whose aqueous solutions have been employed in the oil-field fluid industry [11].

9.2.1 Time–temperature superposition

The use of time–temperature superposition (TTS) to extend the effective frequency or time-scale of viscoelastic measurements for polymer melts and glasses is very widely employed [1]; the method is described in some detail in Section 8.4. However, one clear *caveat* is that the method should not be used if a sample is undergoing any significant physical change (e.g., phase transition) within the temperature range under study. This would suggest that the method cannot be used at all for most physical gels since, as outlined above, the gelation mechanism usually involves some degree of ordering. This is still the overriding rule, but it has been demonstrated in a number of cases that TTS can still be used, and that it provides a valuable insight into long-time relaxation effects, by extending the effective frequency range much below that which it is convenient to study in real time. Successful examples include application of the method to poly(vinyl chloride) (PVC) [7] and globular-protein gels [12].

9.2.2 Strong physical gels

Many, but by no means all, physical gels are thermoreversible, but unfortunately the term physical gel has, in some articles, been confused with thermoreversibility. In this chapter we shall not restrict ourselves in this way. However, for thermoreversible gels, Ferry and Eldridge [13] showed, in their work on gelatin gels, that the concentration C and (weight average) molecular weight (M) dependence of the gel melting-temperature could be described by a simple model, essentially that of Flory for the crystalline melting of a polymer. They established that a plot of $\ln C$ versus $1/T_m$ (gel melting-temperature in Kelvin) was almost linear, as was $\ln M$ versus $1/T_m$. This implied

$$-\left[\frac{\mathrm{d}\ln C}{\mathrm{d}T_m}\right]_M \approx \frac{\Delta H^0}{RT^2} \tag{9.1}$$

where $\mathrm{d}T_m$ was the change in melting temperature. This is of van't Hoff form, and suggested that gel formation was governed by an equilibrium between

$$2 \text{ crosslinking sites} \rightleftharpoons 1 \text{ crosslink}$$

with ΔH^0 the enthalpy for the process.

In much of the published work on cold-set, thermoreversible gels, the very first experiments tried have been carried out to establish how T_m varies with polymer concentration. T_m itself can often be measured by very simple techniques, such as establishing when a suspended ball bearing is first able to fall through the melting gel. Despite the simplicity of the method and the equation, there appear to be few, if any, cases where equation (9.1) does not apply. On the contrary, linear plots are obtained in the great majority of systems investigated. The underlying binary-site equilibrium model has been employed in a number of gel-network models (Section 9.5.2.1).

Nevertheless, it is clear that the binary junction model is oversimplified. Recent work by Tanaka and co-workers [14–16] has been investigating many aspects of this equilibrium model, and trying to make progress in putting it onto a firmer thermodynamic basis. For example, Tanaka and Stockmayer [15] treated the binary model just described. However, the latest paper by Tanaka and Nishinari [16] considers the treatment of junction multiplicity. In this newer version they include parameters for multiplicity and junction-zone length and are able to fit these parameters to data for various synthetic physical gels, including polystyrene–CS_2 and poly(vinyl alcohol) (PVA)–water. There seems no doubt this approach will prove of major importance in the future.

9.3 Preparation and structure of physical gels and networks

As already mentioned, the majority of the gels we shall describe will be of the Flory III type. In other words, the aggregation of heterostructural chains is often a common feature of such materials. If we consider a general -A-B-A- type block structure, where block A is more compatible with the solvent and block B is less compatible, the B-B interactions between units on adjacent chains in solution form the physical crosslinks. This relative compatibility simply suggests that B-B interactions are slightly favoured energetically over other possible pair interactions, with A or with solvent molecules. Alternatively, there may be crystallization or helix formation between B blocks. We will use the term 'frustrated crystallization' to describe this general phenomenon. The presence of the A blocks is crucial here since otherwise junction zones could extend along whole chain lengths.

A particularly clear example of this type of behaviour is the early work by Rees and co-workers on the seaweed polysaccharide, carrageenan, where gel junction zones are formed by intermolecular carrageenan double helices [17]. The existence of 'rogue residues' was established chemically. In these, certain saccharide (sugar) repeat units in the carrageenan chain backbone are in the wrong conformation to allow the helix to propagate. This ensures that, rather than isolated helix pairs being formed, each chain can share portions of ordered, in this case helical, structure with at least two other chains, an essential condition for branching and subsequent gel formation. Although this simple model has subsequently proved not to be totally correct [18] it clearly serves as a model of more general applicability.

As far as synthetic polymers are concerned, the major interest is in PVC, PVA, tactic polystyrene (PS) and poly(methyl methacrylate) (PMMA) systems. One way of categorizing such systems is to divide the gel-network systems into two classes, one in which the nature of the solvent is not significant, and where gelation seems to be an example of a 'frustrated crystallization' phenomenon, and the other in which the nature of the solvent is much more important, so that gelation will occur in some solvents but not in others. In the first category we can place the PVC, PVA and other gels, whereas the second, solvent-specific gels, are exemplified by tactic PS and PMMA systems. Individual systems are discussed only briefly here; fuller details are given in the extended reviews by Guenet [6] and te Nijenhuis [7].

9.3.1 Gelation of synthetic polymers – partial crystallization

9.3.1.1 PVC–plasticizer gels

Much of the detailed work in this area has been carried out on PVC gels, consisting of around 20% of polymer dissolved in a typical plasticizer, for example dioctylphthalate (DOP) or diethyl malonate. There is obviously great technological interest in these systems because of the plastisol process for casting PVC articles, but much of the work has been concerned with the ageing behaviour of the gels. The work by te Nijenhuis and co-workers [19, 20] is particularly noteworthy, since they carried out a series of extremely careful gelation measurements in a home-built rheometer. These experiments clearly established that history, time and temperature effects are extremely significant, suggesting that subsequent efforts to examine properties in terms of equilibrium phase-behaviour, as has been proposed by some authors, are liable to be frustrated. There are two major complications in the work. First, PVC is never totally atactic, and the template for crystallization may be drawn from small isotactic regions. Second, the synthesis of realistic quantities of PVCs of well-defined molecular weights and molecular-weight distributions has proved difficult, so most measurements have been performed on technical-grade samples. In this respect, the work shares much with work on biopolymer systems (discussed later). Indeed, te Nijenhuis' thesis [21] draws together work on PVC and gelatin gels, one of the first attempts to draw parallels between synthetic-gel and biopolymer--gel research.

Several X-ray studies have been carried out on PVC–plastisol gels, with the aim of establishing the precise gelation mechanism. The early work by Alfrey *et al.* [22] established that the network junctions involve crystallites, and subsequent work using SAXS showed that interlamellar spacing increased with plasticizer concentration. More detailed work has been concerned to establish how the junction-zone concentration and the degree of crystallization depend upon tacticity and primary chain length [21]. In this there appears to be less agreement. Many of the uncertainties in this area are associated with the length and number of cooperative junction zones per chain (this clearly determines the network functionality), which can only be ascertained indirectly, and in relating such factors to the gel modulus.

As far as the gel point itself is concerned, there has been particular interest in investigating the applicability of the Chambon–Winter viscoelastic criteria for gelation [21] (at the gel point $\tan\delta$ is constant, and G' and G'' are $\sim\omega^n$, $0 < n < 1$) to systems with non-covalent junctions. A paper by te Nijenhuis and Winter [25] has reanalyzed te Nijenhuis' earlier PVC/DOP data [21] in terms of this model. The slope n at the gelation time was ≈ 0.77, in agreement with results for covalent networks with a deficiency of crosslinking agent. Results for gelatin [26] suggest $n \approx 0.68$, but recent results for polysaccharide gels give a much wider range of values [27].

9.3.1.2 Poly(vinyl alcohol) gels

Poly(vinyl alcohol) (PVA), typically produced by hydrolysing >80% of the ester groups of polyvinyl acetate, is an extremely important synthetic, water-soluble polymer. It can gel by a large number of different routes, including the freeze–thaw method, and in combination with chemical complexing agents, particularly those known to foster hydrogen bonds, such as water itself, or added resorcinol. The PVA–borate-ion system [28] has been employed in the oil industry as both a secondary and a tertiary recovery fluid. However, it is clear that, in terms of our earlier definition, the borate systems are usually weak self-healing gels; indeed, if this were not the case, they would be unsuitable in these applications. Other agents employed include the addition of congo red [29], more widely known as a dyeing and fixing agent in microscopy, and the use of solvents less polar than water, such as ethylene glycol or DMSO [30].

Work by Prins and co-workers in the 1970s [31] established that, under certain circumstances, gel formation is induced by spinodal decomposition (Section 9.4.1), followed by the partial or, to use our terminology, frustrated crystallization of PVA chains. The spinodal locus was established by light-scattering experiments. This mechanism has subsequently proved so attractive to other workers that it has been employed as an explanation, in other quite different physical-gel systems, where there is little or no evidence for phase separation, and where no serious efforts have been made to locate the spinodal or even the binodal (Section 9.4.1). This is not to criticize the Prins work; however, even here, subsequent work by Komatsu and co-workers [32] has made it clear that the sol–gel boundary may be crossed well above the binodal, i.e. in the one-phase region.

The work by Watase and Nishinari in the 1980s [33] has used the attractively simple freeze–thaw method. In this a solution of PVA (say >10% w/w) is frozen (~ −25 °C), and then just allowed to thaw to 0 °C before being refrozen, and the cycle repeated. Rubbery gels are formed, and the gel modulus increases both with initial freezing time, and with the number of freezing cycles. They also investigated the effect of degree of hydrolysis (often referred to more traditionally as degree of saponification, *DS*) on gel modulus *E* and melting behaviour, for samples of *DS* from 96% to 99.9% [34]. *E* increases steadily with *DS* in this range, as does the slope of the Eldridge–Ferry plot. Freeze–thaw PVA gels have been employed in a number of recent studies on 'intelligent gels', so-called because they swell or deswell under external stimuli [35].

9.3.1.3 Polyethylene gels

High-density polyethylene (HDPE) has a great propensity to crystallize, but the first systematic report [36] suggested that stirring solutions of high-molecular-weight HDPE of concentrations as low as 1% w/v in *p*-xylene produced gel-like materials. Subsequent work has shown that stirring, followed by cooling without further stirring, can also produce gels, provided the temperature is high enough (>127 °C) [37]. Other work has concentrated on solutions in decalin [38]. This work is an extension of the work by Smith and Lemstra on the successful gel-process for high-modulus fibres.

9.3.2 Gelation of synthetic polymers – solvent-specific gels

These include physical gels formed from synthetic polymers, including certain isotactic (i) and syndiotactic (s) polymers in organic solvents, such as i-PS and s-PS in decalin, ionomer systems in solvents of low dielectric constant, and a number of block copolymers [39]. Obviously the most studied are the tactic polystyrenes, particularly i-PS, because of the limited availability of s-PS. Other solvents where gelation can take place include 1-chlorodecane and diethyl malonate, while significant differences are observed for gels made in the *cis* and *trans* isomers of decalin. Much of this work has been performed by Guenet and co-workers [40, 41]. Techniques employed include differential scanning calorimetry (DSC) and small-angle neutron scattering (SANS). In some of this work, the term phase diagram is used when referring to data extrapolated from DSC at quite high heating rates (say 5 °C min^{-1}). In this case, we prefer the less contentious term 'state diagram', i.e. a term which does not imply that thermodynamic equilibria are being investigated.

Of course, i-PS crystallization from the melt has been widely investigated, and some of the temperatures and conditions extrapolated from work in solution correspond quite closely to those measured in the melt, so that there is some evidence for the usual chain-folded crystals being formed.

Indeed, early work by Mandelkern and co-workers [42] on low-density PE concluded that the crystallites formed in these gels were the same as those found in dilute solution. In other experiments, more unusual structures appeared to be observed, including helical and so-called ladder-like structures [43]. The latter borrows extensively from the egg-box concept proposed for pectin gels (Section 9.3.3.3) [44]. The evidence for helical structures is not widely accepted, and it seems surprising that unequivocal chiroptical techniques such as circular dichroism (CD) do not appear to have been employed.

The gelation of atactic PS, for example from decalin (*cis* and *trans* forms) and CS_2 solutions, has been known since the mid-1960s, although only as an indirect consequence of light-scattering anomalies [45]. Since atactic PS does not crystallize, the preceding model, suggested for tactic PS, cannot be correct. The mechanism remained a subject of controversy for some years until the suggestion of Arnauts and Berghmans was published in 1987 [46]. They have shown how, in this case, the measured glass-transition temperature T_g versus composition locus crosses the binodal line (Section 9.4.1), i.e., that which separates the single-phase from the two-phase region (Figure 9.2). They suggest that subsequently the more concentrated of the two concomitant phases becomes 'frozen' below its T_g, giving a gel with glassy junctions. This suggestion is now accepted by the vast majority of workers in the field.

There is a considerable body of work on the gelation of i-PMMA and s-PMMA. Although the atactic form also forms gels under more extreme conditions, and generally at higher concentrations. Perhaps most interesting is the work performed on so-called stereo-complexes, mixtures of i-PMMA and s-PMMA [47]. This work has been extended to a wide variety of solvents,

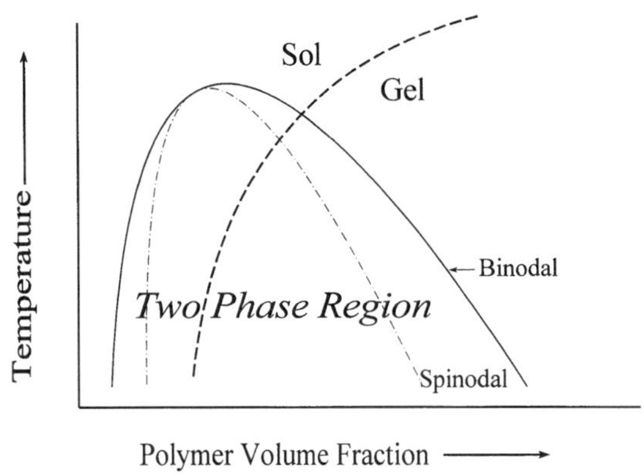

Figure 9.2 Idealized phase diagram for a conventional (upper critical solution temperature, UCST) polymer 'solution' system, crossed by a sol–gel boundary line. Certain heat-set gels (including protein gels) are now believed to have a phase diagram of the lower (L)CST form.

including CCl_4 and toluene. In all of these, it is observed that optimum gel characteristics are observed when the composition is ≈ 2 parts of s-PMMA to 1 part of i-PMMA. The minimum s-sequence length has been reported to be around 3–10 units, again suggesting that cooperativity is a crucial factor.

9.3.3 *Gelation in biopolymer systems – networks from disordered biopolymers*

A fundamental distinction can be made between those systems which, in forming gels, are transformed from essentially disordered random-coil biopolymers (although, even in disordered conformations, in terms of say persistence length, they are sometimes stiff compared with synthetic polymers) into a partly ordered state (e.g., by undergoing a coil–helix transition), and those systems which form, and are maintained in, an essentially ordered state, such as gels formed from globular proteins [8].

9.3.3.1 *Gelatin gels*

Although gelatin has been known for its adhesive properties since ancient Egyptian times [6], like most of the systems described in this chapter, the precise gel structure and gelation mechanism are still not completely established. Gelatin(e) is a proteinaceous material derived by the hydrolytic degradation (by either acid or alkaline treatment) of collagen, the principal protein component of white, fibrous connective tissue (skin, tendon, bone etc.).

Gelatins normally dissolve in warm water ($\gtrsim 40\,^{\circ}C$). Above this temperature the polypeptide exists as flexible, single coils. On recooling, transparent gels are formed (provided the concentration is greater than some critical concentration C_0, typically 0.4–1.0%). The gels contain extended physical junction-zones, formed by a partial reversion to ordered, triple-helical collagen-like sequences, separated along the chain contour by peptide residues in the disordered conformation [48, 49].

It is thought that the coil-to-triple-helix propagation rate is limited by the presence of *cis*-proline residues in the backbone. The subsequent reversion of these to the *trans* form allows the helix to propagate only slowly; the overall growth rate is typically 4–6 orders of magnitude less than for double-helical systems, such as the carrageenans (discussed later). A recent model has suggested that the triple helix is formed from two chains, one of which has an intramolecular β chain, rather than from three separate chains [50].

In the past it was assumed that initial helix formation was followed by substantial lateral aggregation, leading to extended quasi-crystalline junctions, as described historically in the fringed micelle model of polymer crystallization. However, since long-time measurements of OR increase slowly, but apparently without limit (even when plotted on a log–time axis), the proportion of residues in the ordered, helical conformation must also be increasing. This suggests a considerable degree of conformational flexibility, even post-gel, that is rather unlikely to occur if the junction zones are formed of rigid crystallites. More direct evidence comes from SANS by Djabourov and co-workers [51]. They found that in the sol state the cross-sectional chain radius R_c was 0.32 ± 0.1 nm, a value in good agreement with the calculated side-group extension of a collagen chain, whereas in a relatively concentrated gel (5%) R_c was 0.43 ± 0.1 nm, not much greater than this. Now, it is widely accepted that the junction zones consist of (largely) isolated triple helices. This, albeit simpler, hypothesis helps to explain how, and why, of all the physical gels, gelatin behaves, at least isothermally, most like an ideal rubber (Sections 9.5.3.2 and 9.5.4).

9.3.3.2 Marine polysaccharides

The most important of these forming gels are (ι- and κ-) carrageenan, agar(ose), and the alginates. All three are extracted from different species of seaweed. Much evidence suggests that the first two form thermoreversible gels by an extension of the gelatin mechanism and, although some details are still disputed, the general principles are as follows. On heating above the helix–coil transition temperature (for the charged carrageenans this depends crucially on ionic strength and cation species but lies typically in the range 20–50 °C), they apparently undergo an order-to-disorder (helix-to-coil) transition. On recooling they revert, at least partly, to a double helix [17]. Compared with that for gelatin, this is a very fast process, taking only a second or two under most conditions. More recently, it has been suggested that a super-helical structure is involved [18]. However, definitive evidence, for example employing analysis of the angular dependence of light scattering, has yet to be employed to examine this hypothesis.

Subsequently, at lower temperatures, the behaviour becomes less ideal; so for agarose there is then substantial side-by-side aggregation (confirmed by the measurement of R_c by SAXS) [52]. For the carrageenans, gelation is known to depend crucially on the cations present; for Na^+ little is seen, whereas high-modulus gels are formed, for example, with K^+ and Ca^{2+}. The precise details of network formation in these systems is still being actively researched [18].

Alginate gels, by contrast, are not thermoreversible; in fact they appear heat stable up to >100 °C, and their formation can only be induced in the presence of certain, specifically divalent, cations. Alginates are anionic block copolymers composed of two very similar saccharide units, guluronate (G) and mannuronate (M). If Ca^{2+} ions are introduced into a solution of sodium alginate, gelation occurs extremely rapidly. Gelation is induced by specific ion-binding, accompanied by conformational changes, with evidence for Ca^{2+} ions binding cooperatively to G blocks. In one model, the junction zones involve two chains and chelated ions giving the so-called 'egg-box' structure [44].

9.3.3.3 Plant polysaccharides

Pectin, extracted from plant cell walls, particularly from apples and citrus fruits, and starch, extracted, for example, from cereals and tubers (potato), are the most important members of this group. The former consist predominantly of sequences of galacturonic acid residues (which are quite similar to the G units in alginate), with occasional interruptions by rhamnose residues. At least some of the carboxyl groups are methyl esterified, the precise distribution depending upon the plant source and age. Reasonably, in view of their structural similarity, pectins of low degree of esterification behave like alginates, and gel with divalent ions, and the structure has been termed the egg-box model. The more highly esterified materials gel under conditions of low pH and decreased water activity, i.e., where intermolecular electrostatic repulsions are reduced; in this case the junction zones are thermoreversible at, say, 40 °C [8].

Starch, by contrast, consists of two chemically very similar polysaccharides, one predominantly linear and of relatively low molecular weight ($<2 \times 10^6$), amylose, and the other, branched and of much higher M_w ($>5 \times 10^7$), amylopectin. Amylopectin usually predominates (>70% by weight), but the precise proportions of the two polymers, their chain length and the branching frequency depends upon the source (potato, wheat, rice, tapioca, etc.). The polymers themselves are laid down in a highly structured semicrystalline supramolecular granule (≈2–100 μm diameter). On heating, the granules swell and rupture (at temperatures > ≈60 °C). After this, it is believed that the amylose leaches out, and at concentrations greater than, say, 20%, the mixture of granular residues and amylose forms a viscoelastic paste. On cooling, this sets, and the result has been described as a composite of amylose gel filled with swollen granules [53]. This is, however, by no means the whole picture since amylopectin solutions can also gel, albeit extremely slowly [54], and much of the subtlety of starch behaviour is undoubtedly related to the limited compatibility, and mutual gelation of the two polymeric components.

9.3.3.4 Microbial polysaccharides

A number of polysaccharides of interest occur outside the cells of certain cultured microbes, either covalently attached to the cells or secreted into the growth media. They are the microbial exopolysaccharides and, over the past few years, a great number of them have been described. At the moment, on a volume-production basis, the two major members of this group are gellan, an anionic polysaccharide produced by *Auromonas elodea*, and xanthan, also anionic, from *Xanthomonas campestris*. Gellan has a complex, tetrasaccharide repeat unit, and gels in the presence of multivalent cations, via a double-helical intermediate, in a way analogous to the gelling carrageenans. The bulk, mechanical properties are sensitive to the degree of acylation of the chain [55, 56]. There have been quite a number of published viscoelastic data for gellan, but many of the published observations appear rather obvious; quantitative conclusions based on, for example, results for different molecular-weight samples appear lacking. Xanthan, which has a pentasaccharide sequence, forms strong gels only under extreme conditions. It is, however, the archetypal weak-gel structurant and, as mentioned earlier, has been employed in a number of industries because of its unusual rheological properties [57, 58].

9.3.4 Gelation in biopolymer systems – networks from globular and rod-like biopolymers

Almost all of this group of materials are formed from animal and vegetable proteins. In some cases, the resultant gels involve at least partial denaturation, which does not occur *in vivo* (heat-set proteins), whereas, in others, the biological function of the protein is to gel under certain physiological conditions (blood clotting≡fibrin network formation). Distinction must also be made between established gel systems and solutions of entangled rods; there are several cases in the literature where the latter are referred to as gels.

9.3.4.1 Globular proteins forming branched networks

Many globular proteins can form gels, above, say, 5% concentration, just by heating [59]. Perhaps the most familiar example of this is the boiling of an egg (essentially the gelation of ovalbumin), but similar gels can be formed by heating, e.g., serum albumins (SAs), chymotrypsin, globins, whey and vegetable (soy) proteins. In most of these, the fundamental unit is still a globular particle, so they can be regarded as Flory IV gels. However, under extreme conditions, for example, of pH, the globular protein can unfold totally. Much of the published work has been restricted to bovine (B)SA and β-lactoglobulin from milk whey, since pure samples can be obtained relatively cheaply. There are many published data on the rheological properties of all the systems mentioned, but in many cases comparatively crude samples have been used. For example, there is much data on whey-protein gels, but whey protein consists of mixtures of lactalbumin, lactoglobulin and smaller amounts of caseins.

During gelation induced by heating, the conformation of the corpuscular protein is only slightly perturbed, as confirmed by electron microscopy, X-ray scattering and spectroscopic techniques [59]. It appears that denaturation partially disrupts the protein without modifying the overall shape very significantly. Consequently it remains as a 'molten globule', but with some intraglobular hydrophobic residues exposed. At low enough concentrations, these can re-fold all but reversibly, but above a certain concentration there is competition between intramolecular and intermolecular β-sheet formation. If the latter predominates, gels are formed which are fibrillar, with fibrils approximately 1–2 times the width of the original globule. The balance between linear- and branched-chain growth depends upon both pH and ionic strength, *I*, and a range of gels can be prepared, ranging from transparent through translucent, to biphasic gels and turbid coagulates without macroscopic strength [60]. In most cases, prolonged heating at higher temperatures produces a more drastic change, and some intermolecular, covalent, disulfide bonds are formed.

Protein gels can also be produced by means of denaturation, other than simply heat, including treatment with non-solvents (alcohols), or 'hydrogen-bond breakers' (urea). Subjecting ovalbumin solutions to both urea and heat (>80 °C) apparently produces very substantial peptide unfolding [61]. Gels can also be produced by enzymic means, crucially important in the antibody–antigen reaction, and also in the process of red blood-cell agglutination.

9.3.4.2 Globular proteins forming ordered fibrous assemblies

In this category, we consider specifically actin, tubulin and fibrin networks. The first two of these are globular proteins (≈6 nm diameter) that can form very specific, rod-like assemblies by, respectively, the addition of salt and the action of other proteins. The polymeric actin fibril is composed of two helical strands wound around one another, whereas that formed from tubulin 'monomers' is a hollow cylinder [62]. The fibril length can be very large, and these fibrils then form very weak networks. Treatment with a so-called actin-binding protein (ABP) provides covalent junctions, whereas another protein, gelsolin, cuts actin filaments, producing a drastic reduction in gel modulus [63]. Such actin networks are crucial to the self-regulating mechanical behaviour of the intracellular cytoplasm.

Fibrinogen is a relatively compact protein, which can also polymerize under physiological conditions to form very fine stranded, fibrillar networks [64]. The rheological properties of these have been extensively investigated by Ferry and co-workers, and they have demonstrated how the mesh size of the network can be altered by changes in pH and *I*. Ca^{2+}, in the presence of the serum factor, fibrinoligase, produces covalent bonds.

9.3.4.3 Casein gels

Casein (milk) gels are important technologically because they form the basis of cheese, yoghurt and other similar products. The term casein, itself, describes a number of different proteins that occur in milk as roughly spherical but highly voluminous micelles, partially stabilized by colloidal calcium phosphate. Treating whole milk with an enzyme, chymosin (rennet), is believed to cleave away part of the outside of the micelle, producing a coagulate (curd), which is separated from the remaining liquor (whey), a solution of whey proteins. Many detailed studies of casein aggregation have been carried out, exemplified by the work by Niki and co-workers [65].

Casein aggregation can also be induced by heat (>140 °C), Ca^{2+} ions, and acid treatment. Recent work by Walstra, van Vliet and co-workers has employed a fractal aggregation model to describe the mechanical properties of gels formed both by rennet and by acid treatment [66].

9.4 Physicochemical aspects of physical gelation and networks

Real progress in the study of the physical gelation of synthetic and biological macromolecules has so far been limited by a number of factors, but recently there has been more of an appreciation of some of the common features of synthetic and natural gels. From previous remarks, it should be clear that any model study of the phenomenon of physical gelation would require that the molecular weight M (chain length) and polydispersity were known and could be controlled, as could f, the functionality (Section 9.1.2), and the junction-zone length M_j (clearly $fM_j \leq M$). However, the majority of measurements reported in the literature have been obtained for samples of polydispersities which are either unknown or are rather broad ($DP_w/DP_n > 2$), compared with accepted standards in polymer research. This applies not only to the biopolymers, but also to their synthetic polymer analogues. There is also the uncertainty in the lateral extent of the junction zones, and in the distribution of substituent groups along the chain backbones.

At least one attractive way of surmounting these problems has already been mentioned [10]. Model polybutadiene chains were treated to give 'point' hydrogen-bonded junctions, and measurements made of the mechanical spectra of the modified and unmodified melts. One interesting outcome of this work is that point junctions alone were not sufficient to produce a physical gel. This seems to add weight to the concept that a degree of chain-contour cooperativity is required, i.e. that a critical number of adjacent segments needs to be involved in an effective junction zone. (One might even speculate that since most covalent-bond energies are around 10 times as great as the non-covalent-bond energies, this factor could be used to estimate the minimum length of an effective junction zone.)

9.4.1 Phase behaviour

One view of physical gelation is that it implicitly involves a degree of heterogeneity. This is undoubtedly true at the level of detail of the chain backbone. One example of this is the heterostructure in the type of block-copolymer gel-system already mentioned. This degree of microscopic heterogeneity, or phase incompatibility, has led some workers to suggest that physical-gel formation intrinsically involves macroscopic incompatibility. This implies that gelation must occur, either within the metastable (binodal) region, when a more concentrated phase exceeds a certain critical concentration, or even within the unstable region, by the process of non-nucleated, spinodal decomposition. Although there are undoubtedly some cases where one or other of these effects does drive the gelation, and where the phase and gel loci have been carefully mapped, as in the work by Prins on PVA gels [31], it is certainly not, in the author's view, true in every case.

The interplay between liquid–liquid phase separation and gelation has been treated classically by Dušek [67] and using a site-correlated lattice model by Coniglio *et al.* [68]. Figure 9.2 illustrates an idealized phase diagram for a sol–gel transition close to a binodal. The asymmetry of the binodal (and spinodal) suggests one very important aspect – both loci will depend on polymer or aggregate molecular weight. As the gel point is approached and the molecular weight increases, the binodal, for example, becomes so asymmetric that for infinite molecular weight it actually becomes triangular, with effectively no branch to the low concentration side of the maximum. This may explain why there is some controversy between those workers who consider that 'phase separation drives gelation' and those who believe the converse. It may be that when phase separation is seen, it follows chain–chain aggregation, and in these cases the pre-gel polymer–solvent system is close to or even below the theta point. Phase separation is then a consequence of the molecular weight increase at gelation.

In some cases, however, the position of the phase boundary has proved crucial. For example, as already discussed in Section 9.3.2, atactic polystyrene, which unlike the tactic forms cannot crystallize, forms amorphous, thermoreversible gels in both *cis-* and *trans-*decalin [46]. Miller *et al.* [69] have suggested that the synthetic, and in suitable solvents helical, polypeptide poly(γ-benzyl-L-glutamate) (PBLG) forms gels by spinodal decomposition, but other workers have challenged this view. In this case the exact mechanism is complicated by PBLGs tendency to form both liquid- and solid-crystals.

9.5 Mechanical properties of physical networks

9.5.1 Large deformations – experimental aspects

Large-deformation testing involves examining a sample either in tension, compression or shear, up to and sometimes beyond the yield or failure point. Although measurements for elastomeric networks (Chapter 7) are usually quite simple to perform, for some weaker physical gels and networks there are experimental difficulties that have to be overcome if meaningful data are to be obtained. Consequently, we digress briefly to address practical issues.

Many workers have taken the easy option, by preparing and then compressing cylindrical samples. However, in such an experiment the response is often dominated by friction between the sample surface and the compressing plates and the sample begins to barrel or bulge. If Poisson's ratio of the gel is close to

0.5 (totally incompressible), uniaxial compression in one plane is equal to biaxial extension in the other two. Such biaxial extension can occur, only if the surfaces of the specimen are very well lubricated. Tensile measurements are just as difficult to perform, because merely gripping weaker-gel samples with pneumatic or screw clamps will induce preferential failure at the clamping point. Large-deformation measurements in shear are also difficult to perform because the debonding of a sample from the test geometry is very difficult to avoid. It can be reduced by roughening the fixture surfaces, but then the calculations of stresses and strains is made difficult.

Bagley and co-workers [70] have compared the large-deformation behaviour of gels under both lubricated and unlubricated conditions in uniaxial compression. The lubricated surfaces typically require paraffin oil and Teflon-covered platens, whereas the unlubricated experiment uses cyanoacrylate adhesive. Alternative experimental methods that have proved valuable include extending either rings or, even better, 'hippodrome'- or 'racetrack'-shape gel samples over dowel pins [71] so that tensile measurements can be made without the need to use clamps, or measuring cylindrical gel-samples in tension, by fastening the sample ends with 'superglue' onto roughened supports, themselves attached to clamped metal plates [72].

9.5.2 Large deformations – theory

It might seem unusual to discuss the theory of large deformation measurements after, and with little reference to, experiment, but this is a fair assessment of the current level of understanding. There have been remarkably few reliable measurements of large-deformation behaviour, and only a fraction of these have been considered everything beyond the overall shape of the stress–strain curve. We might expect that the simplest form of stress–strain behaviour is that due to classical rubber elasticity, for example, the simplest form (compare equations (7.37)–(7.46)) is

$$\sigma = \frac{E}{3}(\lambda - \lambda^{-2}) \tag{9.2}$$

where σ is the nominal tensile stress, E is Young's modulus and λ is the extension ratio. Although it might be thought that this equation was rather poorly followed, because non-Gaussian and finite-chain extensibility effects would soon dominate, the λ at failure is often <2 (i.e. a failure strain of $<100\%$), so that the curvature in plots of G versus $(\lambda - \lambda^{-2})$ is rather slight. Indeed, some tensile measurements for gelatin gels suggest that $\sigma \propto \varepsilon = \lambda - 1$, i.e. the gels are Hookean, almost up to failure. In order to account, at least phenomenologically, for non-Gaussian behaviour, McEvoy *et al.* [73] used equation (9.3), due to Blatz, Sharga and Tschoegl (BST) [74]:

$$\sigma = \frac{2E}{3n}(\lambda^{n-1} - \lambda^{-(n+2)/2}) \tag{9.3}$$

The BST equation is the classical theory, but with a general exponent n. When $n = 2$, equation (9.3) reduces to equation (9.2), but exponents other than 2 describe non-classical behaviour. McEvoy *et al.* were able to describe their tensile measurements on gelatin, agarose and their mixtures with $n \approx 2.5$–3.5. More recently, Bot *et al.* [75] have employed the same approach for extensive measurements on gelatin gels, in both shear and compression, and obtained values of n ranging from 2.5 up to 5.5. There is some indication from their work that the exponent deviates more at lower concentrations, and after longer times.

As far as more rigorous approaches are concerned, Doi and Kuzuu [76] calculated the stress–strain profile for isotropic glasses of rod elements. In their model, based upon the macroscopic elasticity due to the energetics of rod bending, the profile is S-shaped with initial stress hardening, i.e., behaviour opposite to that suggested by equation (9.2). However, evidence for this class of behaviour for physical gels is still rather limited.

9.5.3 *Small deformations*

Small-deformation measurements are concerned essentially with estimating the zero-strain, limiting shear modulus G or Young's modulus E. For an ideal entropic rubber (Chapter 7), G or E is proportional to absolute temperature T. For thermoreversible gels, the decrease in modulus with increase in temperature is usually much more pronounced than any apparent increase from the chain-entropic effect. Even for non-thermoreversible gels, G tends to decrease as T increases and, typically, albeit over a small range of temperatures, $G \propto T^{-b}$, where $b \approx 3$–5.8. This suggests that the predominant (isothermal) energetic contribution to network elasticity is enthalpic. For thermoreversible gels it is obviously more difficult to argue this, and there is some evidence that gelatin gels are indeed entropic. This is consistent with the modern view of large regions of flexible polypeptide chain between comparatively small triple-helical regions. For other thermoreversible gels, such as PVC and agarose gels, one might argue for a predominant enthalpic contribution to elasticity.

For thermoreversible gels, Nishinari *et al.* [77] have formulated the so-called 'reel-chain' model for the modulus-temperature dependence. In this model, multi-stranded chains are viewed as being wound-up, as if on a fishing reel. If a certain critical stress is applied, the reel begins to unravel to give untwisted, flexible chains. If it is assumed that the junction zones of (finitely extensible, Langevin) chains can be unzipped by a process of sublimation, the resultant modulus appears to be a parabolic function of T. This model has been made more rigorous by Higgs and Ball [78]. They also calculate a parameter equal to the entropic enhancement of the front factor for a stiffened chain.

9.5.3.1 Combinatoric network theories

Although most physical gels are enthalpic, it can still be argued that the number of elastically active network chains (EANCs) (see Section 7.4) *at a given temperature* is independent of the contribution per chain to the gel modulus, so that combinatoric network theories can still be employed. A number of approaches have been made in this area, and some are now described.

The requirement of any theory of the modulus of physical gels is that it must reflect almost universal observations, namely that, at high polymer concentrations (say $>5C_0$), where C_0 is the minimum gelation concentration, the modulus should be proportional to C^n, where n is around 2–2.5, and that, at lower concentrations, as $C \to C_0$ from above, the exponent n becomes significantly greater. (Such an increasing exponent is, of course, a necessary consequence of the assumption that $G \approx 0$ for $C < C_0$.) C_0 itself varies from <0.05% for some microbial polysaccharides to >10% for certain more particulate gel systems, but it is almost always observed that the shape of the scaled $\log G$ versus $\log(C/C_0)$ curve remains the same.

In previous work, we have developed a model using branching or cascade theory [79] to count EANCs and formulated the modulus–concentration relationship in a universal way [8, 80]. To do this, we abandoned the conventional (ideal entropy) assumption of a contribution to the modulus of RT per mole of EANC, and introduced a non-ideality parameter a such that aRT units would be the actual contribution. This factorization into a and RT reflects a particular choice of energy units and (unlike the case of ideal rubbers) is *not* a statement about the expected temperature dependence of the gel modulus. If the assumption is made that the making and breaking of junctions occurs slower than the time-window in a rheological experiment, an equilibrium constant can be defined for the reversible conversion of a junction point into two free sites. The other parameters of the model are f, the network-junction functionality, and K, an association constant. The equilibrium shear modulus G is written in the form [8]

$$G = N_e(\alpha, t)aRT \tag{9.4}$$

where N_e, the number of moles of EANCs per unit volume, is a function of the junction-point density in the system. The parameter a is synonymous with g, the rubber-elasticity 'front factor' of Chapter 7, and R and T have their usual meanings; N_e is a function of f and C/C_0, where C/C_0 is the ratio of polymer concentration to the critical gelation concentration.

The value of f can be estimated from a consideration of the junction-zone structure. For example, for multiple-helical junction-zones, if the molecularity of the helix is n (i.e., $n = 3$ for a triple helix), and the number of such helical junction-zones per chain is x, then f is given by

$$f = (n - 1)(x - 1) + 1 \tag{9.5}$$

and, in general,

$$C_0 = \frac{M(f-1)}{Kf(f-2)^2} \tag{9.6}$$

Best values of the three parameters a, f and K can be determined *provided sufficient data are available.*

The preceding, so-called Clark–Ross-Murphy (CRM) model has proved to be quite successful, and has been applied to a number of different physical-gel systems, including agarose, amylose, PVC and gelatin [80–83]. In all the systems, the parameter a appears to increase with increasing network-strand thickness. The apparent success requires one to examine the assumptions and conclusions very carefully. Since two (or three if f is unconstrained) parameters are employed, it is not surprising that a reasonable fit can be obtained, especially when relatively few modulus-concentration data points exist (typically 5–10). One assumption which may be challenged is that of a junction-zone formation equilibrium, since many physical gels are known to be under kinetic rather than thermodynamic control. However, calculations in which network formation progresses by a pairwise *irreversible* association mechanism have been performed, and have shown that if network formation occurs against the background of a competing wastage process (cyclization, for example), the same treatment may be applied [84].

Despite our own repeated reservations, the method has been criticized by others for being too parametrized [6]. Because we have always been aware of this problem, we have set out to reduce the number of independent parameters. The succeeding text illustrates different approaches. The first is to estimate K (or C_0), a and f from other experiments, the second is to use further information, perhaps from samples of different primary-chain molecular weight M, and to find values of the parameters which are either independent of M, or which change in a reasonable and predictable manner.

9.5.3.2 Application of the Clark–Ross-Murphy (CRM) model to gelatin gels

In one exercise [85], a series of points of G versus C were obtained using a good quality gelatin sample and a, K and f were estimated from other, non-rheological, experiments. Although f is the most fundamental parameter of the model, unlike the case of covalent networks, it is not easy to estimate f *a priori*. Two different arguments justify a final choice of $f = 15$ for the sample. First, renaturation experiments give estimates of the minimum, stable helix length, and this can be compared with the overall chain length. Second, there is work by Djabourov and Leblond [86] that estimated the critical helix content at the gel point h_c from OR experiments. Their value of h_c was ≈ 0.07. From the Flory gel-point equation (and appreciating the

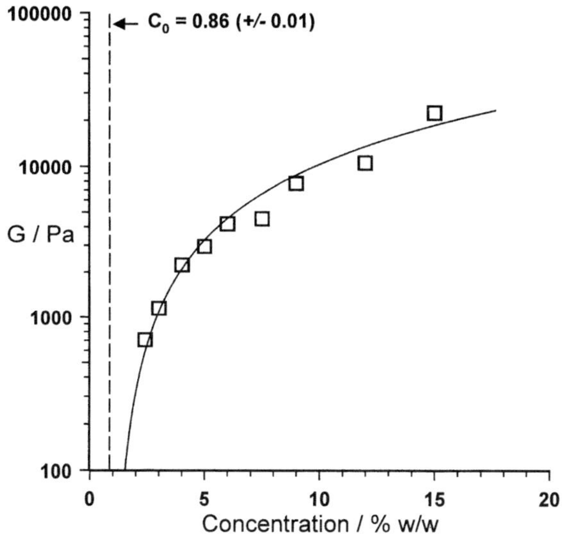

Figure 9.3 Branching-theory fit to gelatin data.

problems in such an approximation), this gives a functionality of $f \approx 15$, close to that ($f \approx 15$–20) estimated from renaturation.

We can also argue using the theory of rubber elasticity (Chapter 7) that a could never be less than ~0.5 (phantom-chain model with tetrafunctional junction points) nor *for rubberlike chains* should it be greater than 1 (affine-chain model). We therefore anticipate a value $0.5 < a < 1$. To maintain consistency with the original affine model, with a front factor of unity (i.e., no correction is applied) we opt here to use the value 1.0. To estimate K we use published results for C_0 obtained using light-scattering measurements on the same sample, and employ equation (9.6) to calculate the value of K. Figure 9.3 shows the result, for the (essentially zero free-parameter) fit with the estimated values. The result is suitably gratifying in view of the various approximations involved.

9.5.3.3 Extension of the CRM model to modulus versus molecular weight data

Another approach is to test the model using different molecular-weight samples at a few discrete concentrations. A recent paper has applied the model to high-quality data on κ-carrageenan [87]. In order to carry out this calculation, it is useful to assume that f is proportional to chain length (or molecular weight) and so define a new parameter k_M which relates f to M:

$$f = \frac{M}{k_M} \tag{9.7}$$

The model now predicts [88] that G remains zero until a critical molecular weight M_c is reached ($M_c \geq 2k_M$) and then G increases, before finally reaching a constant limit value at high M. This behaviour is precisely that observed experimentally. A very acceptable fit was obtained by assuming that $k_M \approx 1000$ (a reasonable value that corresponds to one bonding unit per three disaccharide units).

9.5.3.4 Other related approaches

A broadly similar model, by Oakenfull [89], includes a parameter M_j, the molecular weight of the junction-zone region, and n, the molecularity of the junction zones, i.e., $n = 3$ for intermolecular triple-helical gelatin gelation. (n is always 2 in the model above, since it assumes that binary chain–chain interactions are involved.) M/M_j then corresponds to x in equation (9.5). By allowing $n > 2$, an element of cooperativity is included; there is, however, no parameter corresponding to a in equation (9.4). In practice, the behaviour of the two models is quite similar, and the Oakenfull approach has been applied to gelatin and a number of polysaccharide gels, including high-methoxy pectin.

A newer model for modulus-versus-concentration data is due to te Nijenhuis [7]. He assumed that there is an equilibrium between the concentration of junction zones and that of potential junction sites, in a similar manner to the approach of Tanaka and Nishinari, already described [16]. According to this model, for f-functional junctions

$$K_{tN} = \frac{[\text{junction zones}]}{[\text{potential junction sites}]^{f/2}} \tag{9.8}$$

and

$$[\text{junction zones}] = \frac{\gamma_w}{f/2}\frac{C}{M} \tag{9.9}$$

where γ_w is the number of junction zones per primary chain. From this, for large γ_w, he derives

$$G = \frac{CRT}{M}\frac{4}{f}\left[\frac{C}{C_0}\right]^{f/2-1} \tag{9.10}$$

The method has been applied with some success to the data of Peniche-Covas *et al.* for gelatin [90], adding support to the model of triple-helical junction zones.

9.5.3.5 Non-combinatoric theories

It would not reflect the current state of understanding to give the impression that combinatoric network theories are universally accepted. There is still some controversy in the literature, mainly addressing two points. The first is that some physical gels may be more like colloidal particulate systems (Flory IV) than 'junction-zone', (Flory III)-like. This is undoubtedly true, for example, for casein and globular protein gels prepared under relatively mild conditions of pH or temperature. For these systems a fractal aggregate model has been suggested by Bremer *et al.* [91]. They have proposed that fractal (Hausdorff) exponents be deduced from the $G \propto C^n$ exponent, and compared with those found by, for example, direct structural probes. Agreement is indeed good, but two points need to be made. First, it is a direct assumption of this type of model that the critical concentration C_0 is zero – something quite inconsistent with most experimental results, and, secondly, data for BSA and other globular-protein gels are very well fitted by the model of Section 9.5.2, over a very wide range of conditions.

Another model, that of Jones and Marques [92], addresses the opposite extreme, by considering the properties of a rigid-rod network. This very physical model also incorporates some fractal ideas but, although it is no doubt rigorous, its final conclusions are slightly disappointing. For example, it suggests that, for rods, the shear modulus is $\approx E_r C^2$, where E_r is the modulus of the rod material. As we have already mentioned that C^2 dependencies are commonly seen, this may be regarded as a limited success, since other, much simpler, theories can do as well if not better. The theory does not appear to have been extensively tested, and it is not clear that it will prove very useful.

9.5.4 Permanence of physical networks

The long-time creep behaviour of physical gels is particularly valuable, since it can help one to understand the overall behaviour of such systems, and establish if they are really permanent or if, at very long times, they flow, i.e., whether they have a true equilibrium modulus. Of course, there are many experimental problems, particularly in reducing experimental artefacts due to slippage [93], or loss of solvent (i.e., moisture for hydrogels). However, many early experiments were carried out by Braudo and co-workers [94] of the long-time, creep-compliance behaviour of a range of polysaccharide gels. The systems included pectin, furcellaran and carrageenans. By including the creep-recovery data (the decrease in strain when a constant stress is suddenly removed), the effective viscosity of the gels was estimated to be $\approx 10^8$–10^9 Pa s. More recently Higgs and Ross-Murphy [95] made such measurements on gelatin gels, concluding that the creep-phase viscosity is $\approx 10^8$ Pa s. Very recent calculations by te Nijenhuis confirm that these values are quite consistent with the relaxation seen in his own very precise frequency-domain measurements [7], and he concludes reasonably that, despite other evidence, 'gelatin gels show no rubberlike behaviour up to very low frequencies'.

9.6 Conclusion

The area of physical gels is more than sufficient for a complete volume. Consequently, this chapter can cover the topic only cursorily. Unlike other aspects of polymer networks, the area is still by no means mature: indeed there are still many workers who do not appreciate the subtleties of physical networks. In the future, we hope this chapter will encourage others to study non-covalent as well as covalent gels and networks.

Acknowledgements

The author is very much indebted to Dr. Ir. K. te Nijenhuis for an invaluable pre-publication copy of his review *Viscoelastic Properties of Thermoreversible Gels*, cited as Ref. 7, and to Prof. A.H. Clark for many stimulating discussions.

Glossary of symbols and abbreviations

C	concentration
C_0	critical concentration for physical-gel formation
E	gel modulus; Young's modulus
E_r	modulus of the rods constituting a rigid-rod network
G	shear modulus
G'	real part of the complex shear-modulus
G''	imaginary part of the complex shear-modulus
I	ionic strength
K	(EANC) association constant
M	primary-chain molecular weight
M_e	critical molecular weight for topological entanglements
M_j	junction-zone length (molecular weight)
N_e	moles of EANCs per unit volume
R_c	cross-sectional chain radius
T_g	glass-transition temperature
a	non-ideality parameter associated with concentration of EANCs
f	network-junction functionality; number of junction sites per chain
h_c	critical helix content at the gel point
k_M	constant of proportionality between network-junction functionality (f) and M
n	exponent in the BST equation; junction-zone molecularity
x	number of helical junction-zones per chain
γ_0	maximum strain for linear viscoelastic behaviour

γ_w	number of junction zones per primary chain
λ	extension ratio
σ	nominal tensile stress
ω	frequency of shear strain
BST	Blatz, Sharga and Tschoegl (equation)
CD	circular dichroism
DMSO	dimethyl sulfoxide
DOP	dioctyl phthalate
DS	degree of saponification
DSC	differential scanning calorimetry
EANC	elastically active network chain
G	guluronate (alginate) unit
HDPE	high-density polyethylene
M	mannuronate (alginate) unit
OR	optical rotation
PBLG	poly(γ-benzyl-L-glutamate)
PE	polyethylene
PMMA	poly(methyl methacrylate)
PS	polystyrene
PVA	poly(vinyl alcohol)
PVC	poly(vinyl chloride)
TTS	time-temperature superposition
LCST	lower critical-solution temperature
UCST	upper critical-solution temperature
i-PS	isotactic polystyrene
s-PS	syndiotactic polystyrene

References

1. Ferry, J.D. (1980) *Viscoelastic Properties of Polymers*, 3rd edn, John Wiley, New York.
2. Doi, M. and Edwards, S.F. (1986) *The Theory of Polymer Dynamics*, Clarendon, Oxford.
3. Flory, P.J. (1975) *Faraday Discuss. Chem. Soc.*, **57**, 1.
4. Jordan-Lloyd, D. (1926) in *Colloid Chemistry I* (ed. J. Alexander), Chemical Catalog Co., New York, p. 767.
5. Almdal, K., Dyre, J., Hvidt, S. and Kramer, O. (1993) *Polym. Gels & Networks*, **1**, 5.
6. Guenet, J.-M. (1992) *Thermoreversible Gelation of Polymers and Biopolymers*, Academic Press, London.
7. te Nijenhuis, K. (1997) *Adv. Polym. Sci.* (in press).
8. Clark, A.H. and Ross-Murphy, S.B. (1987) *Adv. Polym. Sci.*, **83**, 57.
9. Ross-Murphy, S.B. and Shatwell, K.P. (1993) *Biorheology*, **30**, 217.
10. Stadler, R. and de Lucca Freitas, L. (1986) *Colloid Polym. Sci.*, **264**, 773.
11. Lapasin, R. and Pricl, S. (1995) *Rheology of Industrial Polysaccharides*, Chapman and Hall, London.
12. Richardson, R.K. and Ross-Murphy, S.B. (1981) *Int. J. Biol. Macromol.*, **3**, 315.
13. Ferry, J.D. and Eldridge, J.E. (1949) *J. Phys. Chem.*, **53**, 184.
14. Tanaka, F. (1989) *Macromolecules*, **22**, 1988.
15. Tanaka, F. and Stockmayer, W.H. (1994) *Macromolecules*, **24**, 6367.

16. Tanaka, F. and Nishinari, K. (1996) *Macromolecules*, **29**, 3625.
17. Rees, D.A. (1969) *Adv. Carbohydr. Chem. Biochem.*, **24**, 267.
18. Chronakis, I.S., Piculell, L. and Borgström, J. (1996) *Carbohydr. Polym.*, **31**, 215.
19. te Nijenhuis, K. and Dijkstra, H. (1975) *Rheol. Acta*, **14**, 71.
20. te Nijenhuis, K. (1981) *Colloid & Polym. Sci.*, **259**, 522.
21. te Nijenhuis, H. (1979) PhD Thesis, University of Delft, The Netherlands.
22. Alfrey, T., Widerhorn, N., Stein, R. and Tobolsky, A.V. (1949) *J. Colloid Sci.*, **4**, 211.
23. Gert, C. and Meyer, H.M. (1978) *Angew. Makromol. Chem.*, **78**, 81.
24. Winter, H.H. and Chambon, F. (1986) *J. Rheol.*, **30**, 411.
25. te Nijenhuis, K. and Winter, H.H. (1989) *Macromolecules*, **22**, 411.
26. Michon, C., Cuvalier, G. and Launay, B. (1993) *Rheol. Acta*, **32**, 94.
27. Michon, C., Cuvalier, G., Launay, B. and Parker, A. (1996) *J. Chim. Phys. et Physico-Chim. Biol.*, **93**, 819, 828.
28. Shibayama, M., Sato, M., Kimura, Y. and Fujiwara, H. (1988) *Polymer*, **29**, 336.
29. Beltman, H. and Lyklema, J. (1975) *Faraday Discuss. Chem. Soc.*, **57**, 92.
30. Hyon, S.-H., Cha, W.-I. and Ikada, Y. (1986) *Polym. Bull.*, **22**, 119.
31. Pines, E. and Prins, W. (1973) *Macromolecules*, **6**, 888.
32. Komatsu, M., Inoue, T. and Miyasaka, K. (1986) *J. Polym. Sci. (Physics)*, **24**, 303.
33. Watase, M. and Nishinari, K. (1985) *J. Polym. Sci. (Physics)*, **23**, 1803.
34. Watase, M. and Nishinari, K. (1989) *Makromol. Chem.*, **190**, 155.
35. Shiga, T., Hirose, Y., Okada, A. and Kurauchi, T. (1993) *J. Appl. Polym. Sci.*, **47**, 113.
36. Pennings, A.J. (1977) *J. Polym. Sci. (Symposia)*, **59**, 55.
37. Barham, P.J., Hill, M.J. and Keller, A. (1980) *Colloid. Polym. Sci.*, **258**, 899.
38. Smith, P. and Lemstra, P.J. (1977) *Makromol. Chem.*, **180**, 2893.
39. Burchard, W. and Ross-Murphy, S.B. (eds) (1990) *Physical Networks – Polymers and Gels*, Elsevier Applied Science, London.
40. Guenet, J.-M., Lotz, B. and Wittmann, J.C. (1985) *Macromolecules*, **18**, 420.
41. Guenet, J.-M. and McKenna, G. (1988) *Macromolecules*, **21**, 1952.
42. Edwards, C.O. and Mandelkern, L. (1982) *J. Polym. Sci. (Lett.)*, **20**, 355.
43. Guenet, J.-M. (1986) *Macromolecules*, **19**, 1960.
44. Grant, G.T., Morris, E.R., Rees, D.A., Smith, P.J.C. and Thom, D. (1973) *FEBS Lett.*, **32**, 195.
45. Benoit, H. and Picot, C. (1966) *Pure Appl. Chem.*, **12**, 545.
46. Arnauts, J. and Berghmans, H. (1987) *Polymer Commun.*, **28**, 66.
47. Vorenkamp, E.J., Bosscher, F. and Challa, G. (1979) *Polymer*, **20**, 59.
48. Djabourov, M., Maquet, J., Theveneau, H., Leblond, J. and Papon, P. (1985) *Brit. Polym. J.*, **17**, 169.
49. Ross-Murphy, S.B. (1992) *Polymer*, **33**, 2622.
50. Busnel, J.P., Morris, E.R. and Ross-Murphy, S.B. (1989) *Int. J. Biol. Macromol.*, **11**, 119.
51. Pezron, I., Herning, T., Djabourov, M. and Leblond, J. (1990) in *Physical Networks – Polymers and Gels* (eds W. Burchard and S.B. Ross-Murphy), Elsevier Applied Science, London, p. 231.
52. Djabourov, M., Clark, A.H., Rowlands, D.J. and Ross-Murphy, S.B. (1989) *Macromolecules*, **22**, 180.
53. Miles, M.J., Morris, V.J., Orford, P.D. and Ring, S.G. (1985) *Carbohydr. Res.*, **135**, 257.
54. Durrani, C.M. and Donald, A.M. (1995) *Polymer Gels and Networks*, **3**, 1.
55. Moorhouse, R., Colegrove, G.T., Sandford, P.A., Baird, J.K. and Kang, K.S. (1981) in *Solution Properties of Polysaccharides* (ed. D.A. Brant), American Chemical Society, Washington, DC, p. 111.
56. Robinson, G., Manning, C.A. and Morris, E.R. (1991) in *Food Polymers, Gels and Colloids* (ed. E. Dickinson), Royal Society of Chemistry, London, p. 22.
57. Milas, M., Rinaudo, M., Knipper, M. and Schuppiser, J.L. (1990), *Macromolecules*, **23**, 2506.
58. Ross-Murphy, S.B., Morris, V.J. and Morris, E.R. (1983) *Faraday Discuss. Chem. Soc.*, **18**, 115.
59. Clark, A.H. and Lee-Tuffnell, C.D. (1986) in *Functional Properties of Food Macromolecules* (eds J.R. Mitchell and D.A. Ledward), Elsevier Applied Science, Barking, UK, p. 203.
60. Richardson, R.K. and Ross-Murphy, S.B. (1981) *British Polymer J.*, **3**, 315.

61. van Kleef, F.S.M., Boskamp, J.V. and van den Tempel, M. (1978) *Biopolymers*, **17**, 225.
62. Janmey, P.A., Hvidt, S., Lamb, J. and Stossel, T.P. (1990) *Nature*, **345**, 89.
63. Janmey, P.A., Hvidt, S., Kas, J., Lerche, D., Maggs, A., Sackmann, E., Schliwa, M. and Stossel, T.P. (1994) *J. Biol. Chem.*, **269**, 32503.
64. Janmey, P.A., Amis, E.J. and Ferry, J.D. (1983) *J. Rheol.*, **27**, 135.
65. Niki, R., Kohyama, K., Sano, Y. and Nishinari, K. (1996) *Polymer Gels and Networks*, **2**, 105.
66. Walstra, P. and van Vliet, T. (1971) in *Food Polymers, Gels and Colloids* (ed. E. Dickinson), Royal Society of Chemistry, London, p. 369.
67. Dušek, K. (1972) *J. Polym. Sci. C*, **39**, 83.
68. Coniglio, A., Stanley, H. and Klein, W. (1982) *Phys. Rev. B*, **25**, 6805.
69. Miller, W.G., Kou, L., Tokyama, K. and Voltaggio, V. (1978) *J. Polym. Sci. (Symp.)*, **C65**, 91.
70. Navickis, L. and Bagley, E.B. (1983) *J. Rheol.*, **27**, 519.
71. Myers, F.S. and Wenrick, J.D. (1974) *Rubber Chem. Technol.*, **47**, 1213.
72. Stading, M. and Hermansson, A.M. (1991) *Food Hydrocoll.*, **5**, 339.
73. McEvoy, H., Ross-Murphy, S.B. and Clark, A.H. (1985) *Polymer*, **26**, 1483.
74. Blatz, P.I., Sharga, S.C. and Tschoegl, N.W. (1976) *Trans. Soc. Rheol.*, **18**, 145.
75. Bot, A., van Amerongen, I.A., Groot, R.D., Hoekstra, L.L. and Agterof, W.G.M. (1996) *Polymer Gels and Networks*, **4**, 189.
76. Doi, M. and Kuzuu, N.Y. (1980) *J. Polym. Sci. Polym. Phys. Ed.*, **18**, 409.
77. Nishinari, K., Koide, S. and Ogino, K. (1985) *J. Phys. (France)*, **46**, 793.
78. Higgs, P.G. and Ball, R.C. (1989) *Macromolecules*, **22**, 2432.
79. Gordon, M. and Ross-Murphy, S.B. (1975) *Pure Applied Chem.*, **43**, 1.
80. Clark, A.H. and Ross-Murphy, S.B. (1985) *Brit. Polymer J.*, **17**, 164.
81. Clark, A.H., Gidley, M.J., Richardson, R.K. and Ross-Murphy, S.B. (1989) *Macromolecules*, **22**, 346.
82. Garcia, A.E., Munoz, M.E., Pena, J.J. and Santamaria, A. (1990) *Macromolecules*, **23**, 5251.
83. Clark, A.H., Ross-Murphy, S.B., Nishinari, K. and Watase, M. (1990) in *Physical Networks – Polymers and Gels* (eds W. Burchard and S.B. Ross-Murphy), Elsevier Applied Science, London, p. 209.
84. Clark, A.H. (1993) *Polymer Gels and Networks*, **1**, 139.
85. Ross-Murphy, S.B. (1994) in *Gums and Stabilizers for the Food Industry 7* (eds G.O. Phillips, D.J. Wedlock and P.A. Williams), IRL Press, Oxford, p. 243.
86. Djabourov, M. and Leblond, J. (1986) in *Reversible Polymeric Gels and Related Systems* (ed. P.S. Russo), American Chemical Society, Washington, DC, p. 211.
87. Clark, A.H. (1994) *Carbohydr. Polym.*, **23**, 247.
88. Clark, A.H., Nishinari, K., Ross-Murphy, S.B. and Watase, M. (1995) *Macromolecular Symposia*, **93**, 187.
89. Oakenfull, D.G. (1984) *J. Food Sci.*, **49**, 1103.
90. Peniche-Covas, C.A.L., Dev, S.B., Gordon, M., Judd, M. and Kajiwara, K. (1974) *Faraday Discuss Chem. Soc.*, **57**, 165.
91. Bremer, L.G.B., van Vliet, T. and Walstra, P. (1989) *J. Chem. Soc., Faraday Trans. 1*, **85**, 3359.
92. Jones, J.L. and Marques, C.M. (1990) *J. Physique*, **51**, 113.
93. Richardson, R.K. and Goycoolea, F.M. (1994) *Carbohydr. Polym.*, **24**, 223.
94. Braudo, E.E., Plashcina, I. and Tolstoguzov, V.B. (1984) *Carbohydr. Polym.*, **4**, 23.
95. Higgs, P.G. and Ross-Murphy, S.B. (1990) *Int. J. Biol. Macromol.*, **12**, 233.

Appendix International Union of Pure and Applied Chemistry Macromolecular Division Commission on Macromolecular Nomenclature

Source-based nomenclature for non-linear macromolecules and macromolecular assemblies

(IUPAC RECOMMENDATIONS 1997)

Synopsis

A source-based nomenclature for non-linear macromolecules and macromolecular assemblies is proposed. Source-based or structure-based names for *linear* regular (single-strand and double-strand) and irregular macromolecules can be derived using rules published previously. In contrast, the present document deals with the source-based naming of non-linear macromolecules and assemblies of macromolecules. The former comprise branched, graft, comb, star, cyclic and network macromolecules; the latter comprise polymer blends, interpenetrating polymer networks and polymer–polymer complexes. The names of non-linear macromolecules are formed by citing source-based name(s) of the constituent linear chain(s) and denoting their mode(s) of combination by prefixes and connectives. Specification of molecular weight, composition or branch-point functionality(ies) of the macromolecule or its constituents may be given in parentheses.

Introduction

The recommendations for macromolecular nomenclature published so far [1] have dealt with regular single-strand organic macromolecules [2, 3], regular single-strand and quasi-single-strand inorganic and coordination macromolecules [4], copolymer macromolecules [5], regular double-strand organic macromolecules [6] and irregular single-strand organic macromolecules [7]. As far as possible, the established principles of organic and inorganic nomenclature have been followed [8–10]. The present document is intended to extend the system to non-linear macromolecules such as branched and cyclic macromolecules, micronetworks, polymer networks, and to macromolecular assemblies held together by non-covalent bonds or forces, such as polymer blends, interpenetrating polymer networks and polymer complexes.

Conventionally, the word 'polymer' as a noun is ambiguous in meaning. It is commonly used to refer to both polymer substances and polymer molecules. Henceforth, 'macromolecule' is used for individual molecules and 'polymer' is used to denote a substance composed of macromolecules [11]. 'Polymer' may also be employed unambiguously as an adjective according to accepted usage, e.g. 'polymer blend', 'polymer molecule'.

For non-linear macromolecules, the skeletal structure should be reflected in the name. Non-linear macromolecular structures and macromolecular assemblies are classified using the following terms:

cyclic
branched
  short-branched
  long-branched
micronetwork
network
blend
semi-interpenetrating polymer network
interpenetrating polymer network
polymer–polymer complex.

Definitions of some of the terms used in this document can be found in previous nomenclature reports [11, 12].

A.1 Definitions

A.1.1 chain

The whole or part of a macromolecule, an oligomer molecule or a block, comprising a linear or branched sequence of constitutional units between two boundary constitutional units, each of which may be either an end group, a branch point, or an otherwise-designated characteristic feature of the macromolecule.

Note 1 Except in linear single-strand macromolecules, the definition of the chain may be somewhat arbitrary.

Note 2 A cyclic macromolecule has no end groups but may nevertheless be regarded as a chain.

Note 3 Any number of branch points may be present between the boundary units.

Note 4 Where appropriate, definitions relating to 'macromolecule' may also be applied to 'chain'.

A.1.2 linear chain

A chain with no branch points intermediate between the boundary units.

A.1.3 branched chain

A chain with at least one branch point intermediate between the boundary units.

A.1.4 main chain, backbone

That linear chain to which all other chains, long or short or both, may be regarded as being pendant.

Note Where two or more chains could equally be considered to be the main chain, that one is selected which leads to the simplest representation of the molecule.

A.1.5 subchain

An arbitrarily chosen contiguous sequence of constitutional units in a chain.

Note The term ‘subchain’ may be used to define designated subsets of the constitutional units in a chain.

A.1.6 side chain, branch, pendant chain

An oligomeric or polymeric offshoot from a macromolecular chain.

Note 1 An oligomeric side chain may be termed a short-chain branch.

Note 2 A polymeric side chain may be termed a long-chain branch.

A.1.7 branch unit

A constitutional unit containing a branch point.

Note A branch unit from which f linear chains emanate may be termed an f-functional branch unit, e.g. five-functional branch unit. Alternatively, the terms trifunctional, tetrafunctional, pentafunctional, etc., may be used, e.g. pentafunctional branch unit.

A.1.8 branch point

A point on a chain at which a branch is attached.

Note 1 A branch point from which f linear chains emanate may be termed an f-functional branch point, e.g. five-functional branch point. Alternatively, the terms trifunctional, tetrafunctional, pentafunctional etc., may be used, e.g., pentafunctional branch point.

Note 2 A branch point in a network may be termed a junction point.

A.1.9 crosslink

A small region in a macromolecule from which at least four chains emanate, and which is formed by reactions involving sites or groups on *existing* macromolecules or by interactions between *existing* macromolecules.

Note 1 The small region can be an atom, a group of atoms, or a number of branch points connected by bonds, groups of atoms or oligomeric chains.

Note 2 In the majority of cases, a crosslink is a covalent structure but the term is also used to describe sites of weaker chemical interactions, portions of crystallites, and even physical interactions and entanglements.

A.1.10 network

A highly ramified macromolecule in which essentially each constitutional unit is connected to each other constitutional unit and to the macroscopic phase boundary by many permanent paths through the macromolecule, the number of such paths increasing with the average number of intervening bonds; the paths must, *on the average*, be coextensive with the macromolecule.

Note 1 Usually, and in all systems that exhibit rubber elasticity, the number of distinct paths is very high but, in most cases, some constitutional units exist which are connected by a single path only.

Note 2 If the permanent paths through the structure of a network are all formed by covalent bonds, the term covalent network may be used.

Note 3 The term physical network may be used if the permanent paths through the structure of a network are not all formed by covalent bonds but, at least in part, by physical interactions, such that removal of the interactions leaves individual molecules or macromolecules or a non-network macromolecule.

A.1.11 micronetwork

A highly ramified macromolecule containing cyclic structures and of colloidal dimensions.

A.1.12 macrocycle

A cyclic macromolecule or a macromolecular cyclic portion of a macromolecule.

Note 1 See Note 2 to Definition A.1.1.

Note 2 In the literature, the term 'macrocycle' is sometimes used for molecules of low relative molecular mass that would not be considered macromolecules.

A.1.13 comb macromolecule

A macromolecule comprising a main chain with multiple trifunctional branch points from each of which a linear side chain emanates.

Note 1 If the subchains between the branch points of the main chain and the terminal subchains of the main chain are identical with respect to constitution and degree of polymerization, *and* the side chains are identical with respect to constitution and degree of polymerization, the macromolecule is termed a regular comb macromolecule.

Note 2 If at least some of the branch points are of functionalities greater than three, the macromolecule may be termed a brush macromolecule.

A.1.14 star macromolecule

A macromolecule containing a single branch point from which linear chains (arms) emanate.

Note 1 A star macromolecule with *n* linear chains (arms) attached to the branch point is termed an *n*-star macromolecule, e.g. five-star macromolecule.

Note 2 If the arms of a star macromolecule are identical with respect to constitution and degree of polymerization, the macromolecule is termed a regular star macromolecule.

Note 3 If different arms of a star macromolecule are composed of different monomeric units, the macromolecule is termed a variegated star macromolecule.

A.1.15 polymer blend

A macroscopically homogeneous mixture of two or more different species of polymer.

Note 1 See Figure A.1(a).

Note 2 In most cases, blends are homogeneous on scales larger than several times visual, optical wavelengths.

(a) (b) (c)

Figure A.1 Schematic representation of macromolecular assemblies: (——) polymer chain 1; (——) polymer chain 2; (•, ●) branch points; (a) polymer blend; (b) semi-interpenetrating polymer network; (c) interpenetrating polymer network.

Note 3 For polymer blends, no account is taken of the miscibility or immiscibility of the constituent polymers, i.e., no assumption is made regarding the number of phases present.

Note 4 The general use of the term 'polymer alloy' as a synonym for a polymer blend is discouraged.

A.1.16 semi-interpenetrating polymer network (SIPN)

A polymer comprising one or more networks and one or more linear or branched polymers characterized by the penetration on a molecular scale of at least one of the networks by at least some of the linear or branched macromolecules.

Note 1 See Figure A.1(b).

Note 2 Semi-interpenetrating polymer networks are distinguished from interpenetrating polymer networks because the constituent linear or branched polymers can, in principle, be separated from the constituent polymer network(s) without breaking chemical bonds; they are polymer blends.

A.1.17 network polymer

A polymer composed of one or more networks.

A.1.18 interpenetrating polymer network (IPN)

A polymer comprising two or more networks which are at least partially interlaced on a molecular scale but not covalently bonded to each other and cannot be separated unless chemical bonds are broken.

Note 1 See Figure A.1(c).

Note 2 A mixture of two or more pre-formed polymer networks is not an IPN.

A.1.19 polymer–polymer complex

A complex, at least two components of which are different polymers.

A.2 General principles

The following procedure is employed to name both non-linear macromolecules and macromolecular assemblies:

1. The topology of the structure of the non-linear macromolecule or modes of combination of the constituent species of macromolecules are ascertained.

2. For each constituent linear subchain in a non-linear macromolecule or for each species of macromolecule, a source-based name is assigned according to the rules for naming homopolymer or copolymer molecules [2, 5].
3. The names of the constituent subchains in a non-linear macromolecule or those of the macromolecules are then combined together with the appropriate prefix or connective(s), or both.

The following italicized qualifiers can be used as both prefixes (e.g. *blend*-, *net*-) and connectives (e.g. -*blend*-, -*net*-), separated by hyphen(s) from the constituent name(s), to designate the skeletal structure of non-linear macromolecules or macromolecular assemblies:

cyclic	*cyclo*
branched, unspecified	*branch*
short-chain-branched	*sh-branch*
long-chain-branched	*l-branch*
branched with branch points of functionality *f*	*f-branch* (*f* is given a numerical value)
comb(like)	*comb*
star	*star*
star with *f* arms	*f-star* (*f* is given a numerical value)
network	*net*
micronetwork	*μ-net*
polymer blend	*blend*
interpenetrating polymer network	*ipn*
semi-interpenetrating polymer network	*sipn*
polymer–polymer complex	*compl*

A.3 Non-linear homopolymer molecules

Rule 1. In naming non-linear homopolymer molecules, the italicized prefix for the skeletal structure of the macromolecule is placed before the source-based name of the constituent linear chain.

Examples:

1.1 *sh-branch*-polyethylene
1.2 4-*star*-polystyrene
1.3 *cyclo*-poly(dimethylsiloxane)
1.4 *net*-polybutadiene

Rule 2. Crosslinks, the branch units of star macromolecules and other junction units [5] are optionally specified by their source-based names after the name of the macromolecule with the connective (Greek) ν, separated by hyphens.

Note When the content of the crosslinking monomer is high, the macromolecule is treated as a copolymer molecule.

Examples:

2.1 *net*-polybutadiene-ν-sulfur
(polybutadiene vulcanized with sulfur)
2.2 *net*-polystyrene-ν-divinylbenzene
(polystyrene crosslinked with divinylbenzene to form a network)
2.3 *branch*-polystyrene-ν-divinylbenzene
(polystyrene crosslinked with divinylbenzene, insufficient to cause a network to form)
2.4 *l-branch*-poly(ethyl acrylate)-ν-(ethylene dimethacrylate)
2.5 3-*star*-polystyrene-ν-methyltrichlorosilane

A.4 Non-linear copolymer molecules

A.4.1 Copolymer molecules comprising a single species of linear chain

Rule 3. In naming non-linear copolymer molecules comprising linear subchains of the same monomeric units in a single type of skeletal structure, the italicized prefix for the skeletal structure is placed before the source-based name of the constituent linear subchains. In the case of star macromolecules with block copolymer arms, the block named first after the prefix emanates from the branch point.

Examples:

3.1 *cyclo*-poly(styrene-*stat*-α-methylstyrene)
3.2 *star*-[polystyrene-*block*-poly(methyl methacrylate)]
(each arm of the star macromolecule is a block copolymer chain with a polystyrene block attached to the central unit)
3.3 μ-*net*-poly[styrene-*stat*-(vinyl cinnamate)]
(crosslinked copolymeric micronetwork)
3.4 *net*-poly(phenol-*co*-formaldehyde)
3.5 *comb*-poly(styrene-*stat*-acrylonitrile)
(both the main chain and side chains are statistical copolymer chains of styrene and acrylonitrile)
3.6 *net*-poly[(hexamethylene diisocyanate)-*alt*-glycerol]
3.7 *net*-poly[styrene-*alt*-(maleic anhydride)]-ν-(ethylene glycol)
(alternating copolymer of styrene and maleic anhydride crosslinked with ethylene glycol to form a network)

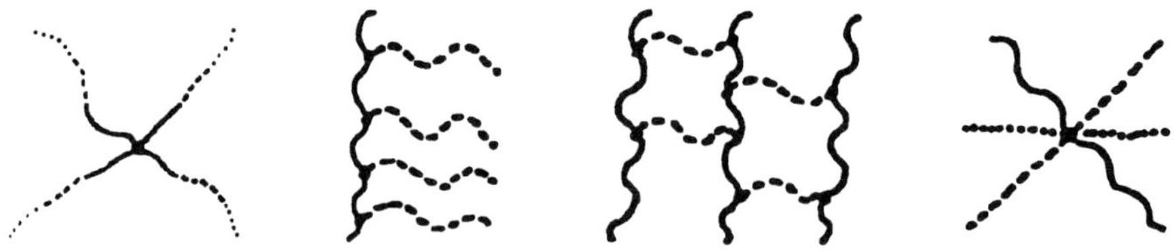

Figure A.2 Schematic representation of non-linear copolymer molecules: (——) polymer chain 1; (– – –) polymer chain 2; (· · ·) polymer chain 3; (●) branch point; (a), (d) star macromolecules; (b) comb macromolecule; (c) network macromolecule.

3.8 *star*-(polyA-*block*-polyB-*block*-polyC)
(each arm of the star macromolecule is a block copolymer chain with a polyA block attached to the central unit, Figure A.2(a)).

A.4.2 Copolymer molecules comprising a variety of species of chains

Rule 4. In naming non-linear copolymer molecules having linear subchains of two or more types, the italicized connective for the skeletal structure is placed between the source-based names of the types of constituent linear subchains. In the case of branched and comb-like macromolecules, the linear chain named before the connective is that which forms the main chain, whereas that (those) named after the connective forms (form) the side chain(s). The names of different species of side chain are separated by semicolons. In the case of variegated star macromolecules the prefix is placed before the name of the macromolecule with the different species of arms separated by semicolons.

Examples:

4.1 polystyrene-*comb*-polyacrylonitrile
(equivalent to polystyrene-*graft*-polyacrylonitrile (Figure A.2(b)); -*graft*- is recommended for these cases. However, *comb* cannot be replaced by *graft* if the former is a prefix; compare Example 3.5)

4.2 polystyrene-*comb*-[polyacrylonitrile; poly(methyl methacrylate)]
(a comb macromolecule with polyacrylonitrile and poly(methyl methacrylate) side chains)

4.3 poly[(ethylene glycol)-*alt*-(maleic anhydride)]-*net*-oligostyrene
(unsaturated polyester cured with styrene; Figure A.2(c))

4.4 poly(methyl methacrylate)-*net*-poly(ethylene oxide)

4.5 *star*-(polyA; polyB; polyC)
(a variegated star copolymer molecule consisting of arm(s) of polyA, arm(s) of polyB and arm(s) of polyC; Figure A.2(d))

Rule 5. The name of a crosslinking monomer molecule having two or more different types of polymerizable groups, each serving as a monomeric unit for one of two or more different linear chains, is cited with the name of each of the chains with the symbol *ν*.

Note When the proportion of the crosslinking monomer having two or more different polymerizable groups is a significant fraction of the total, the constituent chains are treated as copolymer chains.

Examples:

5.1 [poly(butyl vinyl ether)-*ν*-(2-vinyloxyethyl methacrylate)]-*net*-[poly(methyl methacrylate)-*ν*-(2-vinyloxyethyl methacrylate)]
(poly(butyl vinyl ether) and poly(methyl methacrylate) chains crosslinked mutually with 2-vinyloxyethyl methacrylate to form a network)

5.2 poly[(butyl vinyl ether)-*co*-(2-vinyloxyethyl methacrylate)]-*net*-poly[(methyl methacrylate)-*co*-(2-vinyloxyethyl methacrylate)]
(a copolymer of butyl vinyl ether and 2-vinyloxyethyl methacrylate formed into a network with a copolymer of methyl methacrylate and 2-vinyloxyethyl methacrylate)

5.3 [*branch*-poly(butyl vinyl ether)-*ν*-(2-vinyloxyethyl methacrylate)]-*ν*-[*branch*-poly(methyl methacrylate)-*ν*-(2-vinyloxyethyl methacrylate)]
(poly(butyl vinyl ether) and poly(methyl methacrylate) chains crosslinked mutually with 2-vinyloxyethyl methacrylate, insufficient to cause a network to form)

A.5 Macromolecular assemblies

Assemblies of macromolecules include polymer blends, semi-interpenetrating polymer networks, network polymers, interpenetrating polymer networks, and polymer–polymer complexes.

Rule 6. Assemblies of macromolecules held together by non-covalent bonds are named by combination of the names of the constituent macromolecules together with an italicized connective between them.

Examples:

6.1 polystyrene-*blend*-poly(2,6-dimethylphenol)

6.2 (*net*-polystyrene)-*blend*-(*net*-polybutadiene)
(blend of two networks)

6.3 (*net*-polystyrene)-*sipn*-poly(vinyl chloride)
(SIPN of a polystyrene network and a linear poly(vinyl chloride); Figure A.1(b))

6.4 (*net*-polybutadiene)-*ipn*-(*net*-polystyrene)
(IPN of two networks; Figure A.1(c))

6.5 [*net*-poly(styrene-*stat*-butadiene)]-*ipn*-[*net*-poly(ethyl acrylate)]
(IPN of two networks)
6.6 poly(acrylic acid)-*compl*-poly(4-vinylpyridine)
(complex of two species of linear macromolecules)
6.7 poly[(methyl methacrylate)-*stat*-(methacrylic acid)]-*compl*-poly(*N*-vinyl-2-pyrrolidone)
(complex of statistical copolymer and homopolymer molecules)

A.6 Quantitative specifications

Quantitative characteristics of a macromolecule or an assembly of macromolecules, such as mass and mole fractions or percentages as well as the degrees of polymerization and molar masses, may be expressed by placing corresponding figures after the complete name. The order of citation is the same as for the monomer species in the name. Some characteristics cannot be defined for all types of macromolecules and assemblies dealt with in this document, e.g., molar mass of a network.

Rule 7. Mass or mole fractions or percentages of the monomeric units are placed in parentheses after the name of the macromolecule or asssembly of macromolecules with the symbol w or x, respectively, followed by the mass or mole fractions or percentages. The order of citation in the parentheses is the same as that of the monomers in the name; the individual values are separated by colons. Unknown quantities can be designated by a, b etc., or by hyphens.

Note For examples, see Rule 8.1 of Ref. 5.

Rule 8. For simple systems, the molar mass, relative molecular mass or degree of polymerization may be included in the scheme of Rule 7 by the symbol M, M_r or DP, respectively, followed by the corresponding numerical values, separated by colons. Symbols are qualified if quantities refer to parts of macromolecules or assemblies, e.g. M (block), M_r (arm), M (network chain) etc.

Note 1 Average quantities are denoted by the symbol $\bar{M}$, $\bar{M}_r$ or $\overline{\text{DP}}$.

Note 2 M (network chain) is often denoted M_c. Similarly, M_r (network chain) would be denoted $M_{r,c}$.

Examples:

8.1 *net*-polystyrene-*ν*-divinylbenzene (w 98:2%)
(polystyrene crosslinked with 2% divinylbenzene)
8.2 *star*-(polyacrylonitrile; polystyrene) (M_r 100 000:20 000)
(star macromolecule consisting of arms of polyacrylonitrile of a total $M_r = 100\,000$ and arms of polystyrene of a total $M_r = 20\,000$)

8.3 *star*-(polyacrylonitrile; polystyrene) (M_r (arm) 50 000 : 10 000)
(star macromolecule consisting of arms of polyacrylonitrile each of $M_r = 50\,000$ and arms of polystyrene each of $M_r = 10\,000$; the macromolecule could be identical to that of Example 8.2 with two arms of polyacrylonitrile and two arms of polystyrene)

Rule 9. The functionality of a monomer, oligomer or polymer molecule (or their mixtures) or of a branch point may be included in the scheme of Rule 7 by adding the symbol '*f*' and the figure corresponding to the functionality in parentheses after the name of the monomer, oligomer or polymer molecule or branch point.

Note If a mixture of monomer, oligomer or polymer molecules is present with components of the mixture bearing the same type of chemical functionality, an average value of *f* for the mixture may be quoted in parentheses following the list of components in the mixture.

Examples:

9.1 *net*-poly[glycerol; pentaerythritol (*f* 3.5)-*co*-(tolylene diisocyanate)]
(network formed by the reaction of the glycerol/pentaerythritol mixture of average functionality 3.5 with tolylene diisocyanate)

9.2 *star*-polystyrene-*ν*-methyltrichlorosilane (*f* 2.8)

9.3 6-*star*-(polyacrylonitrile (*f* 3); polystyrene (*f* 3)) (M_r (arm) 50 000 : 10 000)
(a six-armed star macromolecule consisting of three arms of polyacrylonitrile, each of $M_r = 50\,000$, and three arms of polystyrene, each of $M_r = 10\,000$)

9.4 4-*star*-(polyacrylonitrile-*block*-polystyrene) (M_r (arm) 50 000)
(a four-armed star macromolecule, each arm of which consists of the same copolymeric block of $M_r = 50\,000$)

9.5 *net*-polystyrene ($\bar{M}_{r,c}$ 10 000)
(network chains of average $M_r = 10\,000$)

References

1. IUPAC (1991) *Compendium of Macromolecular Nomenclature*, Blackwell Scientific Publications, Oxford.
2. IUPAC (1976) Nomenclature of regular single-strand organic polymers. *Pure Appl. Chem.*, **48**, 373–385. (Reprinted as Chapter 5 in Ref. 1.)
3. IUPAC (1989) A classification of linear single-strand polymers. *Pure Appl. Chem.*, **61**, 243–254. (Reprinted as Chapter 8 in Ref. 1.)
4. IUPAC (1985) Nomenclature for regular single-strand and quasi-single-strand inorganic and coordination polymers. *Pure Appl. Chem.*, **57**, 149–168. (Reprinted as Chapter 6 in Ref. 1.)
5. IUPAC (1985) Source-based nomenclature for copolymers. *Pure Appl. Chem.*, **57**, 1427–1440. (Reprinted as Chapter 7 in Ref. 1.)

6. IUPAC (1993) Nomenclature of regular double-strand (ladder and spiro) organic polymers. *Pure Appl. Chem.*, **65**, 1561–1580.
7. IUPAC (1994) Structure-based nomenclature for irregular single-strand organic polymers. *Pure Appl. Chem.*, **66**, 873–889.
8. IUPAC (1979) *Nomenclature of Organic Chemistry*, Sections A, B, C, D, E, F and H, Pergamon Press, Oxford.
9. IUPAC (1994) *A Guide to IUPAC Nomenclature of Organic Compounds*, Blackwell Scientific Publications, Oxford.
10. IUPAC (1990) *Nomenclature of Inorganic Chemistry*, 3rd edn, Blackwell Scientific Publications, Oxford.
11. IUPAC (1996) Glossary of terms in polymer science. *Pure Appl. Chem.*, **68**, 2287–2311.
12. IUPAC (1989) Definitions of terms relating to individual macromolecules, their assemblies, and dilute polymer solutions. *Pure Appl. Chem.*, **61**, 211–241. (Reprinted as Chapter 3 in Ref. 1.)

Prepared by a Working Group consisting of J. KAHOVEC (Czech Republic), P. KRATOCHVÍL (Czech Republic), A.D. JENKINS (UK), I. MITA (Japan), I.M. PAPISOV (Russia), L.H. SPERLING (USA), and R.F.T. STEPTO (UK). Membership of the Commission during the preparation of this report (1983–1994) was as follows: *Titular Members:* G. Allegra (Italy, to 1989); R.E. Bareiss (Germany, to 1993); N.M. Bikales (USA, *Secretary* to 1987); K. Hatada (Japan, Associate Member from 1987, Titular Member from 1989); A.D. Jenkins (UK, *Chairman* to 1985, Associate Member to 1987); J. Kahovec (Czech Republic, Associate Member from 1987, Titular Member from 1991); P. Kratochvíl (Czech Republic, *Chairman* to 1991); E. Maréchal (France, Associate Member from 1991, Titular Member from 1993); W.V. Metanomski (USA, Associate Member from 1987, Titular Member from 1991); I. Mita (Japan, to 1989, Associate Member to 1991); C. Noël (France, from 1985 to 1993); I.M. Papisov (Russia, to 1987, Associate Member to 1991); V.P. Shibaev (Russia, from 1987); R.F.T. Stepto (UK, Associate Member from 1987, Titular Member from 1989, *Chairman* from 1991); U.W. Suter (Switzerland, to 1991); W.J. Work (USA, Associate Member from 1985, *Secretary* from 1987). *Others contributing to this report:* R.B. Fox (USA); J.I. Kroschwitz (USA).

Index